T3-BSC-236

THEORETICAL ASPECTS OF BAND STRUCTURES AND ELECTRONIC PROPERTIES OF PSEUDO-ONE-DIMENSIONAL SOLIDS

PHYSICS AND CHEMISTRY OF MATERIALS WITH LOW-DIMENSIONAL STRUCTURES

Series B: Quasi-One-Dimensional Materials

THEORETICAL ASPECTS OF BAND STRUCTURES AND ELECTRONIC PROPERTIES OF PSEUDO-ONE-DIMENSIONAL SOLIDS

Edited by

HIROSHI KAMIMURA

Department of Physics, Faculty of Science,
University of Tokyo, Bunkyo-ku, Tokyo, Japan

D. REIDEL PUBLISHING COMPANY

DORDRECHT / BOSTON / LANCASTER / TOKYO

Library of Congress Cataloging-in-Publication Data
Main entry under title:

Theoretical aspects of band structures and electronic properties of
pseudo-one-dimensional solids.

(Physics and chemistry of materials with low-dimensional
structures. Series B, Quasi-one-dimensional materials)
Includes bibliographies and indexes.
1. One-dimensional conductors. 2. Conduction band.
3. Quantum chemistry. I. Kamimura, Hiroshi, 1930–
II. Series.
QC176.8.E4T446 1985 530.4'1 85-19401
ISBN 90-277-1927-6

Published by D. Reidel Publishing Company,
P.O. Box 17, 3300 AA Dordrecht, Holland

Sold and distributed in the U.S.A. and Canada
by Kluwer Academic Publishers,
190 Old Derby Street, Hingham, MA 02043, U.S.A.

In all other countries, sold and distributed
by Kluwer Academic Publishers Group,
P.O. Box 322, 3300 AH Dordrecht, Holland

Printed in The Netherlands

TABLE OF CONTENTS

PREFACE

This volume presents a sequence of articles which describe the theoretical treatments of investigating the fundamental features in the electronic structures and properties of typical quasi-one-dimensional solids; organic conductor TTF-TCNQ, polyacetylene, metallic and superconducting polymer $(SN)_x$, and linear chain chalcogenides and halides of transition elements including $NbSe_3$. The aim of this volume is not to present an exhaustive review but rather to touch on a selective class of problems which appear to be fundamental for typical quasi-one-dimensional solids. Thus the topics in this volume are rather confined to the key basic properties of quasi-one-dimensional systems.

The quasi-one-dimensional solids are one of the most extensively investigated subjects in current physics, chemistry and materials science. These materials are unique in attracting a broad range of scientists, chemists, experimental and theoretical physicists, materials scientists and engineers. In 1954 Fröhlich constructed a theory of superconductivity based on a one-dimensional model of moving charge density waves. In 1955 Peierls predicted that any one-dimensional metal is unstable against the distortion of a periodic lattice so that a metal-nonmetal transition occurs at a certain temperature for a one-dimensional metal. According to these theories a gap is opened at the Fermi surfaces of one-dimensional conductors at low temperatures and the charge density wave is created in connection with the occurrence of the gap. Besides this characteristic the prediction of high temperature superconductivity for a certain kind of one-dimensional metal by Little in 1964 and some criticism by Ginzburg in 1970 further stimulated the research of quasi-one-dimensional conductors.

Much of current activity in the research of one-dimensional conductors started with the synthesis of TTF-TCNQ in 1970. In particular, observations of a high conductivity peak just above the phase transition in this compound, and attempts to understand this phenomenon, sharply accelerated activities, although it is now known that the peak is not so large that a special mechanism of conductivity is necessary. At nearly the same time $(SN)_x$ also aroused a great deal of interest because superconductivity was observed for the first time for metallic polymers and their band structures were intensively investigated by a large number of theoretical groups.

Linear chain chalcogenides and halides of transition elements are one of a recently discovered class of quasi-one-dimensional solids. These materials show a variety of electronic states depending on a combination of elements. For example, $TaSe_3$ remains metallic and becomes superconducting below 2.3 K, while TaS_3 undergoes a commensurate Peierls transition from a metallic to a semiconducting phase. Among the compounds in this class $NbSe_3$ attracted special attention because it shows remarkable transport properties including two types of charge density wave phase transition.

Polyacetylene is also a newly discovered class of quasi-one-dimensional solids. Recent reports suggesting the possible existence of neutral and/or charged solitons in polyacetylene gave impetus to the research in one-dimensional systems. It is striking that it became possible to add dopants to polyacetylene and as a result the electrical conductivity increased enormously.

Looking at the remarkable development of activity in the field of one-dimensional conductors mentioned above, the following five kinds of quasi-one-dimensional solids have been chosen as the object of topics in this volume; TTF-TCNQ, $(SN)_x$, $NbSe_3$ and linear chain chalcogenides and halides of transition elements, and conducting polymers such as polyacetylene. The topics associated with the above materials cover the most active areas in the field of quasi-one-dimensional solids except the topic of platinum cyanide complexes represented by KCP. The reason why KCP has not been included in this volume in spite of the fact that it has been studied quite earlier is that KCP is now being reinvestigated as a mixed valence system and a clear description as to the status of KCP is still premature.

This volume consists of five articles. Three articles are devoted to the description of the band structures of $(SN)_x$, $NbSe_3$ and linear chain chalcogenides and halides of transition elements. In particular, in articles of $(SN)_x$ and $NbSe_3$ the methods of the first principles self-consistent band structure calculations suitable for the quasi-one-dimensional solids are presented. In the quasi-one-dimensional systems large fluctuations can occur in connection with phase transitions. Further, many different types of ordering, such as charge density waves, spin density waves, superconducting pairing and etc. may exist at low temperatures. Furthermore, one might want to know how charge transfer and charge transport occur in such systems, including the role of Fröhlich conduction, how three-dimensional effects such as interchain interactions are important, etc. In order to clarify these problems and to obtain considerable insight into them, precise information on a band structure of a real material is needed. For this purpose the articles on $(SN)_x$, $NbSe_3$ and linear chain chalcogenides and halides of transition elements will be useful.

In TTF-TCNQ, on the other hand, the electron–electron interaction plays an important role, as compared with electron–phonon interactions, which are primarily important in other classes of materials included in this volume. That is, among quasi-one-dimensional solids, TTF-TCNQ is the most typical material in which many-body effects are essential. With this in view TTF-TCNQ has been chosen as one of the topics, and one article is devoted to a description of a theoretical method to treat the effects of the Coulomb interactions in TTF-TCNQ. In that article it will be also discussed how the Coulomb interactions govern the equilibrium properties of TTF-TCNQ.

The linear conjugated polymer, polyacetylene $(CH)_x$, is one of recently discovered one-dimensional organic materials. In particular, a new technique of doping other elements or molecules into polyacetylene has led to a number of interesting physical and chemical phenomena in this system. In this view one article is devoted to a theoretical description of conducting polymers, in particular the π-electronic structure and related bond alternation patterns. Various defect states such as kinks, polarons, bipolarons, and excitons are described and the physics related to these defect states is discussed.

Finally I wish to thank the authors for the thought and care they have put into their material. I would also like to express my sincere gratitude to Prof. F. Lévy, the managing editor, for his valuable suggestions and his constant encouragement.

Department of Physics
University of Tokyo

Hiroshi Kamimura

THEORETICAL ASPECTS OF CONDUCTING POLYMERS: ELECTRONIC STRUCTURE AND DEFECT STATES

D. BAERISWYL*

Physics Laboratory I, H. C. Ørsted Institute
DK-2100 Copenhagen Ø, Denmark

1. Introduction

The successful synthesis of polyacetylene films by Itô *et al.* (1974) and the subsequent discovery by Chiang *et al.* (1977) that adding dopants to the polymer can lead to an enormous increase of electrical conductivity have stimulated a great deal of interest both among chemists and physicists by opening a large new field of research. This is well documented by recent conference proceedings (Epstein and Conwell, 1981/82; Comès *et al.*, 1983) and review papers (Baeriswyl *et al.*, 1982; Etemad *et al.*, 1982). Examples of conducting polymers are shown in Figure 1. The common feature of these compounds is their planar geometry which leads to 'conjugation', i.e. delocalization of π orbitals along the chain, as illustrated by the sequences of single and double bonds. Notice that only in *trans*-polyacetylene the C—C and C=C bonds can be interchanged without changing their respective environment. *Trans*-$(CH)_x$ is therefore twofold degenerate with respect to the bond alternation sequence while the other compounds are not.

This article is devoted to the theoretical description of conducting polymers, in particular the π-electron structure and the related bond alternation patterns. The aim is not to present an exhaustive review but rather to touch on a selected class of problems which appear to be fundamental such as the question of the applicability of a single-electron picture (as compared to a many-body approach including electronic correlation), the nature of 'polaronic' states (local defect structures which are sometimes associated with solitons) and the effect of disorder (which appears to be unavoidable within the present generation of materials). Detailed derivations are given for a few points (where I felt it might be helpful), for others only the main results are mentioned together with appropriate references. Experiments are often quoted (to keep the feet on earth) but the selection is more or less arbitrary. More information about experimental aspects can be found in the reviews and proceedings mentioned above.

The article is organized as follows. In Section 2 the Hückel theory for a conjugate chain is described and the concepts of bond order and bond–bond polarizability are introduced. The system is shown to be unstable with respect to bond alternation leading to a semiconducting ground state. A critical analysis of the different choices of parameters is also given. In Section 3 the continuum limit of the Hückel model is

*Permanent Address: Institute for Theoretical Physics, ETH-Hönggerberg, CH-8093 Zürich, Switzerland.

H. Kamimura (ed.), Theoretical Aspects of Band Structures and Electronic Properties of Pseudo-One-Dimensional Solids, 1–48.

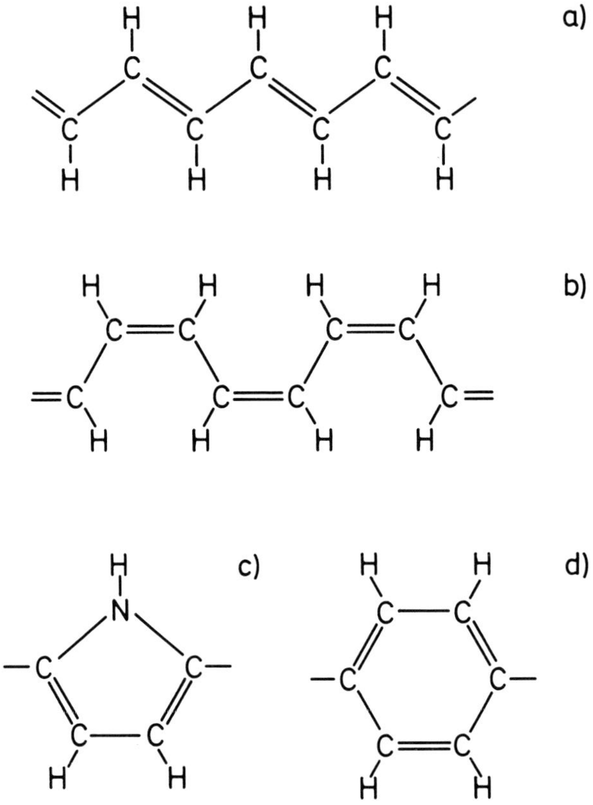

Fig. 1. Chemical structure of conducting polymers: (a) *trans*-polyacetylene, (b) *cis*-polyacetylene, (c) polypyrrole, (d) polyparaphenylene.

presented. Both the properties of the homogeneous ground state and the spectrum of small amplitude fluctuations are derived. Section 4 contains a rather detailed discussion of polaronic effects produced by the electron–phonon coupling. It is shown that various localized defect structures (kinks, polarons, bipolarons) are exact solutions of the continuum model. Furthermore, some difficulties with the soliton concept and the interpretation of experiments are mentioned. Section 5 is concerned with complications arising from various effects which are not taken into account in the Hückel model such as electron correlation, disorder and interchain coupling. It is shown that they can lead to profound changes in the ground-state configuration as well as in the defect structures.

2. Ground State Configuration (Hückel Theory)

In this section the Hückel theory for conjugated polymers is presented. It is shown how the parameters can be determined on the basis of properties of small molecules. The concepts of bond order and bond–bond polarizability play an important role. The slow spatial decay of the nonlocal polarizability, which is a typical one-dimensional effect, is responsible for the bond alternation in long chains.

2.1. MODEL AND BASIC ASSUMPTIONS

The basic assumptions of the molecular-orbital approach for conjugated hydrocarbon systems introduced by Hückel (1931, 1932) and reviewed by Salem (1966) are:

(1) to consider only the π electrons explicitly and to take into acocunt the effect of σ electrons in terms of local bond-stretching forces;
(2) to evaluate the π-electron states relative to fixed atomic configurations (adiabatic approximation) and, furthermore, to treat the atomic displacements as classical variables;
(3) to replace the many-body forces among the π electrons by an effective single-particle Hamiltonian;
(4) to choose the eigenfunctions of this Hamiltonian as linear combinations of atomic orbitals which are assumed to be orthonormal (Wannier-type) states;
(5) to neglect all matrix elements of the effective Hamiltonian except those between states associated with bonded sites (tight-binding approximation).

Introducing creation and annihilation operators c_{ns}^{+} and c_{ns}, respectively, for π electrons at site n and with spin projection s the effective single-particle Hamiltonian is

$$H = \sum_{n} f(r_n) - \sum_{ns} t(r_n)(c_{ns}^{+}c_{n+1s} + c_{n+1s}^{+}c_{ns}), \tag{1}$$

where the first term describes the elastic energy associated with σ-bond compression and the second term is the tight-binding π-electron Hamiltonian. Both the elastic energy per σ bond $f(r_n)$ and the resonance (or hopping) integral $t(r_n)$ depend on the bond length r_n.

An important general property of the π-electron Hamiltonian is its symmetry with respect to the zero of energy. Indeed it changes sign under the canonical transformation $c_{ns} \to (-1)^n c_{ns}$ and therefore, given an electronic level at ε_ν, there is necessarily also a level at $-\varepsilon_\nu$. For an even-numbered chain with one electron per site this corresponds to full electron-hole symmetry. In the case of an odd-numbered chain (with free ends) there is necessarily a level at energy zero.

2.2. BOND ORDER

It follows immediately from Equation (1) that the bond-order operator

$$p_n = \tfrac{1}{2}\sum_{s}(c_{ns}^{+}c_{n+1s} + c_{n+1s}^{+}c_{ns}) \tag{2}$$

will play a primordial role in determining the electronic structure of π-electron systems. Consider the simplest cases of ethane and ethylene (Figure 2) which have

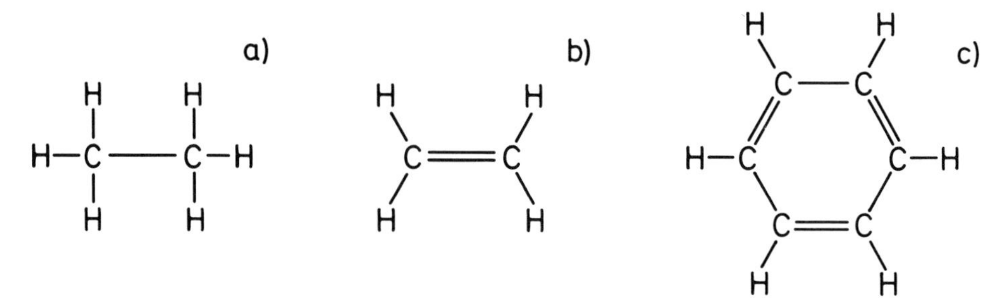

Fig. 2. Chemical structure of small organic molecules: (a) ethane, (b) ethylene, (c) benzene.

only two sites. Both the Hamiltonian and the bond order are diagonalized by the canonical transformation

$$c_{ns} = 2^{-1/2}(c_{bs} - (-1)^n c_{as}), \qquad n = 1, 2, \tag{3}$$

giving

$$p = \tfrac{1}{2}\sum_s (c_{bs}^+ c_{bs} - c_{as}^+ c_{as}) \tag{4}$$

in terms of operators associated with bonding and antibonding states. The energy levels are $\varepsilon_b = -t$ and $\varepsilon_a = +t$, respectively, which justifies *a posteriori* the choice of the minus sign in front of the π-electron Hamiltonian, Equation (1) (see also Salem, 1966, 1.5). In the case of ethane there is no π electron and therefore $p = 0$, whereas ethylene has two π electrons in the bonding state giving $p = 1$. Notice that the hypothetical case with four π electrons would again correspond to $p = 0$. Therefore, the bond order is not a monotonic function of the total number of π electrons. We obtain more insight into the meaning of the bond order upon examining its connection to the electron density described by the operator

$$\rho(\mathbf{x}) = \sum_{nm} \phi_n^*(\mathbf{x})\,\phi_m(\mathbf{x}) c_{ns}^+ c_{ms}, \tag{5}$$

where $\phi_n(\mathbf{x})$ is the Wannier (or atomic) orbital at site n. In the case of two sites the expectation value is obtained as

$$\langle \rho(\mathbf{x}) \rangle = \tfrac{1}{2} N_\pi (|\phi_1|^2 + |\phi_2|^2) + \langle p \rangle (\phi_1^* \phi_2 + \phi_2^* \phi_1), \tag{6}$$

where N_π is the total number of π electrons. In the case of ethylene this corresponds to the usual expression for the density associated with a bonding state. Equation (6) indicates that the bond order measures the degree of quantum interference in the many-particle state. There is no interference if all orbitals are empty (or filled), whereas there is maximum interference in the case of a half-filled band.

The bond order decreases slightly with the number of sites N due to the increasing delocalization of molecular orbitals. To show this we consider the case of cyclic chains with one π electron per site and equal bond lengths $r_n = r_0$. The Hamiltonian (1) is easily diagonalized by the transformation

$$c_{ns} = \frac{1}{\sqrt{N}} \sum_k \mathrm{e}^{ikna} c_{ks}, \qquad k = 2\pi\nu/(Na), \quad -N/2 < \nu \leq N/2, \tag{7}$$

where we have introduced the lattice constant $a = L/N$, L being the chain length (this is convenient since it is in accordance with the concept of a Brillouin zone, yet the result will not depend on a). The π-electron part of the Hamiltonian is

$$H_\pi = \sum_{ks} \varepsilon_k^0 c_{ks}^+ c_{ks} \quad \text{with} \quad \varepsilon_k^0 = -2t(r_0)\cos ka. \tag{8}$$

In the ground state of an even-numbered ring the levels with negative energy are doubly occupied and the positive energy levels are empty. The expectation value for the bond order is

$$\langle p_n \rangle = \frac{1}{N}\sum_{ks} \cos \text{ka} \, \langle c_{ks}^+ c_{ks} \rangle . \tag{9}$$

In the case of benzene, the levels $k = 0$ and $ka = \pm\pi/3$ are occupied, which yields $\langle p_n \rangle = 2/3$. In the limit $N \to \infty$, which corresponds to the case of an ideal polyacetylene chain, the k sum is replaced by an integral over the half-filled band $-\pi/2 < ka \leq \pi/2$, yielding $\langle p_n \rangle = 2/\pi$.

2.3. Equilibrium configuration and bond–bond polarizability

According to the adiabatic approximation and the classical treatment of the lattice, the variables r_n correspond to parameters which are determined by minimalizing the total energy. Let $\mathbf{r} = (r_1, r_2, \ldots, r_N)$ denote a given lattice configuration and $|\psi_0(\mathbf{r})\rangle$ the electronic ground state associated with this configuration. The total energy is given by

$$E(\mathbf{r}) = \langle \psi_0(\mathbf{r})| H |\psi_0(\mathbf{r})\rangle \tag{10}$$

and extrema are found from the equation

$$\frac{\partial E(\mathbf{r})}{\partial r_n} = \left\langle \psi_0(\mathbf{r}) \left| \frac{\partial H}{\partial r_n} \right| \psi_0(\mathbf{r}) \right\rangle = 0, \tag{11}$$

where the Hellmann–Feynman theorem has been used (e.g. Landau and Lifshitz, 1958a). For the Hückel model, Equations (1) and (2), this yields the equation of state

$$\frac{\mathrm{d}f(r_n)}{\mathrm{d}r_n} - 2\frac{\mathrm{d}t(r_n)}{\mathrm{d}r_n} p_n(\mathbf{r}) = 0, \tag{12}$$

where we have introduced the notation $p_n(\mathbf{r}) = \langle \psi_0(\mathbf{r})| p_n |\psi_0(\mathbf{r})\rangle$ for the expectation value of the bond order. The dynamical matrix which determines the force constants is obtained as

$$\begin{aligned} \frac{\partial^2 E}{\partial r_n \, \partial r_m} = \delta_{nm}\left[\frac{\mathrm{d}^2 f(r_n)}{\mathrm{d}r_n^2} - 2\frac{\mathrm{d}^2 t(r_n)}{\mathrm{d}r_n^2} p_n(\mathbf{r})\right] - \\ - 2\frac{\mathrm{d}t(r_n)}{\mathrm{d}r_n}\frac{\partial}{\partial r_m} p_n(\mathbf{r}). \end{aligned} \tag{13}$$

The diagonal term is a localized force constant which, in general, contains contributions from both σ- and π electrons. The nonlocal contributions are induced by the extended π-electron system and are proportional to the bond–bond polarizability (Salem, 1966)

$$\pi_{nm}(\mathbf{r}) = -\frac{\partial p_n(\mathbf{r})}{\partial t(r_m)}. \tag{14}$$

In the ground-state configuration the dynamical matrix has only positive eigenvalues. We shall see that the long-range nature of the bond–bond polarizability leads to a diverging negative contribution to the dynamical matrix in the case of equal bond lengths and for an infinitely long chain. This is the origin of the instability with respect to bond alternation.

2.4. Parametrization

There have been many different proposals how to parametrize the functions $f(r)$ and $t(r)$ which represent, respectively, the elastic energy of a σ bond and the hopping matrix element for π electrons. Due to the quite drastic approximations of the Hückel theory we cannot expect any choice to be consistent with all the empirical data. The most simple forms would be a harmonic potential about the single bond length for $f(r)$ and a linear function for $t(r)$. Equation (12) would then imply a

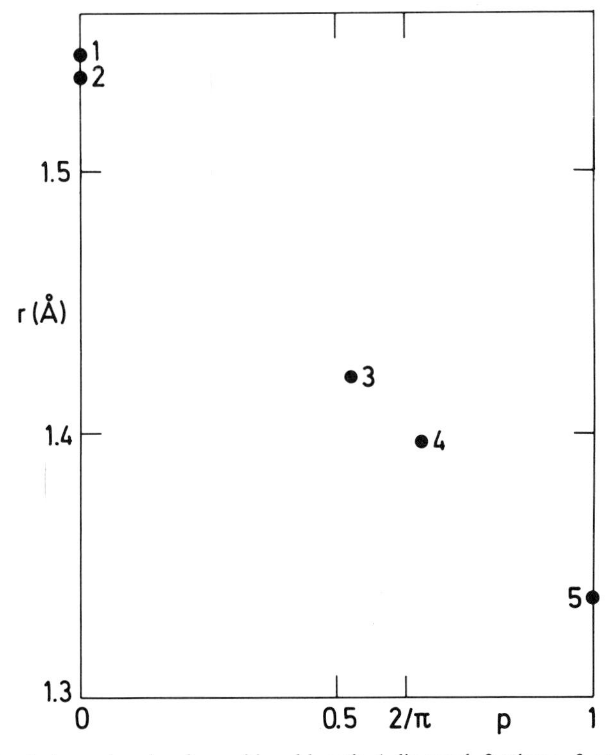

Fig. 3. Correlation between bond order and bond length: 1 diamond, 2 ethane, 3 graphite, 4 benzene, 5 ethylene.

linear relationship between bond order and equilibrium bond length. This is indeed approximately observed, as shown in Figure 3. However, such a parametrization would predict that the force constants associated with the fully symmetrical vibrational modes (where only the diagonal terms in Equation (13) are involved) should be the same for different molecules, whereas the experiments imply that these force constants vary appreciably. A more elaborate representation has been proposed by Kakitani (1974) in terms of an expansion about the single bond length by including terms up to the fourth order for $f(r)$ and up to the third order for $t(r)$ and by determining the seven coefficients from measured vibrational and electronic energies of small molecules and from the empirical bond order–bond length relationship. Using Kakitani's analysis for polyacetylene (i.e. for a bond order of $2/\pi$) we obtain an equilibrium bond length $r_0 = 1.40$ Å. Furthermore, expanding up to second order in $(r - r_0)$ we find

$$\begin{aligned} f(r) &= -\frac{4}{\pi}\alpha(r - r_0) + \tfrac{1}{2}K_0(r - r_0)^2 \\ t(r) &= t_0 - \alpha(r - r_0), \end{aligned} \tag{15}$$

where we have absorbed the second-order term in $t(r)$ within the elastic constant K_0 by replacing the bond-order operator for this contribution by its average value $2/\pi$ (this is legitimate since, according to Equation (12), the fluctuations in bond order are proportional to the fluctuations in bond length and therefore the term we have neglected is of higher order). This yields a reduction of the force constant by about 30% relative to the bare value of the σ bond. The linear term in $f(r)$ is essential for preserving the stability of the chain with respect to an overall contraction, in particular in the case of local defect structures (Vanderbilt and Mele, 1980; Su and Schrieffer, 1980).

The parameters obtained from Kakitani's analysis are given in Table I. They are consistent with values deduced by Misurkin and Ovchinnikov (1964) from the vibrational spectrum of benzene. Since the bond orders of benzene and polyacetylene are very similar, the respective Hückel parameters should indeed be essentially the same. The high-order terms in Kakitani's expansion give appreciable contribution to the force constant K_0 and thus one may wonder if the resulting parameters are not sensitive to the Ansatz chosen for $f(r)$ and $t(r)$. However, as shown be the following argument, the parameter set of Kakitani appears in fact to be very rea-

TABLE I

Hückel parameters for polyacetylene

K_0 (eV Å^{-2})	t_0 (eV)	α (eV Å^{-1})	Reference
46	2.9	4.5	Kakitani (1974)
28	2.5	4.7	Su *et al.* (1980)
90	3	9	Vanderbilt and Mele (1980)
45	2.4	6.3	Pietronero and Strässler (1981)

sonable for polyacetylene. Using the representation (15) together with Equation (12) and assuming equal bond lengths we find

$$K_0(r - r_0) + 2\alpha[p(r) - p_0] = 0. \tag{16}$$

Comparing this expression with the empirical bond order–bond length relationship (Figure 3) yields the ratio $\alpha/K_0 \approx 0.1\,\text{Å}$. On the other hand, the force constant associated with the fully symmetric vibrational mode in benzene is $K_0 = 47.5\,\text{eV}\,\text{Å}^{-2}$ (e.g. Salem, 1966). This is consistent with Kakitani's parameters but it disagrees with the values proposed by other authors (Table I).

This discrepancy originates in the different ways of determining the parameter values. Su *et al.* (1979, 1980) use the force constant of ethane for K_0 which is much smaller than the bond-stretching force constant of benzene. The coupling constant α is then determined by identifying the theoretical gap induced by bond alternation (Peierls gap) with the observed optical absorption edge. Vanderbilt and Mele (1980) determine both K_0 and α from optical properties of polyacetylene (Raman frequency and optical gap) by relating them to the Peierls mechanism. (Notice that in the SSH Hamiltonian (Su *et al.*, 1979, 1980) the parameters K_0 and α are reduced by factors of $\frac{3}{4}$ and $(\frac{3}{4})^{1/2}$, respectively, due to the zigzag structure of *trans*-polyacetylene). We will argue later that such a procedure amounts to incorporate electron–electron interactions in an average way within the single-particle Hamiltonian. The parameter set of Pietronero and Strässler (1981a) is intended to be representative for graphite. The value for K_0 is consistent both with the analysis of Kakitani and with the elastic properties of graphite. The determination of α by using simple Slater orbitals for atomic wavefunctions may be questionable and rather approximate. Nevertheless, this parameter set yields good results both for the bond-length change observed in intercalated graphite and the electrical conductivity (Pietronero and Strässler, 1981a, b).

2.5. Peierls Instability and Bond Alternation in Polyacetylene

The question of bond alternation in polymeric chains has been studied theoretically by Lennard-Jones (1937) and Coulson (1938) and advanced as a hypothesis for explaining optical spectra as a function of chain length by Kuhn (1948, 1949). But a rigorous proof that bond alternation must occur within Hückel theory has only been produced by Longuet-Higgins and Salem (1959). It is closely related to the 'Peierls theorem' (Fröhlich, 1954; Peierls, 1955) which states that a one-dimensional metal is unstable with respect to a periodic lattice distortion. In order to describe this instability we consider the dynamical matrix (13) and show that it has negative eigenvalues in the case of equal bond lengths. For $r_n = r_0$, $n = 1, \ldots, N$, the bond–bond polarizability, Equation (14), is found to be

$$\pi_{nm} = \frac{2}{N^2}\sum_{kk'} \exp[i(k - k')(n - m)a] \\ [1 + \cos(k + k')a]\frac{n_k - n_{k'}}{\varepsilon_k^0 - \varepsilon_{k'}^0}, \tag{17}$$

where ε_k^0 is the spectrum of Equation (8) and $n_k = 1$ for $\varepsilon_k^0 < 0$ and $n_k = 0$ for $\varepsilon_k^0 > 0$. In the infinite chain limit the diagonal term is obtained as

$$\pi_{nn} = -\frac{2}{\pi^2 t_0}\int_0^{\pi/2} \mathrm{d}x \int_0^{\pi/2} \mathrm{d}y \frac{1 - \cos x \cos y}{\cos x + \cos y} = -\frac{0.491}{t_0}. \tag{18}$$

Using Equation (13) and (15), this yields an effective local force constant

$$\frac{\partial^2 E}{\partial r_n^2} = K_0 - 0.982\frac{\alpha^2}{t_0} \tag{19}$$

which is positive for small coupling. Introducing the characteristic parameter

$$\lambda = \frac{2\alpha^2}{\pi t_0 K_0} \tag{20}$$

we find that the local potential is stable for $\lambda < \lambda_c = 0.648$. The critical parameter λ_c may be related to the peculiar phase transition 'by breaking of analyticity' found recently by Le Daëron and Aubry (1983) for incommensurate band filling. Notice that the parameter sets of Table I give values $0.1 \lesssim \lambda \lesssim 0.2$ for which the local potential is stable. This is in contrast to the canonical model of structural phase transitions where an unstable local potential is responsible for the lattice distortion in the low-temperature phase (e.g. Bruce, 1980). In the present case it is the non-diagonal part of the dynamical matrix which is responsible for the instability. The Fourier transform of the bond–bond polarizability (Equation (17)) can be calculated explicitly:

$$\begin{aligned}\pi(q) &= \frac{2}{N}\sum_k [1 + \cos(2k + q)a]\frac{n_k - n_{k+q}}{\varepsilon_k^0 - \varepsilon_{k+q}^0} \\ &= -\frac{2}{\pi t_0}\left(\frac{1}{\sin(qa/2)} \ln\left(\frac{1 + \sin(qa/2)}{\cos(qa/2)}\right) - 1\right].\end{aligned} \tag{21}$$

This function is shown in Figure 4. It diverges logarithmically for $q \to 2k_F = \pi/a$, which shows that the dynamical matrix (13) has negative eigenvalues for $q \to \pi/a$ however small the coupling constant α. Therefore the lattice is unstable with respect to bond alternation.

The procedure for determining the distorted ground state has been described many times (e.g. Longuet-Higgins and Salem, 1959; Su *et al.*, 1980). Inserting the Ansatz $r_n = r_0 + (-1)^n y$ into the π-electron Hamiltonian, Equation (1), and using the parametrization (15) the electronic spectrum is obtained as

$$\varepsilon_k = \operatorname{sign} \varepsilon_k^0[(2t_0)^2 \cos^2 ka + \Delta_0^2 \sin^2 ka]^{1/2}, \tag{22}$$

where $\Delta_0 = 2\alpha y$. It exhibits a gap $2\Delta_0$ at the two Fermi points, the size of which is determined by minimalizing the total energy

$$E_0 = \tfrac{1}{2}NK_0 y^2 + 2\sum_{|k|<k_F} \varepsilon_k = \frac{4N}{\pi}t_0\left[\frac{\delta^2}{4\lambda} - E(1 - \delta^2)\right], \tag{23}$$

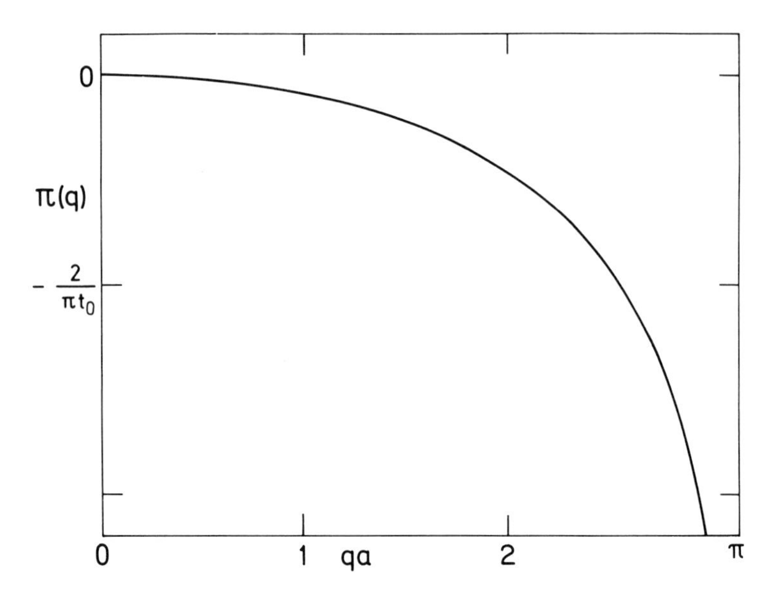

Fig. 4. Bond–bond polarizability $\pi(q)$.

where $\delta = \Delta_0/(2t_0)$ is the ratio between bandgap $E_g = 2\Delta_0$ and bandwidth $W = 4t_0$, λ is given by Equation (20) and $E(1 - \delta^2)$ is a complete elliptic integral. For small coupling (i.e. $\delta^2 \ll 1$), we can use the asymptotic expansion

$$E(1 - \delta^2) \approx 1 + \frac{\delta^2}{4}\left(2 \ln \frac{4}{\delta} - 1\right). \tag{24}$$

We notice that, however small λ, the decrease in electronic energy due to the opening of a gap always becomes larger than the increase in elastic energy for small enough δ. Minimalizing the energy with respect to δ we find the gap parameter in the weak-coupling limit

$$\Delta_0 = 2\alpha y \approx 2W/\exp\left(1 + \frac{1}{2\lambda}\right). \tag{25}$$

Up to this point the dimerized state simply represents a variational state and not necessarily a true minimum of the energy. But, as we shall now see, and as has been pointed out by Shastry (1983), this state represents indeed a solution of Equation (12) and thus an extremum of the energy. With the parametrization of Equation (15) the equation of state is given by

$$-4\alpha/\pi + K_0(r_n - r_0) + 2\alpha\langle p_n\rangle = 0. \tag{26}$$

For the dimerized state, where $(r_n - r_0) = (-1)^n\Delta_0/(2\alpha)$, the bond order is obtained as $\langle p_n\rangle = p_0 + (-1)^n p_1$, where $p_0 = 2[E(1 - \delta^2) - \delta^2/(2\lambda)]/\pi$ and $p_1 = -\delta/(\pi\lambda) = -\Delta_0 K_0/(4\alpha^2)$. Therefore, this state exhibits an alternating bond-order component, whereas the local density remains unchanged relative to the undimerized state,

$\langle \rho_n \rangle = 1$ (see Appendix A for details). This has to be contrasted to the Fröhlich model (Fröhlich, 1952, 1954; Peierls, 1955) where the electronic density is coupled to the phonons and thus the ground state exhibits an oscillating charge–density component (charge–density wave). The alternating terms in Equation (26) exactly compensate each other, whereas the constant terms do not. However, the discrepancy is very small and could easily have been eliminated by starting from a slightly different bond length r_0 which would modify the coefficient in front of the linear term in $f(r)$ (cf. Equation (15)). We conclude that, indeed, the dimerized state represents an extremum of the energy. Numerical calculations for weak couplings (Schulz, 1978; Mele and Rice, 1980a) show, furthermore, that the dynamical matrix has no negative eigenvalues in the dimerized state, which is thus a true minimum. Whether or not it represents an absolute minimum appears still to be open; on the other hand, it is hard to imagine why it should not do so in the weak-coupling limit.

One of the key questions concerning polyacetylene is if the Peierls instability is indeed the dominant mechanism leading to the observed semiconducting behaviour. The optical gap E_g in *trans*-polyacetylene has been measured by several groups. Depending on how the experiments are analysed it varies between 1.4 and 1.8 eV (e.g. Fincher *et al.*, 1979; Moses *et al.*, 1982). The bond alternation parameter y has been determined both using X-ray diffraction (Fincher *et al.*, 1982) and NMR nutation experiments (Clarke *et al.*, 1983), yielding $y \approx 0.05$ Å. This value agrees with earlier diffraction data obtained on β-carotene (Bart and MacGillavry, 1968). Using these experimental results, together with Equation (25) and a bandwidth W of 10 eV, we would obtain empirical values for the parameters α and λ, $\alpha \approx$ 7–9 eV Å^{-1}, $\lambda \gtrsim 0.2$. These are appreciably higher than the 'unbiased' parameters obtained from Kakitani's analysis (Table I). We shall later attribute this discrepancy to correlation effects. On the other hand, the parameters can easily be adjusted to reproduce the observed data. In particular, the values proposed by Vanderbilt and Mele (Table I) not only yield an electronic gap and a bond alternation parameter in agreement with experiment, but also reproduce the frequencies of the observed Raman modes (Mele and Rice, 1980a). We shall thus take the Hückel theory together with these adjusted parameters as an effective single-particle model which contains, in an averaged way, correlation effects. Such a procedure is very useful if one wants to describe complicated defect structures. But one has always to keep in mind that a coherent theory will have to treat consistently the problem of electronic correlation.

2.6. Phonons

The Hückel model as presented here describes a conjugated polymer in terms of bond length changes since, in a tight-binding scheme, these are the relevant coordinates which are coupled strongly to the π electrons. The vibrations of the lattice involve naturally all the other internal coordinates, in particular the bond angles and the CH bond length. Also, in order to calculate vibrational spectra, we have to add the kinetic energy associated with the ionic motion. This term is most simply represented in terms of Cartesian coordinates. Accordingly, one has to represent also the dynamical matrix in this basis. This transformation from internal to Cartesian coordinates which is widely used in lattice dynamics of moleules (Wilson *et al.*, 1955) and polymers (Zannoni and Zerbi, 1983a) depends on the geometrical structure of the system. As a consequence, the problem becomes quite complicated in general; moreover, it depends on many parameters (see Zannoni and Zerbi, 1983a, for a recent review). On the other hand, if we are not interested in the detailed structure of dispersion curves but only in the gross behaviour of those modes which are strongly coupled to the π electrons the problem can be considerably simplified, at least in the case of *trans*-polyacetylene. Consider a *trans*-$(CH)_x$ chain and assume that the CH groups move rigidly together. It is easy to convince oneself that displacements of these groups associated with bond length alternation only involve motion parallel to the chain axis if we keep the bond angles fixed. Therefore, the ground-state configuration which does not deform bond angles can equally well be described in terms of the Cartesian components parallel to the chain axis. This consideration leads to the SSH model (Su *et al.*, 1979, 1980) defined by the Hamiltonian

$$H = \tfrac{1}{2}M\sum_n \dot{u}_n^2 + \tfrac{1}{2}K_0'\sum_n (u_{n+1} - u_n)^2 - \\ - \sum_{ns} t_{n,n+1}(c_{ns}^+ c_{n+1s} + c_{n+1s}^+ c_{ns}), \tag{27}$$

where u_n describes the displacement of the nth CH group parallel to the chain axis and the connection to the Hückel model is given in Table II. The lattice dynamics is much simpler in the SSH model than in the full Hückel theory, but also quite approximate. In fact, this simplification rests on the (quite drastic) assumption that there are eigenvectors which are well described in terms of the displacements u_n and that all the other modes do not involve these coordinates. Consider a vibration where neighbouring atoms move in opposite direction, $\delta u_n = (-1)^n \, \delta u$. This is a simple bond-stretching mode and thus the associated potential energy can easily be calculated from Equation (23). In the weak-coupling limit ($\lambda \ll 1$) one obtains a mode frequency

$$\Omega_0^2 = 2\lambda\omega_0^2, \tag{28}$$

where $\omega_0^2 = 4K_0'/M$ corresponds to the bare frequency of the undistorted system. The explicit numerical calculations involving all in-plane vibrational modes show that the limitation to the coordinates u_n is a crude approximation (Inagaki *et al.*,

TABLE II

Connection between Hückel and SSH parameters

Parameter	Hückel	SSH	Connection
Distortion parameter	$y = (-1)^n(r_n - r_0)$	$u = (-1)^n u_n$	$u = -y/\sqrt{3}$
Elastic energy	$\frac{1}{2}K_0(r_n - r_0)^2$	$\frac{1}{2}K_0'(u_{n+1} - u_n)^2$	$K_0' = 3K_0/4$
Resonance integral	$t(r_n) = t_0 - \alpha(r_n - r_0)$	$t_{n,n+1} = t_0 - \alpha'(u_{n+1} - u_n)$	$\alpha' = \sqrt{3}\,\alpha/2$

1975; Mele and Rice, 1980a). In particular, the stretching mode is split into one involving mostly the C—C bonds and another involving mostly the C=C bonds. This splitting which is actually observed is therefore not a signature for bond alternation (Rabolt *et al.*, 1979). Nevertheless, the softening predicted by Equation (28) is also found in the more complete lattice-dynamical calculations (Mele and Rice, 1980a).

3. The Continuum Limit

The model presented in Section 2 describes a coupled electron–phonon system on a discrete one-dimensional lattice, where the one-dimensionality is not to be taken literally (i.e. in a geometrical sense) but rather dynamically: every lattice point is coupled to two neighbouring points. We have seen that the Peierls mechanism leads to a gap at the Fermi level. For weak coupling, i.e. $\lambda \ll 1$, only states close to $\pm k_F$ are affected, namely within a region $|\hbar v_F \Delta k| \lesssim \Delta_0$, where $\hbar v_F = |d\varepsilon_k^0/dk|_{k_F} = 2t_0 a$. This introduces a characteristic length $\xi = \hbar v_F/\Delta_0 = aW/E_g \approx 7a$. Quite generally we expect, therefore, that electrons and phonons will strongly interact only within fluctuations of length scales $\gtrsim \xi$. For these fluctuations the discreteness of the lattice (with lattice constant $a \ll \xi$) should play a minor role and the continuum description in terms of field variables should be appropriate. The continuum limit has the advantage that many results can be obtained analytically (Brazovskii, 1978; Takayama *et al.*, 1980; Horovitz, 1981a; Brazovskii and Kirova, 1981; Campbell and Bishop, 1981), whereas the discrete model requires numerical computations (Su *et al.*, 1980; Su and Schrieffer, 1980).

3.1. Definition of the Fields and Appropriate Boundary Conditions

The Hückel description in terms of bond lengths r_n and operators c_{ns} is mapped onto a continuous line as follows. We associate with any bond between sites n and $n+1$ a point $x_n = (n + \frac{1}{2})a$ and represent the bond-length changes in terms of the function

$$y(x_n) = (-1)^n (r_n - r_0) \tag{29}$$

which is constant in the ground state and expected to be slowly varying for the relevant fluctuations. Correspondingly we introduce operators

$$\begin{aligned} \psi_{1s}(x_n) &= \frac{1}{\sqrt{L}} \sum_k e^{ikx_n} c_{k_F+ks} \\ \psi_{2s}(x_n) &= \frac{-i}{\sqrt{L}} \sum_k e^{ikx_n} c_{-k_F+ks} \end{aligned} \tag{30}$$

for right- and left-moving electrons, respectively, where $L = Na$ and $-\pi/(2a) < k \leq \pi/(2a)$. For states where the k-values are restricted to the neighbourhood of $\pm k_F$ we can expand the operators c_{ns} and c_{n+1s} with respect to the bond coordinate x_n. Using Equation (A2) of Appendix A we find

$$
\begin{aligned}
c_{ns} &\approx \sqrt{a}\left(1 - \tfrac{1}{2}a\frac{\partial}{\partial x_n}\right)[i^n \psi_{1s}(x_n) + (-i)^{n-1} \psi_{2s}(x_n)] \\
c_{n+1s} &\approx \sqrt{a}\left(1 + \tfrac{1}{2}a\frac{\partial}{\partial x_n}\right)[i^{n+1} \psi_{1s}(x_n) + (-i)^{n} \psi_{2s}(x_n)].
\end{aligned}
\tag{31}
$$

This expansion will enable us to represent observables like the bond order or the current density in terms of local operators. Notice that this mapping does not depend on the geometrical form of the chain. Also the 'lattice constant' a is in principle arbitrary, but we choose it as L/N where L is the length of the system as measured along the chain direction.

The rapidly varying phase factors in Equation (29) and (31) have important consequences on the boundary conditions for the fields as has been pointed out by Kivelson *et al.* (1982). We assume periodic boundary conditions for the bond lengths, $r_N = r_0$, as well as for the electrons, $c_{Ns} = c_{0s}$ and $c_{N+1s} = c_{1s}$. Using Equation (29) this implies

$$
y(x_N) = (-1)^N y(x_0) \tag{32}
$$

which shows that in odd chains a misfit at the chain ends is prevented only if the function $y(x_n)$ changes sign somewhere. Using Equations (31) the boundary conditions for the electron fields are found to be

$$
\begin{aligned}
\psi_{1s}(x_N) &= (-i)^N \cdot \psi_{1s}(x_0) \\
\psi_{2s}(x_N) &= i^N \cdot \psi_{2s}(x_0).
\end{aligned}
\tag{33}
$$

This will have important effects on the extended states of odd chains. Observables like bond order, charge or current density nevertheless satisfy still periodic boundary conditions (Kivelson *et al.*, 1982).

3.2. Bogoliubov–de Gennes Equations and Self-Consistency Condition

Above we defined fields on a discrete set of points x_n. Going over to the continuum limit, $x_n \to x$, corresponds to including more and more Fourier coefficients both for the phonons and the electrons. In order to avoid unphysical divergences due to an infinite Fermi sea we shall have to use a cut-off wavevector Λ which can be associated with an effective bandwidth. This cut-off will be essential for the properties of the ground state, in particular the value of the gap parameter Δ_0 which, according to Equation (25), is proportional to the bandwidth. On the other hand, the fluctuations will be cut-off independent in the weak-coupling limit.

Extending, thus, the domain of k values in Equations (30) to $\pm\infty$ and taking the limit $L \to \infty$ the fields $\psi_{is}(x)$ are found to satisfy the anticommutation relations

$$
\begin{aligned}
\{\psi_{is}(x), \psi_{js'}(x')\} &= \{\psi_{is}^{+}(x), \psi_{js'}^{+}(x')\} = 0 \\
\{\psi_{is}(x), \psi_{js'}^{+}(x')\} &= \delta_{ij}\, \delta_{ss'}\, \delta(x - x'),
\end{aligned}
\tag{34}
$$

where we have assumed that $\{c_{k_F+ks}, c^{+}_{-k_F+k's}\} = 0$ for all k, k', like in the

Tomonaga–Luttinger model (cf. Mattis and Lieb, 1965). Using the parametrization of Equations (15), the Hamiltonian (1) takes the following form in the continuum limit

$$H = \sum_s \int \mathrm{d}x\, \Phi_s^+ \left(-i\hbar v_F \sigma_3 \frac{\mathrm{d}}{\mathrm{d}x} + \Delta \sigma_1 \right) \Phi_s + (2\pi\hbar v_F \lambda)^{-1} \int \mathrm{d}x\, \Delta^2, \tag{35}$$

where we have used spinor notation $\Phi_s = \begin{pmatrix} \psi_{1s} \\ \psi_{2s} \end{pmatrix}$. The gap parameter is defined as $\Delta(x) = 2\alpha y(x)$ and σ_1, σ_3 are Pauli matrices. We have neglected terms of the form $\Sigma_n (-1)^n f(x_n)$ according to our assumption of smooth spatial variations.

Similar Hamiltonians have been studied in the context of inhomogeneous superconducting structures (de Gennes, 1966) and the method used there can be taken over for the present problem. It consists of two steps: first, given a certain field configuration $\Delta(x)$, one determines the electronic eigenstates; second, one has to find that particular configuration which corresponds to a minimum of the total energy. The first step produces the Bogoliubov–de Gennes (BdG) equations, the second step yields the self-consistency condition or gap equation.

The BdG equations are obtained as follows. Labelling the eigenstates of H by ν we expand the fields ψ_{is} in terms of operators $a_{\nu s}$ associated with these states ($a_{\nu s}$ annihilates an electron in the energy level ν and with spin projection s)

$$\begin{aligned} \psi_{1s}(x) &= \sum_\nu u_\nu(x) a_{\nu s} \\ \psi_{2s}(x) &= \sum_\nu v_\nu(x) a_{\nu s}. \end{aligned} \tag{36}$$

The electronic part of the Hamiltonian is given by the first term of Equation (35) or, in terms of operators $a_{\nu s}$, by

$$H_e = \sum_{\nu s} \varepsilon_\nu a_{\nu s}^+ a_{\nu s}. \tag{37}$$

The equations determining, respectively, the wave function amplitudes $u_\nu(x)$, $v_\nu(x)$ and the energy levels ε_ν are obtained by calculating the commutators $[H_e, \psi_{is}(x)]$ for the two forms of the Hamiltonian, Equations (35) and (37), and by identifying the two expressions. This yields the BdG equations

$$\begin{aligned} \varepsilon_\nu u_\nu &= -i\hbar v_F \frac{\mathrm{d}u_\nu}{\mathrm{d}x} + \Delta v_\nu \\ \varepsilon_\nu v_\nu &= i\hbar v_F \frac{\mathrm{d}v_\nu}{\mathrm{d}x} + \Delta u_\nu. \end{aligned} \tag{38}$$

An alternative way is to insert the expansion (36) into Equation (35) and to require the resulting expression to be identical to Equation (37). This yields, in addition to the BdG equations, the normalization condition

$$\int \mathrm{d}x (|u_\nu|^2 + |v_\nu|^2) = 1. \tag{39}$$

In order to derive the self-consistency condition we have to determine the total energy (or the free energy at finite temperatures) associated with the configuration $\Delta(x)$

$$E\{\Delta(x)\} = \sum_{\nu s}{}' \varepsilon_\nu + (2\pi\hbar v_F\lambda)^{-1}\int dx\,\Delta^2, \tag{40}$$

where the prime denotes summation over occupied levels. The spectrum ε_ν is a functional of $\Delta(x)$. Multiplying the BdG equations (38) by u_ν^* and v_ν^*, respectively, integrating and adding, we have

$$\varepsilon_\nu \int dx(|u_\nu|^2 + |v_\nu|^2) =$$
$$\int dx\left[-i\hbar v_F\left(u_\nu^*\frac{du_\nu}{dx} - v_\nu^*\frac{dv_\nu}{dx}\right) + \Delta(u_\nu^* v_\nu + u_\nu v_\nu^*)\right]. \tag{41}$$

A variation of this equation with respect to u_ν (or v_ν) for fixed Δ reproduces the BdG equations provided that $\delta\varepsilon_\nu/\delta u_\nu = 0$. Therefore, ε_ν is stationary with respect to variations of u_ν or v_ν and we find, taking into account the normalization condition (39),

$$\delta\varepsilon_\nu/\delta\Delta = u_\nu^* v_\nu + u_\nu v_\nu^*. \tag{42}$$

The variation of the total energy, Equation (40), then yields the self-consistency condition

$$\Delta(x) = -\pi\hbar v_F\lambda \sum_{\nu s}{}'(u_\nu^* v_\nu + v_\nu^* u_\nu). \tag{43}$$

The same equation is obtained from the continuum limit of Equation (12). Alternatively, one may also derive it by using the Hellmann–Feynman theorem with H given by Equation (35).

3.3. Ground State and Small Amplitude Fluctuations

It is straightforward to study the Peierls distortion in the continuum limit. Assuming a homogeneous order parameter ($\Delta = \Delta_0$) the electronic states derived from Equations (38), normalized according to Equation (39), are plane waves

$$\begin{pmatrix} u_{k\sigma} \\ v_{k\sigma} \end{pmatrix} = (2L)^{-1/2} e^{ikx} \begin{pmatrix} [1 + \sigma\sin\phi_k]^{1/2} \\ \sigma[1 - \sigma\sin\phi_k]^{1/2} \end{pmatrix} \tag{44}$$

with a spectrum $\varepsilon_{k\sigma} = \sigma E_k$, where $\sigma = \pm 1$ discerns conduction and valence band, $E_k = [(\hbar v_F k)^2 + \Delta_0^2]^{1/2}$, $\sin\phi_k = \hbar v_F k/E_k$ and $\cos\phi_k = \Delta_0/E_k$. Inserting Equation (44) into Equation (43) we obtain the gap equation

$$1 = 2\hbar v_F\lambda \int_0^\Lambda dk \frac{1}{E_k} \tag{45}$$

with the solution

$$\Delta_0 = \frac{\hbar v_F \Lambda}{\sinh[(2\lambda)^{-1}]} \approx 2\hbar v_F \Lambda \exp\left(-\frac{1}{2\lambda}\right) \tag{46}$$

in the weak-coupling limit. This is very similar to the corresponding expression for the discrete model, Equation (25). The condensation energy per unit length is found to be

$$[E(\Delta_0) - E(0)]/L \approx -\Delta_0/(2\pi\xi) \tag{47}$$

in terms of the characteristic length $\xi - \hbar v_F/\Delta_0$.

In order to study fluctuations we consider the function

$$D(x - x') = \frac{\delta^2 E}{\delta\Delta(x)\,\delta\Delta(x')} = (\pi\hbar v_F \lambda)^{-1}\,\delta(x - x') +$$
$$+ \frac{\delta^2}{\delta\Delta(x)\,\delta\Delta(x')}\sum_{\nu s}{}' \varepsilon_\nu \tag{48}$$

in close analogy to the dynamical matrix introduced previously (Equation (13)). The electronic contribution which is proportional to the bond–bond polarizability (Equation (14)) provides the screening of the elastic forces. The perturbation expansion yields for the Fourier transform of $D(x - x')$

$$D(q) = (\pi\hbar v_F\lambda)^{-1} -$$
$$- \frac{2i}{\hbar}\sum_{\alpha\beta}\int\frac{dk}{2\pi}\int\frac{d\omega}{2\pi}G_{\alpha\beta}(k, \omega)G_{-\alpha,-\beta}(k + q, \omega), \tag{49}$$

where the indices $\alpha = \pm 1$, $\beta = \pm 1$ in the electron Green's functions $G_{\alpha\beta}(k, \omega)$ label the two types of fermion fields ('plus' for type 1, 'minus' for type 2 particles). Notice that, in contrast to the incommensurate limit which has been studied by Lee *et al.* (1974), we do not have to deal with a matrix $D_{\alpha\beta}$ since the points $2k_F + q$ and $-2k_F + q$ are equivalent for a half-filled band. Therefore, there is no phase mode in the present case. We find

$$D(q) = (\pi\hbar v_F\lambda)^{-1} - \frac{1}{\pi}\int_{-\Lambda}^{\Lambda} dk\frac{1 - \cos(\phi_k + \phi_{k+q})}{E_k + E_{k+q}} \tag{50}$$

which is much simpler than the corresponding expression for the discrete system (cf. Piseri *et al.*, 1982). We now show that $D(q) > 0$, which demonstrates that the Peierls-distorted state is (linearly) stable. Using the gap equation (45) to eliminate the unphysical divergence of the integral for $\Lambda \to \infty$, the remaining expression depends only weakly on the cut-off (provided that $|q| \ll \Lambda$, $\xi^{-1} \ll \Lambda$) and thus we evaluate it for $\Lambda \to \infty$. The integration gives the simple formula

$$D(q) = \frac{2}{\pi\hbar v_F}\eta_q \coth \eta_q, \tag{51}$$

where $\eta_q = \text{Arsh}(\xi q/2)$. It has a minimum for $q = 0$ with $D(0) > 0$ and increases monotonically with $|q|$. This proves that the state with bond alternation corresponds to a minimum of the energy. For $|q\xi| \ll 1$ we obtain the expansion

$$D(q) \sim \frac{2}{\pi\hbar v_F}\left[1 + \frac{1}{12}\xi^2 q^2 - \frac{1}{120}\xi^4 q^4 + O(\xi^6 q^6)\right], \tag{52}$$

which indicates that the characteristic interaction range is ξ. This has some consequences for attempts of constructing effective field theories in which the electrons are 'integrated out'. Gradient expansions based on Equation (52) will give reliable results only for fluctuations with wavelengths longer than ξ. For defects with an extent of the order of ξ, local field theories such as the ϕ^4 model (Rice, 1979) can yield at most a qualitative description.

For $|q\xi| \gg 1$ we find from Equation (51) and the gap equation (46) the asymptotic behaviour

$$D(q) \sim \frac{2}{\pi\hbar v_F}\left(\ln\frac{q}{2\Lambda} + \frac{1}{2\lambda}\right) \tag{53}$$

which is identical to the form obtained in the gapless case. Interestingly enough it also agrees with the corresponding quantity in the discrete case

$$[K_0/(2\alpha^2 a) + \pi(2k_F - q)]/(2a). \tag{54}$$

This expression becomes identical to Equation (53) if we insert the polarizability (21) for wavevectors close to $2k_F$ and if we choose the cut-off as $\Lambda a = 2/e$ (this also brings the gap equations into agreement). We conclude that for weak coupling (i.e. $\xi \gg a$) the continuum limit yields a very good representation of the discrete model, both for the ground state and for fluctuations with $|qa| \ll 1$.

In order to get the spectrum we have to add the kinetic energy of the lattice which, if written in terms of internal coordinates, depends on the geometrical structure of the polymer. For *trans*-$(CH)_x$ we use the continuum limit of the SSH Hamiltonian (27) to get

$$E_{kin} = (2\pi\hbar v_F \lambda\omega_0^2)^{-1}\int dx\,\dot{\Delta}^2, \tag{55}$$

where ω_0 is the bare phonon frequency, as in Equation (28). Together with Equations (48) and (52) this yields a spectrum

$$\begin{aligned}\Omega_q^2 &= \omega_0^2\,\pi\hbar v_F \lambda\, D(q)\\ &\sim 2\lambda\omega_0^2\left(1 + \frac{1}{12}\xi^2 q^2\right) \qquad \text{for} \quad |\xi q| \ll 1\end{aligned} \tag{56}$$

in agreement with Equation (28) and also with the amplitude mode dispersion of Lee *et al.* (1974) (besides a negligible nonadiabatic term in their expression; notice also that 2λ has to be replaced by λ in the incommensurate limit).

Unfortunately the complicated morphology of polyacetylene films makes it very difficult to measure phonon dispersion curves by using inelastic neutron scattering. There is, however, indirect evidence for positive dispersion from Raman experiments. Certain Raman lines have shoulders on the high-frequency side which depend strongly on the wavelength of the incident light. Two different models have been proposed for explaining this effect. In the first model it is traced back to selective resonance enhancement of chains with varying conjugation lengths (Harada *et al.*, 1978; Kuzmany, 1980; Lichtmann *et al.*, 1979/1980). In the second model it is suggested that the dominant process involved may be hot luminescence rather than resonant Raman scattering (Mele, 1982). It seems not yet clear which of the two is the main mechanism responsible for the peculiar line shapes (cf. Lauchlan *et al.*, 1983; Mulazzi *et al.*, 1983) but in both models the electron–phonon coupling and the positive phonon dispersion play an essential role.

4. Intrinsic Defects

The energy associated with a homogeneous bond alternation is an even function of the distortion parameter y (cf. Equation (23)). Therefore, there are two degenerate minima which correspond to the two possible bond alternation sequences: The even-numbered bonds are either longer ($y > 0$) or shorter ($y < 0$) than the odd-numbered bonds. This degeneracy appears, strictly speaking, only in *trans*-$(CH)_x$ since for all the other materials the bonds are not all equivalent. Thus the two possible forms, *cis*-transoid (Figure 1) and *trans*-cisoid (obtained from *cis* by interchanging single and double bonds), have different energy. This can be taken into account by associating different resonance integrals with the inequivalent bonds (cf. Section 4.2). In the exactly degenerate case one expects that domain walls connecting a 'phase' with $y > 0$ to one with $y < 0$ will be the relevant excitations at finite temperatures besides the small amplitude fluctuations. However, the energy of such a domain wall turns out to be too large ($2\Delta/\pi$) for being thermally excited at room temperature. The reason why, nevertheless, the domain walls may play a role in polyacetylene is the defected nature of the material. It appears that during the processes of polymerization and *cis–trans* isomerization various types of defects are introduced such as cross-links, chain-twists or impurities. Some of them interrupt the conjugation, restricting thus the extent of the π orbitals. The conjugation lengths may be still large enough for preserving quasi-continuous bands, but there will be a clear distinction between even- and odd-numbered chains. An odd-numbered chain with strong double bonds at the ends (or one with free boundary conditions, cf. Su, 1980) contains necessarily a domain wall in the inner part. The topological constraints (the number of sites together with the boundary conditions) are necessary prerequisites for the existence of domain walls, their particular internal structure, however, is determined by the unperturbed Hamiltonian. It is in this sense that we are using the term 'intrinsic defects'.

An odd-numbered chain exhibits necessarily an electronic state at midgap which is singly occupied if the chain is neutral. We shall see that this lonely electron is localized within the extent of the domain wall. This combination of self-localization of a carrier and local lattice distortion has much in common with the conventional polaron concept. However, there exists another intrinsic defect to which the term 'polaron' has been applied specifically (Brazovskii and Kirova, 1981; Campbell and Bishop, 1981). It is the collective state which is produced upon adding or subtracting a single electron from an even-numbered chain. To distinguish these two types of intrinsic defects we shall use the term 'kink' for domain walls.

4.1. The kink solution

Bond alternation defects interpolating between the two degenerate ground states (Figure 5) have been discussed a long time ago and several of their particular properties – such as their high mobility, their extent over the characteristic healing length (Longuet-Higgins and Salem, 1959; Ooshika, 1959) and the associated electronic midgap state leading to particular spin–charge relations (Pople and Walmsley, 1962; Kventsel and Kruglyak, 1968) – have been anticipated, at least qualitatively, before the recent excitement about these objects. The renewed interest was stimulated by the observation of highly mobile spins in *trans*-polyacetylene (Goldberg *et al.*, 1979). In the following, detailed theories were developed for describing quantitatively these defects (Rice, 1979; Su *et al.*, 1979).

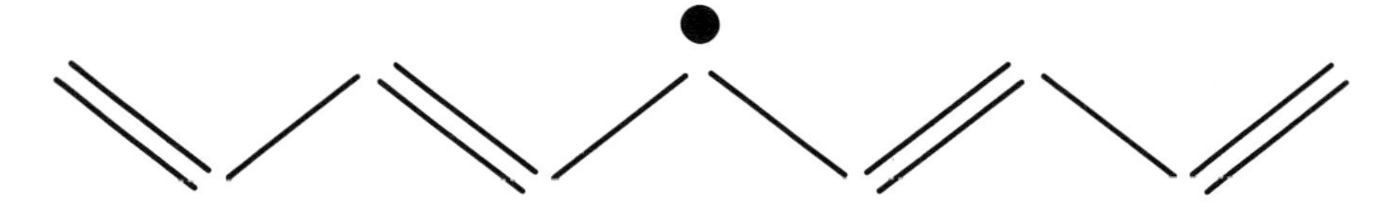

Fig. 5. Bond alternation defect in *trans*-polyacetylene.

Su *et al.* (1979, 1980) determined numerically the stable configuration of the defect by choosing an Ansatz $u_n = (-1)^n u_0 \tanh(n/l)$ and by varying the length l. They obtained an extent $l \approx 7$ and an energy $E_K \approx 0.6\Delta_0$. One of the remarkable properties of the continuum model is its ability to produce analytical results for the intrinsic defect structures. For the kink defect this has been independently shown by Brazovskii (1978), Takayama *et al.* (1980) and Rice *et al.* (1980). Kivelson *et al.* (1982) showed how to choose the boundary conditions in order to avoid divergences in band-to-band transitions; the derivation given below is based on their work.

We want to show that the Ansatz

$$\Delta(x) = \Delta_0 \tanh(x/\xi), \tag{57}$$

where $\xi = \hbar v_F/\Delta_0$, yields an exact solution of the continuum model. Since this is an odd function, i.e. $\Delta(-L/2) = -\Delta(L/2)$, we choose, in agreement with Equation (32) and (33), the following boundary conditions for the wavefunctions

$$\begin{aligned} u(L/2) &= -iu(-L/2) \\ v(L/2) &= iv(-L/2). \end{aligned} \tag{58}$$

Following Takayama *et al.* (1980) we introduce the auxiliary functions $f_\pm = u_\nu \pm iv_\nu$. Omitting the index ν the BdG equations (38) become

$$\begin{aligned} \varepsilon f_+ &= \left(-i\hbar v_\mathrm{F}\frac{\mathrm{d}}{\mathrm{d}x} + i\Delta\right) f_- \\ \varepsilon f_- &= \left(-i\hbar v_\mathrm{F}\frac{\mathrm{d}}{\mathrm{d}x} - i\Delta\right) f_+. \end{aligned} \tag{59}$$

Differentiating and rearranging we obtain the second-order differential equation

$$\left[(\hbar v_\mathrm{F})^2\frac{\mathrm{d}^2}{\mathrm{d}x^2} + \varepsilon^2 - \Delta^2 \pm \hbar v_\mathrm{F}\frac{\mathrm{d}\Delta}{\mathrm{d}x}\right] f_\pm = 0. \tag{60}$$

Noticing that the Ansatz (57) satisfies the relation $\Delta^2 + \hbar v_\mathrm{F}\,\mathrm{d}\Delta/\mathrm{d}x = \Delta_0^2$ we obtain a simple wave equation for f_- with solutions

$$\begin{aligned} f_- &= A_k\,\mathrm{e}^{ikx} \\ f_+ &= A_k\,\mathrm{e}^{ikx}(\hbar v_\mathrm{F}k + i\Delta)/\varepsilon, \end{aligned} \tag{61}$$

where $A_k = (L - \xi\cos^2\phi_k)^{-1/2}$ according to the normalization (39) and $\varepsilon = \pm E_k = \pm[(\hbar v_\mathrm{F}k)^2 + \Delta_0^2]^{1/2}$. The boundary conditions (58) imply $f_\pm(-L/2) = -if_\mp(L/2)$ and thus

$$kL = \begin{cases} 2n\pi + \phi_k & \text{for } \varepsilon > 0 \\ (2n+1)\pi + \phi_k & \text{for } \varepsilon < 0, \end{cases} \tag{62}$$

where $\phi_k = \tan^{-1}(\hbar v_\mathrm{F}k/\Delta_0)$, $-\pi/2 < \phi_k \leqslant \pi/2$. We have the same energy spectrum as in the homogeneous case, but with a modified density of states

$$\mathrm{d}n/\mathrm{d}k = (2\pi A_k^2)^{-1}. \tag{63}$$

It follows that the number of states has decreased by $\frac{1}{2}$ per band, which is compensated by the midgap state ($\varepsilon = 0$)

$$f_- = 0, \qquad f_+ = \xi^{-1/2}\,\mathrm{sech}(x/\xi), \tag{64}$$

a particular solution of Equations (59) and (60). It is now easy to show that the self-consistency equation (43) is exactly fulfilled for a full valence and an empty conduction band. Rewriting it as

$$\Delta(x) = \tfrac{1}{2}i\pi\hbar v_\mathrm{F}\lambda\sum_{\nu s}{}'(f_+f_-^* - f_+^*f_-)_\nu \tag{65}$$

we notice that the r.h.s. is independent of the occupation of the midgap state and given by

$$\Delta(x)\,2\hbar v_\mathrm{F}\lambda\int_0^\Lambda \mathrm{d}k\frac{1}{E_k} = \Delta(x) \tag{66}$$

in view of the gap equation (45). This completes the proof that the kink Ansatz (57) provides an exact (inhomogeneous) state of the continuum model.

The extent of the kink defect (and of the wave function of the midgap state) is given by $\xi = \hbar v_F/\Delta_0 = aW/E_g \approx 7a$, in agreement with the numerical result of Su *et al.* (1980). In order to determine the kink energy we consider first an even-numbered chain where kinks and antikinks can be created in pairs. For a well-separated kink–antikink pair the valence band has one state (two electrons) less than in the ground-state configuration. In the continuum limit this can be taken into account by choosing slightly different cut-offs for the two cases. For an odd-numbered chain with a single kink it follows that the cut-off Λ' has to be chosen such that one electron is removed from the valence band. Using Equation (63), Λ' is related to Λ by

$$L\Lambda' - \phi(\Lambda') = L\Lambda - \pi/2. \tag{67}$$

With $\phi(\Lambda') \approx \pi/2 - (\Lambda\xi)^{-1}$ we obtain

$$\Lambda' \approx \Lambda - (L\Lambda\xi)^{-1}. \tag{68}$$

It is now straightforward to calculate the kink energy from Equation (40). One finds

$$E_K = \frac{1}{\pi}\Delta_0\left[-\frac{1}{\lambda} + \hbar v_F \int_{-\Lambda}^{\Lambda} \mathrm{d}k \frac{1}{E_k} + 2L\int_{\Lambda'}^{\Lambda} \mathrm{d}k \frac{E_k}{\Delta_0}\right], \tag{69}$$

where the first term represents the decrease in elastic energy and the other terms the increase in electronic energy. The first two terms compensate each other in view of the gap equation (45) and the remaining integral gives $E_K = 2\Delta_0/\pi$, in good agreement with the numerical result of Su *et al.* (1980). Due to the modification of the cut-off, our implicit assumption that Δ_0 is not modified by the presence of a kink is not strictly valid. However, this effect would be of order $(\Lambda\xi)^{-2}$ and can be neglected.

The kink structure is illustrated in Figure 6. The electron–hole symmetry implies that for the neutral kink K^0 the midgap level is singly occupied and thus the defect has a spin $\frac{1}{2}$. On the other hand, the charge defects K^+, K^- do not carry spin. It is these unusual spin–charge assignments which make the kink defect so exotic. It becomes even more exotic in the case of spinless fermions where the energy levels can at most be singly occupied since then charge neutrality requires an occupation number $\frac{1}{2}$ for the midgap level. This problem of charge fractionalization represents

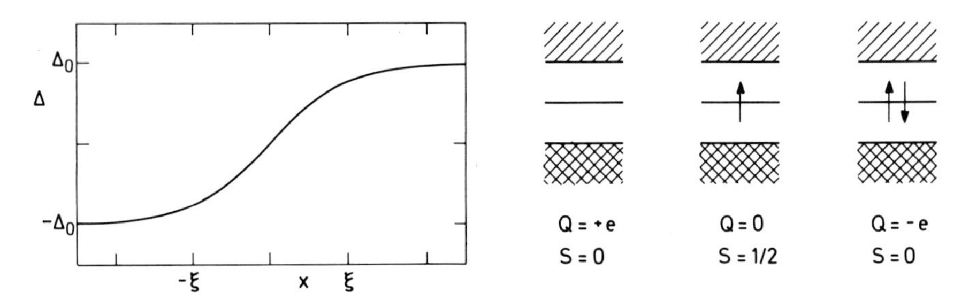

Fig. 6. Kink configuration and allowed level fillings.

a fascinating field in itself and much effort is devoted to it at present (cf. Schrieffer, 1983; Laughlin, 1983).

The kink mass M_K can be calculated from the kinetic energy of the lattice associated with a moving defect, in the spirit of the adiabatic approximation. Using Equation (27) (or Equation (55)) it is related to the mass M of a CH group by $M_K = (4u^2/3a\xi)M$. With the values $u = 0.03\,\text{Å}$, $\xi = 7a$, $a = 1.2\,\text{Å}$ we obtain $M_K \approx 3m_e$, where m_e is the free electron mass, which implies a high mobility. In the continuum model a kink can move freely due to the translational symmetry. This does not remain true in the discrete model where a moving defect has to overcome a small energy barrier. This pinning energy due to lattice discreteness has been estimated to be about 2 meV (Su *et al.*, 1979), and therefore this effect is expected to become relevant only at very low temperatures, $T \lesssim 20\,K$.

4.2. Polarons, bipolarons, and excitons

The kink solution represents a large amplitude defect which satisfies the boundary condition $\Delta(L/2) = -\Delta(-L/2)$. It is natural to look for similar large amplitude defects where the order parameter does not change sign. Such states do indeed exist and an exact analytical representation of their structure can again be given in the continuum limit. The most simple of these defects is the polaron which is created by adding an electron (or a hole) to the ground state and by allowing the system to relax locally. It has been observed in numerical simulations by Su and Schrieffer (1980). Subsequently, the continuum solution has been derived by Brazovskii and Kirova (1981) and by Campbell and Bishop (1981).

Let us first generalize the model to include the nondegenerate *cis* form. We simply represent the inequivalent bonds by different resonance integrals $t_0 \pm t_1$, where t_1 accounts for the modification of π orbitals due to the field produced by the hydrogen atoms. Allowing again for bond alternation the resonance integrals are then given by $t(r_n) = t_0 - (-1)^n(t_1 + \alpha y_n)$. This leads to a continuum order parameter $\Delta(x) = \Delta_i(x) + \Delta_e$, where $\Delta_i(x) = 2\alpha y(x)$ represents the contribution of the lattice distortion and $\Delta_e = 2t_1$ the external field. The BdG equations (38) remain the same, but the self-consistency equation (43) is changed due to the modification of the elastic energy (Δ has to be replaced by $\Delta_i = \Delta - \Delta_e$ in Equation (40)), yielding

$$\Delta(x) = \Delta_e - \pi\hbar v_F\lambda{\sum_{\nu s}}'(u_\nu^* v_\nu + v_\nu^* u_\nu). \tag{70}$$

Using the same procedure as in Section 3.3, the gap equation for the homogeneous ground state $\Delta = \Delta_0$ is readily derived. We obtain

$$1 - \frac{\Delta_e}{\Delta_0} = 2\hbar v_F\lambda\int_0^\Lambda \mathrm{d}k\frac{1}{E_k} \tag{71}$$

with $E_k = [(\hbar v_F k)^2 + \Delta_0^2]^{1/2}$, which has the (implicit) solution

$$\Delta_0 \approx 2\hbar v_F\Lambda\,\mathrm{e}^{-1/(2\lambda)}\,\mathrm{e}^{\gamma} \tag{72}$$

in the weak-coupling limit ($\lambda \ll 1$, $\gamma \ll 1$), where $\gamma = \Delta_e/(2\lambda\Delta_0)$ measures the strength of the external potential. It implies that the electronic gap $2\Delta_0$ should be larger for *cis*- than for *trans*-polyacetylene. This is confirmed by optical absorption experiments (e.g. Fincher *et al.*, 1979) which give $2\Delta_0 = 1.8$ eV for *cis*- and 1.4 eV for *trans*-polyacetylene. It further indicates that $\gamma \ll 1$, in which case we may represent Equation (72) as

$$\Delta_0 \approx 2\hbar v_F \Lambda\, e^{-1/(2\lambda)} + \Delta_e/(2\lambda). \tag{73}$$

There is no particular reason for expecting strong variations in bandwidth or in λ by going from *trans* to *cis*, and therefore we may estimate Δ_e by identifying Δ_e/λ with the change in the optical gap observed upon *cis*/*trans* isomerization. Using $\lambda = 0.2$, this yields $\Delta_e = 0.08$ eV, $t_1 = 0.04$ eV and $\gamma \approx 0.2$.

The defect state proposed by Brazovskii and Kirova (1981) and by Campbell and Bishop (1981) is defined as

$$\Delta(x) = \Delta_0 - \hbar v_F \kappa_0 \{\tanh[\kappa_0(x + x_0)] - \tanh[\kappa_0(x - x_0)]\}, \tag{74}$$

where the parameters κ_0 and x_0 are related by

$$\tanh 2\kappa_0 x_0 = \hbar v_F \kappa_0/\Delta_0. \tag{75}$$

It may be viewed as a superposition of a kink and an antikink, in particular for $\hbar v_F \kappa_0 \to \Delta_0$ where the kink–antikink separation $2x_0$ becomes much larger than their size. In the opposite limit $\hbar v_F \kappa_0 \ll \Delta_0$, Equation (74) describes a small local indentation with a minimum of $\Delta_0[1 - (\hbar v_F \kappa_0/\Delta_0)^2]$ at $x = 0$. The parameter κ_0 is determined from the self-consistency relation (70) and found to depend crucially on both the value of the external potential and the filling of the electronic levels. The proof that the self-consistency relation is exactly fulfilled for particular values of κ_0 is similar to that given in Section 4.1 for the kink and is outlined in Appendix B. Here we quote only the essential results. The defect (74) leads to pronounced changes in the electronic structure: it produces two localized states with levels in the gap (at $\varepsilon = \pm\varepsilon_0$, where $\varepsilon_0 = [\Delta_0^2 - (\hbar v_F \kappa_0)^2]^{1/2}$) and a modified density of band states. The self-consistency condition (70) is exactly satisfied if the following relation is fulfilled

$$\sin\theta\left[\theta - \frac{\pi}{4}(n_+ - n_- + 2) + \gamma\tan\theta\right] = 0, \tag{76}$$

where $\sin\theta = \hbar v_F \kappa_0/\Delta_0$, $0 \leq \theta \leq \pi/2$, and $n_\pm$ is the occupation number of the levels $\pm\varepsilon_0$. The defect energy is found to be

$$E(\theta) = \frac{4}{\pi}\Delta_0\left\{\left[\frac{\pi}{2} - \theta + \frac{\pi}{4}(n_+ - n_-)\right]\cos\theta + \sin\theta + \right.$$
$$\left. + \gamma[\tanh^{-1}(\sin\theta) - \sin\theta]\right\}. \tag{77}$$

One easily verifies that Equation (76) can also be obtained by minimizing $E(\theta)$. Using these two equations which have been first given by Brazovskii and Kirova (1981) and which are derived in Appendix B, it is easy to discuss the equilibrium configurations of various types of defects as functions of both the external potential strength (parameter γ) and the occupation numbers n_+, n_- (Bishop and Campbell, 1982).

We discuss first the degenerate case where $\gamma = 0$. Besides the trivial solution $\theta = 0$ there are two solutions $\theta = \pi/2$ for $n_- - n_+ = 0$ and $\theta = \pi/4$ for $n_- - n_+ = 1$. The former corresponds to a well-separated kink–antikink pair and does not represent a new type of defect. The latter is the polaron solution shown in Figure 7. The order parameter describes a local indentation with a minimum of $\Delta_0(\sqrt{2} - 1)$ at $x = 0$ and an extent which is slightly larger than that of the kink. The two localized electronic states occur at $\pm\Delta_0/\sqrt{2}$ and the filling is either acceptor-like, where the upper level is empty and the lower level singly occupied, or donor-like, where the upper level is singly and the lower one doubly occupied. Therefore, the polaron state exhibits the conventional spin–charge characteristics: it has the charge of a single electron (or hole) and it carries spin. The polaron energy is easily obtained from Equation (77), $E_p = 2\sqrt{2}\,\Delta_0/\pi \approx 0.9\Delta_0$, which is higher than the kink energy but lower than the energy needed to add an electron at the bottom of the conduction band (Δ_0).

For finite γ the polaron solution ($n_- - n_+ = 1$) remains qualitatively the same. The second solution however ($n_+ = n_-$) which corresponds to well-separated kink–antikink pairs for $\gamma = 0$, is strongly modified since the external potential leads to confinement, as illustrated in Figure 8 for the *cis* structure. Electronically, we have to distinguish between the case where both levels in the gap are empty or doubly occupied, which is often referred to as 'bipolaron' (Figure 9), and the 'exciton' (Figure 10) which is a bound state of oppositely charged defects where both levels are singly occupied. In the latter case Coulomb forces would naturally enhance the binding and lead to confinement also for the *trans* isomer (Brazovskii

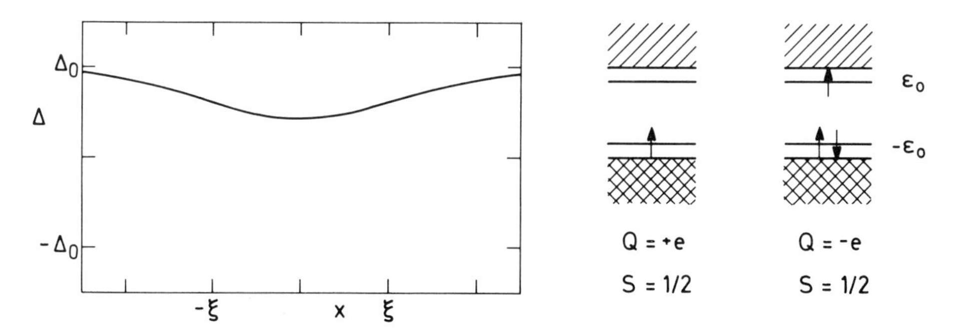

Fig. 7. Polaron configuration for $\gamma = 0$ and allowed fillings of the electronic levels.

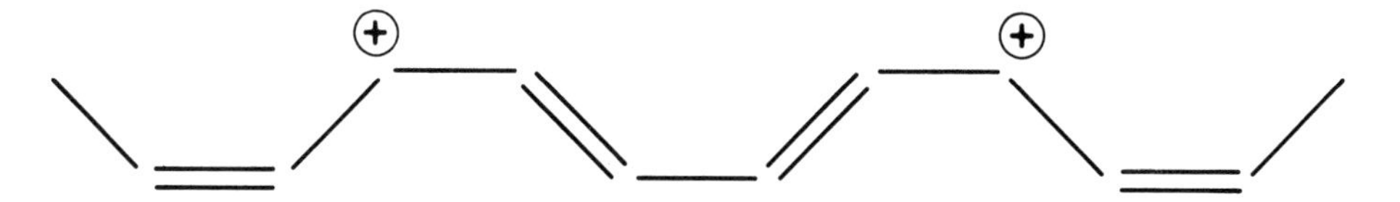

Fig. 8. Bound kink–antikink pair in *cis*-polyacetylene.

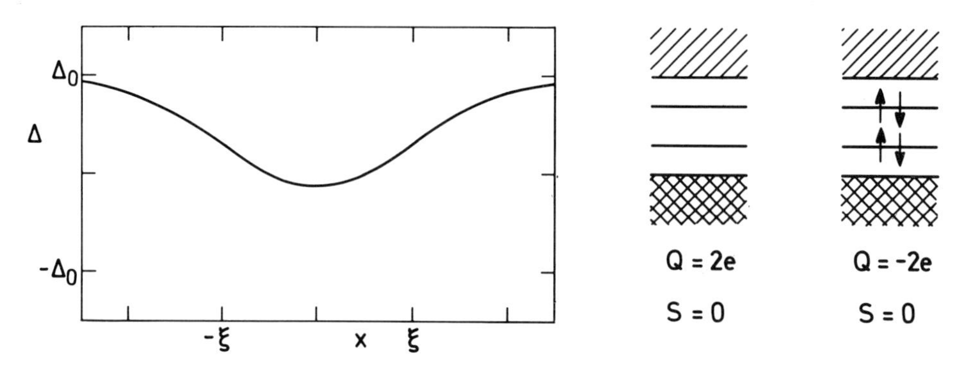

Fig. 9. Bipolaron for $\gamma \approx 0.2$ (corresponding to *cis*-polyacetylene).

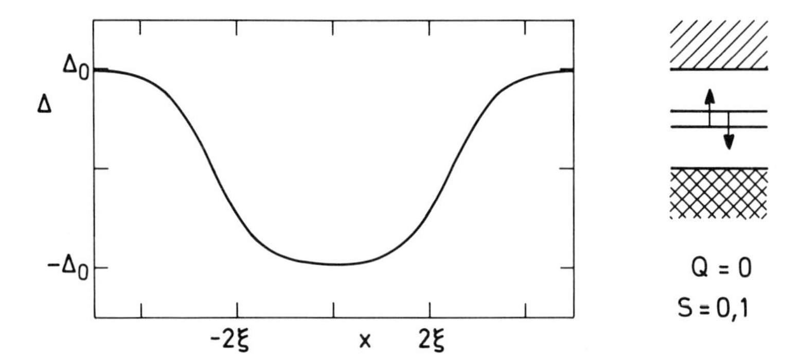

Fig. 10. Exciton in *trans*-polyacetylene (bound due to interchain coupling, cf. Section 5.3).

and Kirova, 1981; Rice, 1983). Consider the limit $\gamma \ll 1$ which seems to be appropriate for *cis*-polyacetylene. For the cases where $n_+ = n_-$ we find from Equation (76) $\theta \approx \pi/2 - \gamma^{1/2}$ and from Equation (75) $2x_0 \approx \frac{1}{2}\xi \ln(4/\gamma)$. Together with the previous estimate $\gamma \approx 0.2$, this corresponds to a distance $2x_0 \approx 1.5\xi \approx 10a$ for the *cis* isomer. Correspondingly the energy levels in the gap occur at $\pm\sqrt{\gamma}\Delta_0 \approx \pm 0.4\Delta_0$. The limit $\gamma \gg 1$ is hard to realize, since necessarily $\gamma < 1/(2\lambda)$ (cf. Equation (71)), but nevertheless interesting since in this case valence and conduction bands are only weakly mixed. Consequently, the polaron is well described in terms of the 'frozen valence band' approximation (Campbell *et al.*, 1982) where it represents a single-electron state and becomes equivalent to the polaron states discussed by Holstein (1959, 1981). Polyacetylene, however, appears to correspond to the opposite limit where the polaron is a collective state involving a strong polarization of the valence band.

The polaron mass can be calculated in the same way as the kink mass, using the adiabatic approximation, and it turns out to be even smaller (Bishop and Campbell, 1982). On the other hand, there is no simple connection to the electron 'band mass' v_F^2/Δ_0, moreover it seems not yet clear how to construct a 'polaron band' (Campbell and Bishop, 1982). The concept of polarons and bipolarons has also been applied to other conjugated polymers like polyparaphenylene or polypyrrole (Brédas *et al.*, 1982, 1983). Bipolarons have recently been proposed as the spinless charge carriers in polypyrrole (Scott *et al.*, 1983).

4.3. Solitons?

Soon after the first experiments on polyacetylene films – in particular, magnetic resonance on pristine (not intentionally doped) material and optical absorption on doped samples – the 'soliton question' has become the focus of debate, as is well documented by the recent conference proceedings. A lot of confusion is related to the lack of a clear, generally accepted soliton concept as applied to solid-state physics. This is in sharp contrast to the well-defined 'mathematical soliton' which corresponds to particular solutions of certain nonlinear differential equations (cf. Scott *et al.*, 1973). For solid-state physics where perturbations of various types (lattice discreteness, impurities, interchain coupling) will inevitably destroy the exact separability of the Hamiltonian into nonlinear and linear modes, one has to relax the requirements somewhat. The difficult question is how and to what extent the mathematical definition has to be modified in order to become a useful physical concept. The proposal made by Bishop (1978) to define 'physical solitons' as 'stable, finite-energy, particle-like field patterns which retain some central physical nonlinear properties' is appealing but not as quantitative as, for example, the quasiparticle concept where the lifetime determines the range of applicability (cf. Pines, 1963). Therefore, it may be premature to refer to the soliton concept as a 'new paradigm' in solid-state physics (Bishop *et al.*, 1980). An additional difficulty is related to the way elementary excitations are probed experimentally. The most complete information is provided by scattering experiments where a specific momentum $\hbar\mathbf{q}$ and a particular energy $\hbar\omega$ are transferred to a sample. This is an excellent method for studying linear modes which are characterized by a spectrum $\varepsilon(\mathbf{q})$ since, in this case, the scattering events are adequately described in terms of creation or annihilation of elementary excitations. There exists, unfortunately, no comparable experimental set-up for nonlinear modes and therefore the evidence for 'solitons' will generally be rather indirect.

The various defect states discussed previously have several 'soliton-like' properties such as stability, transparency to band states and mobility. The stability of a single kink is related to the external constraints (boundary conditions or 'topological charge'), whereas the polaron is expected to be stable due to its charge and spin characteristics. The defects will even survive quantum fluctuations and weak electron–electron interaction since, as long as these effects do not qualitatively change the ground state, they will lead merely to shape deformations and quantitative changes in the electronic structure of the defects. The transparency to band states means that extended electronic states u_k, v_k which are simple plane waves far away from the defects are not reflected by the defect potential but exhibit merely phase shifts. In fact, it is easily seen from Equations (61) and (B5) that $|u_k(\infty)| = |u_k(-\infty)|$, which implies that the reflection coefficient vanishes. The mobility is perfect in the continuum limit where the defect states are infinitely degenerate with respect to their location and thus can move without hindrance. Impurities and other types of disorder will tend to pin the defects. Lattice discreteness induces a small periodic potential with an amplitude of the order of 20 K (Su *et al.*, 1979); furthermore, numerical simulations show that there is an upper limit to the velocity, larger than the velocity of sound (Bishop *et al.*, 1984).

One may then ask to what extent the defects should be stable, transparent and mobile in order to merit the label 'soliton'. There is, unfortunately, no consensus in this respect; moreover, very often the word 'soliton' is used exclusively for the kink state although other defects have similar 'solitonic properties'. It appears that 'solitons' are rather elusive objects in the context of conducting polymers. The important questions are then, on the one hand, how the defect states are modified due to various kinds of interactions and perturbations which are not included in the idealized model, and, on the other hand, to what extent these states are observed experimentally. A short review of these two aspects has been given recently (Baeriswyl, 1983a). In the remaining part of this section we shall briefly summarize the second question, whereas the first aspect is treated in more detail in Section 5.

Let us first describe the possible signatures of kink defects. Neutral kinks are predicted to lead to one-dimensional spin diffusion, as discussed by Maki (1982). Charged kinks which can be generated either optically or upon doping produce additional optical absorption both due to transitions involving the midgap state (Suzuki *et al.*, 1980; Kivelson *et al.*, 1982) and due to infrared active local vibrations of the defect (Mele and Rice, 1980b; Horovitz, 1981b). Direct photoproduction of kink–antikink pairs leads to a pronounced Urbach tail in optical absorption and possibly to a threshold for photoconduction below the interband absorption edge (Su and Lu Yu, 1983). Furthermore, the symmetry of kink–antikink pairs forbids the direct photoproduction of neutral kinks (Ball *et al.*, 1983). A combination of mobile neutral and pinned charged kinks is involved in the hopping mechanism proposed by Kivelson (1982a) for electrical transport in weakly doped ($y < 1\%$) polyacetylene leading to rather specific predictions for the frequency and temperature dependence of the conductivity, whereas diffusion of depinned charged kinks has been proposed to represent the dominant transport mechanism in the intermediate ($1 < y < 6\%$) 'soliton liquid' phase (Su *et al.*, 1981).

The question if the kink signatures have been observed in polyacetylene is still not settled. This is due to experimental discrepancies on the one hand and the possibility of alternative interpretations on the other hand. Therefore, it is not surprising that some views are more optimistic (Heeger and MacDiarmid, 1981; Etemad *et al.*, 1982; Schrieffer, 1983), others are less (Tomkiewicz *et al.*, 1981; Baeriswyl *et al.*, 1982; Zannoni and Zerbi, 1983b; Roth *et al.*, 1983). Let us discuss a few difficulties. A quantitative measure for the one-dimensional spin diffusion has been extracted from the observed $\omega^{-1/2}$ dependence of the proton spin–lattice relaxation time yielding a very high diffusion constant for the neutral kink (Nechtschein *et al.*, 1983). This value disagrees with results obtained from electron spin–echo experiments which give much lower diffusion constants (Shiren *et al.*, 1982; Mehring *et al.*, 1983). In addition, proton NMR experiments on deuterated samples (Clark *et al.*, 1983) and C^{13} NMR measurements (Scott and Clarke, 1983) indicate that the diffusion of electron spins may not be the only relaxation mechanism for nuclear spins but that nuclear spin diffusion may also be involved, producing the same frequency dependence. A series of photogeneration experiments show additional optical absorption slightly below the optical gap (Orenstein and Baker, 1982) as well as near midgap and in the infrared (Vardeny *et al.*, 1983; Blanchet *et al.*, 1983). The photo-induced absorption close to the gap decays extremely rapidly (Shank *et al.*, 1982) as does the luminescence in *cis*-polyacetylene (Hayes *et al.*, 1983), namely at least within 10^{-11} s. This is in agreement with time-resolved photoconductivity experiments from which an upper limit of the order of 10^{-12} s is estimated for the free drift time of photogenerated carriers (Yacoby *et al.*, 1983). These fast time-scales represent a problem since the formation of kinks by relaxing excited electrons and holes through phonon emission is expected to take longer (cf. Mele, 1982). The threshold for photoconduction is appreciably below the optical gap (Lauchlan *et al.*, 1981) and no photo-induced ESR signal has been observed (Flood *et al.*, 1982), in agreement with the kink model. On the other hand, the photoconduction is not simply correlated with photo-induced IR absorption (Kiess *et al.*, 1982), the latter being saturated already at low light intensity (Blanchet *et al.*, 1983). The Kivelson mechanism for 'intersoliton hopping' which has been successfully used to interpret both the frequency and temperature dependence of electrical conductivity (Epstein *et al.*, 1981) is a particular realization of hopping transport in disordered systems. A different realization not requiring mobile neutral kinks can equally well explain the data (Chroboczek and Summerfield, 1983). The proposal of charge transport by kink diffusion is based on experiments showing negligible Pauli susceptibility χ_p in a doping range ($1 < y < 6\%$) where the electrical conductivity has almost reached saturation (Ikehata *et al.*, 1980; Peo, 1982; Chung *et al.*, 1984). According to these experiments χ_p suddenly increases above 6% to its final value. This step-like behaviour which has not been observed in other experiments (Tomkiewicz *et al.*, 1981) seems still not to be fully established (Roth *et al.*, 1983). The semiconductor–metal transition in polyacetylene is far from being understood, in particular since 'soliton signatures' in the optical absorption persist up to the highest dopant levels where no trace of the original optical gap remains (Kiess *et al.*, 1981).

Other defect states like polarons, bipolarons and excitons have been less considered so far although they may be generic excitations for the whole class of conjugate polymers (Brédas *et al.*, 1983). Their signatures, e.g. in optical absorption (Fesser *et al.*, 1983; Brédas *et al.*, 1984) may be less pronounced (as well as less unusual) than those of kinks. Nevertheless, magnetic experiments both on doped polyparaphenylene (Peo *et al.*, 1980) and polypyrrole (Scott *et al.*, 1983) indicate that there is no Pauli susceptibility in the highly conducting regime which has been attributed to bipolarons. Optical absorption experiments on doped polypyrrole (Scott *et al.*, 1984) are in agreement with this interpretation.

5. Towards a More Realistic Theory of Conducting Polymers

The theory developed in the previous sections is largely based on the simple Hückel picture and thus relies on drastic assumptions about the relevant degrees of freedom and the interactions between them. It is therefore important to discuss effects arising from modifications of the Hückel theory. These include quantum fluctuations, explicit treatment of σ electrons, inclusion of other phonon modes, electron–electron interaction, disorder and interchain coupling. The latter three points will be treated in some detail, whereas the former three are briefly discussed below.

The effect of quantum fluctuations on the ground-state properties of a single chain has been studied by Monte Carlo methods (Su, 1982; Fradkin and Hirsch, 1983). The fluctuations in the displacement coordinates are found to be of the order of the distortion amplitude but nevertheless do not destroy long-range order. This is in agreement with calculations by Nakahara and Maki (1982) showing that quantum fluctuations reduce the energy of a kink, but do not destroy its stability. It is conceivable that interchain coupling would strongly reduce the quantum effects.

There exists a vast literature on all-electron calculations for the band structure and ground-state properties of polymers, mostly on the Hartree–Fock level (cf. Kertész, 1982). Local density functional methods have been applied only recently to polyacetylene (Grant and Batra, 1983; Mintmire and White, 1983). It is generally found that π electrons indeed dominate the region around the Fermi energy, in agreement with optical absorption experiments which indicate that the contributions of σ electrons appear at higher energies than the π–π^* transitions (Baeriswyl *et al.*, 1983). For many problems like the structure of defect states, electrical transport or magnetic properties it will therefore be sufficient to consider only the π electrons explicitly.

The discrete model presented in Section 2 exhibits both an acoustic and an optical phonon branch in the dimerized state (Schulz, 1978). In the continuum limit discussed in Section 3 only optical modes have been taken into account. It is, however, straightforward to include the acoustic modes as well. As shown in Appendix C, they do not affect the ground state and induce only a small contraction of the kink structure. On the other hand, it is essential to include the acoustic phonons for describing the low-temperature kink dynamics (Maki, 1982). Furthermore, a realistic lattice dynamical calculation of phonon dispersion curves and eigenvectors has to include all possible atomic displacements and all relevant force constants (Zannoni and Zerbi, 1983a).

5.1. Electron–electron interaction

An 'unbiased' determination of Hückel parameters (Section 2.4) gives $\lambda \approx 0.1$ and therefore a Peierls gap $2\Delta_0 \approx 0.1\,\mathrm{eV}$, which is an order of magnitude smaller than the observed optical gap. Due to this discrepancy Ovchinnikov *et al.* (1972) have proposed that the gap is produced by electron correlation, characterized by the Hubbard parameter U, rather than by electron–phonon interaction. In the strong-coupling limit ($U \gg t_0$) the Hubbard Hamiltonian can be approximated by a spin $\frac{1}{2}$ Heisenberg model

$$H = -\sum_n J(r_n)\mathbf{s}_n \cdot \mathbf{s}_{n+1} \tag{78}$$

with antiferromagnetic exchange coupling $J = -2t^2/U$ (Emery, 1976). Using Equations (15) we find the following bond-length dependence of the exchange integral

$$J(r_n) = -\frac{2t_0^2}{U}[1 - 2\alpha(r_n - r_0)/t_0] \tag{79}$$

which leads to a spin–phonon coupling and consequently to a spin–Peierls transition (for a recent review see Bray *et al.*, 1983). This instability has been proposed as the driving mechanism for bond alternation in polyacetylene by Kondo (1980) and Nakano and Fukuyama (1980). Using the result of the latter authors we find a bond alternation amplitude

$$y(U) = (4/\pi)\alpha^2 t_0(UK_0/2)^{-3/2} \qquad \text{for } U \to \infty, \tag{80}$$

which decreases with increasing U. By representing the electron–electron interaction in terms of a single parameter U one tacitly assumes that it contains 'in some way' the long-range part of the Coulomb potential. In the case of a half-filled band (and in the strong-coupling limit) all sites are singly occupied in the ground state. It is then quite natural to choose U as the energy required to move one electron to a neighbouring site while keeping all the others fixed (Mazumdar and Bloch, 1983). This gives $U = U_0 - V$, where U_0 is the on-site and V the nearest neighbour Coulomb interaction. Various parametrizations of the Coulomb interaction have been given (Ohmine *et al.*, 1978); for the formula proposed by Nishimoto and Nataga (1957) one obtains $U = 5.8\,\text{eV}$. Using this value together with the 'unbiased parameters' deduced from Kakitani's analysis (Table I) we find a spin–Peierls distortion $y = 0.05\,\text{Å}$, in surprisingly good agreement with the measured bond alternation amplitude (Fincher *et al.*, 1982; Clarke *et al.*, 1983). This may be an accident since the value chosen for U does not correspond to the strong-coupling limit, but rather to an intermediate regime. On the other hand, this value of y is an order of magnitude larger than the value obtained from the Peierls instability for $U = 0$ (using the same Hückel parameters). This strongly suggests that small U enhances the bond alternation and at some intermediate value the function $y(U)$ goes through a maximum. Explicit calculations (Ukrainskii, 1979; Horsch, 1981; Kivelson and Heim, 1982) show indeed that in the weak-coupling limit ($U \ll t_0$) the bond alternation increases with increasing U. This effect is at first sight surprising since the Hartree–Fock solution for the bond-order wave does not depend on U (see below). It represents therefore a true correlation effect, as pointed out clearly by Horsch (1981). The increase of bond alternation with U can be traced back to correlation enhancement of the bond–bond polarizability. Indeed the logarithmic divergence of the polarizability for $U = 0$ (cf. Section 2.5) turns into a power law singularity for $U > 0$ with exponents increasing with U in the weak-coupling limit (cf. Sólyom, 1979). Recent Quantum Monte Carlo calculations by Hirsch (1983) embracing both the weak- and strong-coupling regimes fully confirm this picture for the U-dependence of dimerization, showing an initial increase, a maximum at $U \approx 2t_0$ and a subsequent decrease. Similar conclusions have been reached by Mazumdar and Dixit (1983) using finite-chain calculations together with qualitative 'valence-bond' arguments. The electronic gap is then in general a combination of a single-particle gap produced by (correlation enhanced) bond alternation and correlation gap, the latter dominating in the strong-coupling limit where $2\Delta_0 = U$.

The correlation problem becomes more involved if part of the long-range Coulomb interaction is explicitly included as in the extended Hubbard model, where

$$H_{\text{int}} = U_0 \sum_m n_{m\uparrow} n_{m\downarrow} + V \sum_{mss'} n_{ms} n_{m+1s'} \tag{81}$$

with $n_{ms} = c^+_{ms} c_{ms}$. On the basis of the correlation effects described above it is obvious that a 'biased' determination of Hückel parameters related to measured values of the gap and the dimerization has to be modified if H_{int} is taken into account. This is not the case for the 'unbiased' parameter set determined in Section 2.4 since, on the one hand, the terms to be added to the equation of state (12) are expected to be small and, on the other hand, the mean value of the bond order as well as the frequency of the fully symmetric mode in benzene will depend only weakly on U_0. In the following we restrict ourselves to the weak-coupling limit, $U_0 \lesssim t_0$, $V \lesssim t_0$, and therefore use the transformation to continuum fields, Equations (31). Keeping only the relevant terms (those involving both k_{F} and $-k_{\text{F}}$), we find

$$\begin{aligned} H_{\text{int}} = \sum_{ss'} \int \mathrm{d}x \{ & g_1 \psi^+_{1s} \psi^+_{2s'} \psi_{1s'} \psi_{2s} + g_2 \psi^+_{1s} \psi^+_{2s'} \psi_{2s'} \psi_{1s} - \\ & - \tfrac{1}{2} g_3 (\psi^+_{1s} \psi^+_{1s'} \psi_{2s'} \psi_{2s} + \psi^+_{2s} \psi^+_{2s'} \psi_{1s'} \psi_{1s}) \}, \end{aligned} \tag{82}$$

where $g_1 = a(U_0 - 2V)$ is associated with backscattering, $g_2 = a(U_0 + 2V)$ with forward scattering and $g_3 = a(U_0 - 2V)$ with Umklapp scattering processes (cf. Sólyom, 1979). We look for the mean-field (or unrestricted Hartree–Fock) ground state of $H = H_0 + H_{\text{int}}$ where H_0 is given by Equation (35) and thus contains the coupling to the phonon field $y(x)$. We replace the electronic part of the Hamiltonian by the mean-field expression

$$H_m = \sum_s \int \mathrm{d}x \, \Phi^+_s \left(-i\hbar v_{\text{F}} \sigma_3 \frac{\mathrm{d}}{\mathrm{d}x} + \Delta_{1s} \sigma_1 + \Delta_{2s} \sigma_2 \right) \Phi_s \tag{83}$$

in terms of a complex order parameter $\Delta_s = \Delta_{1s} + i\Delta_{2s}$ and take the ground state of H_m as a variational state for the full Hamiltonian. Proceeding as in Section 3.2 we obtain the BdG equations

$$\begin{aligned} \varepsilon u_{\nu s} &= -i\hbar v_{\text{F}} \frac{\mathrm{d}}{\mathrm{d}x} u_{\nu s} + \Delta^*_s v_{\nu s} \\ \varepsilon v_{\nu s} &= i\hbar v_{\text{F}} \frac{\mathrm{d}}{\mathrm{d}x} v_{\nu s} + \Delta_s u_{\nu s} \end{aligned} \tag{84}$$

where the wave functions $u_{\nu s}$ and $v_{\nu s}$ are associated with the fields ψ_{1s} and ψ_{2s}, respectively. Minimalizing the total energy with respect to $y(x)$ and $\Delta_s(x)$ we find the self-consistency equations

$$y(x) = -\frac{2\alpha a}{K_0}\sum_{\nu s}{}' (u^*_{\nu s} v_{\nu s} + u_{\nu s} v^*_{\nu s}) \tag{85}$$

$$\Delta_s(x) = 2\alpha y(x) + \sum_{\nu s'} (g_1 - g_2 \delta_{ss'}) u^*_{\nu s'} v_{\nu s'} - g_3 \sum_{\nu}{}' u_{\nu - s} v^*_{\nu - s}. \tag{86}$$

These equations are easily solved for homogeneous order parameters corresponding to the following phases: bond-order wave (BOW) for $\Delta_s = \Delta_0$, charge-density wave (CDW) for $\Delta_s = i\Delta_0$, spin-density wave (SDW) for $\Delta_s = is\Delta_0$ and spin-bond-order wave (SBOW) for $\Delta_s = s\Delta_0$, where Δ_0 is real (detailed calculations for these different cases have been presented by Fukutome and Sasai, 1982, 1983). In all four cases one obtains the gap equation (46) with λ values as given in Table III, where $\lambda_0 = 2\alpha^2/(\pi t_0 K_0)$. The condensation energy is again given by Equation (47) and therefore, for given coupling strengths, the largest λ value determines the most stable phase. This yields the phase diagram of Figure 11. Notice that only the BOW phase admits bond alternation with an amplitude

$$y = \frac{\hbar v_F \Lambda \lambda_0}{\alpha \lambda} e^{-1/(2\lambda)} \tag{87}$$

which increases as a function of λ (and thus V) up to a maximum for $\lambda = \frac{1}{2}$ and subsequently decreases. Dimerization is therefore enhanced for small V (provided that $\lambda_0 < \frac{1}{2}$), but it is completely independent of U within mean-field theory.

According to the phase diagram of Figure 11 the BOW phase is stable within a relatively narrow stripe. Therefore it is interesting to notice that, using $\lambda_0 = 0.1$ together with the frequently quoted couplings of the Mataga–Nishimoto and Ohno formulae (cf. Nakano and Fukuyama, 1980), we obtain a BOW ground state in the former case, whereas the latter choice would be located close to the BOW–CDW boundary. One may argue that these parametrizations contain long-range interactions as well, which are not included within the extended Hubbard model. However the long-range part would mainly modify the forward scattering (coupling g_2) which affects all instabilities considered here equally (cf. Table III) and therefore

TABLE III

λ values for the mean-field solutions

	$\lambda(g_1, g_2, g_3)$	$\lambda(U_0, V)$
BOW	$\lambda_0 + \dfrac{g_2 - 2g_1 + g_3}{4\pi\hbar v_F}$	$\lambda_0 + \dfrac{V}{2\pi t_0}$
CDW	$\dfrac{g_2 - 2g_1 - g_3}{4\pi\hbar v_F}$	$\dfrac{2V - \frac{1}{2}U_0}{2\pi t_0}$
SDW	$\dfrac{g_2 + g_3}{4\pi\hbar v_F}$	$\dfrac{\frac{1}{2}U_0}{2\pi t_0}$
SBOW	$\dfrac{g_2 - g_3}{4\pi\hbar v_F}$	$\dfrac{V}{2\pi t_0}$

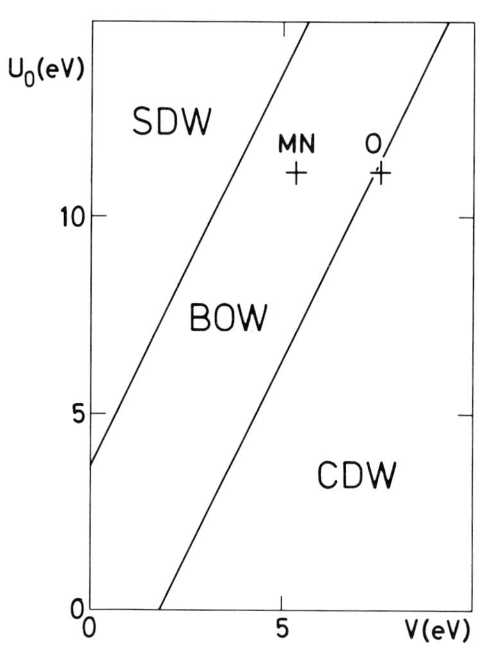

Fig. 11. Phase diagram in mean-field approximation as a function of on site (U_0) and nearest neighbour coupling (V). MN and O denote Mataga-Nishimoto and Ohno parametrization, respectively.

the phase diagram would not be strongly affected. A more serious defect is the use of mean-field theory. For vanishing electron–phonon coupling the mean-field approximation reproduces the exact phase diagram which can be traced back to the underlying symmetry of the Hamiltonian (Baeriswyl and Forney, 1980), but these symmetry arguments fail when electron–phonon interactions are included. Furthermore, the ground state of a one-dimensional chain does not necessarily exhibit long-range order (Hirsch, 1983) and the coupling between chains required to establish long-range order can strongly favour a particular phase at the expense of others (interchain Coulomb forces favour CDWs and tend to suppress SDWs; Lee *et al.*, 1977). Therefore the 'phase diagram' of Figure 11 should not be taken as more than a qualitative representation of the dominant instabilities (in the strictly one-dimensional chain). More quantitative information can only be obtained from detailed calculations taking both the lattice degrees of freedom and the electronic correlation into account.

Keeping these reservations in mind, let us qualitatively discuss the effect of electron–electron interactions on the defect states described in Section 4, using the mean-field equations. We assume that the couplings are weak and do not destroy the BOW state. It is quite clear from Equation (86) that spinless defects such as charged kinks or bipolarons will in general render Δ complex and thus induce a local CDW component. Similarly, defects with spin-like neutral kinks or polarons will induce SDW components since $\Delta_\uparrow \neq \Delta_\downarrow$. This has indeed been found by explicit calculation (Subbaswamy and Grabowski, 1981; Chance *et al.*, 1983) and apparently observed experimentally (Thomann *et al.*, 1983). It is amusing that for particular couplings exact inhomogeneous solutions of Equations (84) to (86) can be found (Baeriswyl, 1983b), but we do not pursue this here. Nakano and Fukuyama (1980) have studied the kink defect in the spin–Peierls regime and obtained similar results as in the absence of interaction both for its energy and its mass. The exciton problem has been addressed by Ukrainskii (1981) who concludes that the triplet exciton is strongly bound due to the on-site interaction U_0, whereas the binding of the singlet exciton depends sensitively on the value of the nearest-neighbour coupling V.

5.2. Disorder

The polymer films used for experimental studies are usually highly disordered materials. There are various possible origins of disorder such as fibrillar morphology, partial crystallinity, cross-linking, short conjugation lengths, incomplete isomerization, impurities, inhomogeneous doping. Within the Hückel framework the effects of randomness on the π electrons can be modelled in essentially two ways: either as fluctuations in local energies (diagonal disorder) or as variations of the resonance integrals (off-diagonal disorder). Diagonal disorder can arise from an external potential (e.g. the Coulomb potential of randomly distributed charged impurities) or from hybridization of π orbitals with other electronic states (of dopant molecules or of sites located on neighbouring chains). Off-diagonal disorder may originate from the presence of *cis* segments within a trans chain or from randomly distributed lattice defects (such as the 'soliton glass' proposed by Su *et al.*, 1981). Weak off-diagonal disorder leads to a pronounced reduction of the square-root singularities at the band edges and a significant broadening of polaronic levels whereas the midgap state associated with a kink is only weakly affected (White *et al.*, 1983). This is a consequence of electron–hole symmetry (cf. Section 2.1) which is not destroyed by off-diagonal randomness. For diagonal disorder where this symmetry is not preserved both kink and polaron states are found to be strongly modified and additional levels associated with localized impurity states appear in the gap (Bryant and Glick, 1982).

Let us discuss the case of diagonal randomness by adding the term

$$H' = \sum_{ns} v_n c_{ns}^{+} c_{ns} \tag{88}$$

to the Hückel Hamiltonian (1). We furthermore assume perfect dimerization ($y_n = y$), leading to a semiconducting ground state for the pure chain with a gap $2\Delta_0 = 4\alpha y$, as shown in Section 2.5. A single impurity at $n = 0$ produces a donor state for $V_0 < 0$ and an acceptor state for $V_0 > 0$ with a level at $\varepsilon = -\Delta_0 \sin \eta$, where $\cot \eta = V_0/(2t_0)$, $-\pi/2 < \eta < \pi/2$, whereas the extent ξ' of the impurity state is related to the (kink) length $\xi = \hbar v_F/\Delta_0$ as $\xi' = \xi/\cos \eta$ (for details see Baeriswyl, 1983c). Both energy and extent of the impurity state tend to the corresponding values of the kink induced midgap state in the limit $|V_0| \to \infty$. It is interesting to note that the wave function amplitude which decays exponentially on both sides of

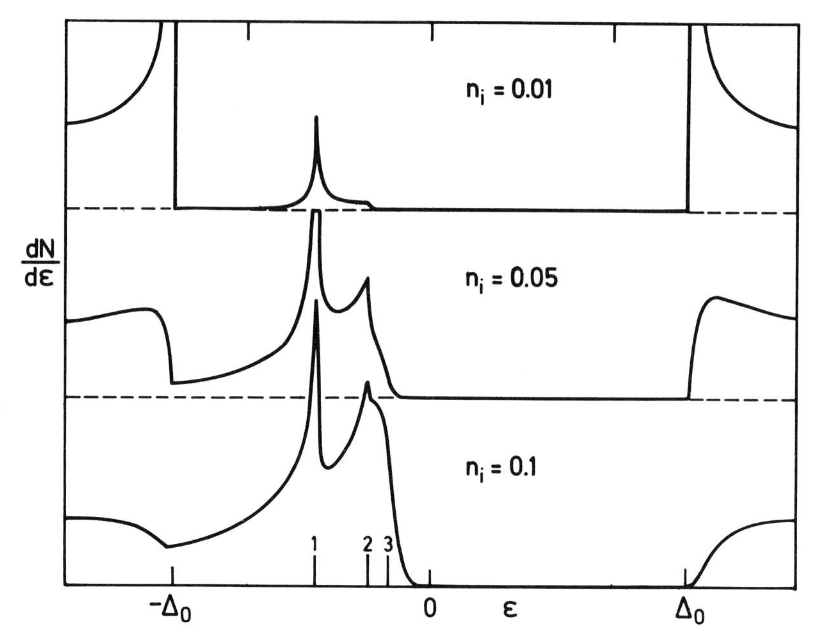

Fig. 12. Electronic density of states in the case of diagonal disorder for different impurity concentrations n_i and a potential strength $V_0 = 4t_0$. The energies marked by 1 to 3 correspond to levels associated with an isolated impurity (1), two close impurities (2), three close impurities (3).

the impurity is not equal on even and odd sites, the ratio being given by tan $\eta/2$. The density of states for a distribution of impurities is shown in Figure 12. It has been calculated using the method of Frisch and Lloyd (1960) and assuming that all impurities have equal strengths V_0 and are randomly distributed over even sites. In this case the continuum limit of the impurity potential is given by

$$V(x) = V_0 a \sum_s \int \mathrm{d}x \sum_i \delta(x - x_i) \Phi_s^+ (1 + \sigma_2) \Phi_s, \tag{89}$$

As before (Equation (60)) the BdG equations can be transformed to a single second-order differential equation

$$\left[(\hbar v_F)^2 \frac{\mathrm{d}^2}{\mathrm{d}x^2} - 2V_0 a \varepsilon \sum_i \delta(x - x_i) \right] f_- = (\Delta_0^2 - \varepsilon^2) f_- \tag{90}$$

which is equivalent to the model of Frisch and Lloyd (1960). Notice that bound states with $\varepsilon^2 < \Delta_0^2$ require $V_0 \varepsilon < 0$. This is the reason why the density of states of Figure 12 vanishes exactly for $0 < \varepsilon < \Delta_0$. It is an artefact of the assumption that impurities are restricted to even sites; for a general distribution there would be a small density of states for $0 < \varepsilon < \Delta_0$. With increasing impurity concentration the acceptor level develops into a structured impurity band with an additional peak from configurations where two impurities are close to each other and even

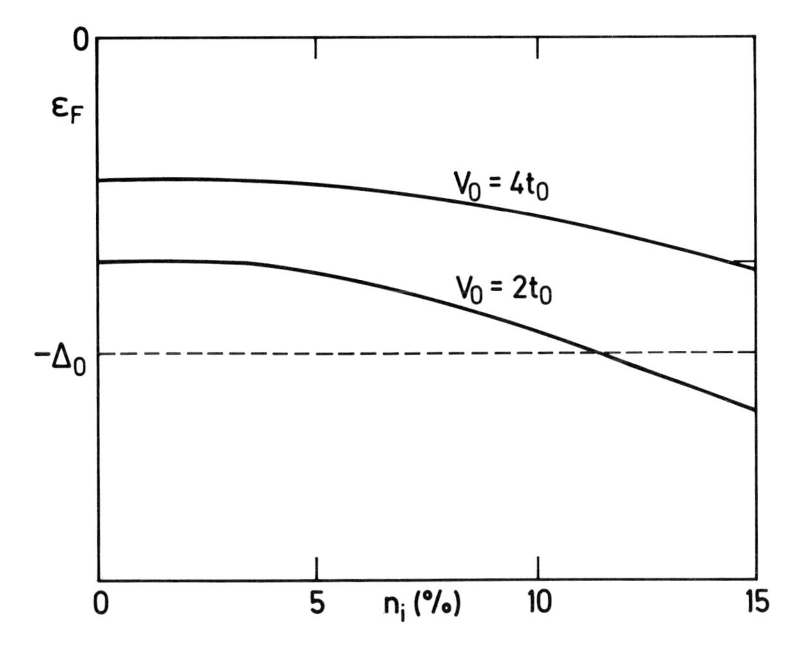

Fig. 13. Fermi energy ε_F as a function of dopant concentration n_i for two different strengths V_0 of the impurity potential.

signatures for three-impurity clusters. At the same time the density of states at the band edges is strongly suppressed. Figure 13 shows the Fermi level as a function of impurity concentration where it has been assumed that every impurity removes one electron from the chain. For low concentration the Fermi level is 'pinned' at the acceptor level. For higher concentrations it decreases and finally dips into the valence band. This represents a very simple model for a semiconductor–metal transition provided that the band states remain quasi-extended.

The model considered above may be far from being realistic and should not be taken too seriously. In particular, the assumption of perfect dimerization cannot be maintained if we allow the lattice to relax. These relaxation effects are expected to be large since, neglecting impurities, the system will exhibit a 'soliton lattice' when electronic charge is removed from the chain (Horovitz, 1981). This configuration of equally spaced kinks and antikinks continuously develops into an incommensurate Peierls distortion for increasing charge removal. The combined effects of impurities and lattice relaxation have been studied numerically by Mele and Rice (1981) with results depending markedly on impurity concentration. Charged kinks pinned to impurity sites are produced for low concentration. Adding more impurities drives the system gradually into a disordered incommensurate phase with 'metallic' behaviour. Finally, for very dense distributions of impurities the lattice distortion is suppressed. This quenching of the Peierls distortion due to strong disorder has also been found by Boyanowsky (1983) in the extreme quantum limit (for the ionic motion).

5.3. Interchain Coupling

Up to now we have been concerned with the dimerization of uncoupled chains at zero temperature. At finite temperatures the bond alternation will not survive thermal fluctuations unless interchain coupling is taken into account which leads to three-dimensional ordering below some critical temperature T_c. There exist various possible mechanisms producing interchain coupling: Coulomb interaction between bond charges, dispersion forces of van der Waals type or interchain tunnelling (Josephson-type coupling). The first two types of coupling have been studied by Baughman and Moss (1982), the third by Baeriswyl and Maki (1983). All these mechanisms favour antiparallel alignment of the dimerization patterns of neighbouring chains, the energy difference δE between parallel and antiparallel alignment being of the order of 1 K per site. This type of ordering appears to have been found experimentally (Fincher *et al.*, 1982).

It is obvious that within such a three-dimensionally ordered system a kink on one chain leads to mismatch with neighbouring chains. The energy required to sustain such a defect increases proportionally to the length of the misaligned segment. A kink is therefore unlikely to be produced except together with an antikink or close to a chain end. This is similar to the case of *cis*-polyacetylene where the inequivalence of neighbouring bonds implies kink–antikink confinement. In order to find the confinement energy due to interchain coupling it is natural to consider the bipolaron (or exciton) state discussed in Section 4.2. Without interchain coupling the energy of a kink–antikink pair decreases (for $\gamma = 0$, i.e. in the *trans* case) according to Equation (77) as

$$E(d) = \frac{4}{\pi}\Delta_0(1 + 2e^{-2d/\xi}) \tag{91}$$

for large distances $d = 2x_0$. Adding the term due to interchain coupling

$$\Delta E(d) = z\,\delta E(d/a), \tag{92}$$

where z is the number of neighbouring chains, we find an energy minimum for

$$d = \tfrac{1}{2}\xi \ln\left(\frac{16\Delta_0 a}{\pi z\,\delta E \xi}\right) \tag{93}$$

which depends only weakly on the precise value of δE or on the coordination number z. In the crystalline regions the packing of the chains seems to form approximately a triangular lattice (cf. Heeger and MacDiarmid, 1981). Therefore the relative ordering of chains can be described in terms of an Ising model with antiferromagnetic exchange constant $J < 0$ on a two-dimensional triangular lattice. This system is frustrated and one may argue that consequently the confinement energy is strongly reduced (Kivelson, 1982b). However, it is easy to visualize that for low-energy configurations the sites with spin up are generally surrounded by four down spins and two up spins. Therefore turning a spin requires for most sites an energy $2|J|$, yielding an effective coordination number $z = 2$. This value, together with $\xi = 7a$, $\Delta_0 = 0.7$ eV and $\delta E = 1$ K, gives an equilibrium distance $d \approx 4\xi$. The corresponding exciton state is illustrated in Figure 10. The energy $E + \Delta E$ can be interpreted as a potential energy for the relative kink–antikink movement. The eigenfrequencies for this motion have been recently calculated by Maki (1984) and associated with the 'pinning mode' observed in photogeneration experiments (Blanchet *et al.*, 1983).

For finite but not too high temperatures, long-range order persists and the kink defects remain confined within pairs. Unbinding will occur as soon as the confinement potential is sufficiently screened due to thermal excitations of additional kink–antikink pairs, namely at the critical temperature T_c where long-range order is lost. In order to estimate T_c we use the coarse-graining method of Bohr and Brazovskii (1983) and subdivide the chains into cells of length d which is defined by Equation (93). The state of the ith cell on the αth chain is described by an Ising spin $s_{i\alpha} = \pm 1$, corresponding to an order parameter $y(x)$ having preferentially the value $\pm y_0$ within this cell. Since a large energy is required to put more than one kind into the same cell (cf. Equation (91)) this procedure takes the relevant low-temperature kink configurations into account. It leads to the Ising Hamiltonian

$$H = -J_{\|} \sum_{\alpha i} s_{i\alpha} s_{i+1\alpha} - J_{\perp} \sum_{\substack{\langle \alpha, \beta \rangle \\ i}} s_{i\alpha} s_{i\beta}, \tag{94}$$

where $J_{\|} = \frac{1}{2} E_K$ and $J_{\perp} = -\frac{1}{2} \delta E(d/a)$. Neglecting the frustration problem, we estimate the critical temperature using the exact solution of the two-dimensional Ising model where

$$\sinh(2\beta_c J_{\|}) \sinh(2\beta_c J_{\perp}) = 1 \tag{95}$$

with $\beta_c = (k_B T_c)^{-1}$. For the above determined parameters and $E_K = 2\Delta_0/\pi \approx 5000\,\text{K}$ we find $T_c \approx 1000\,\text{K}$, which is not accessible experimentally. However, as Bohr and Brazovskii (1983) have pointed out, T_c can be decreased dramatically if a finite density of kinks is created intentionally (e.g. upon doping or by photogeneration). They find that $T_c \rightarrow 0$ for a concentration $\nu_c = 0.1817$, which corresponds to about 0.006 kinks per site in polyacetylene. Neglecting all complications like impurity pinning or short-chain effects, this represents the critical concentration above which interchain coupling is inefficient in restraining kinks from moving along chains and contributing to transport.

6. Concluding Remarks

In this paper it has been shown that the simple Hückel model for a one-dimensional chain represents a rather crude approximation for dealing with conducting polymers and that complications due to electron correlation, disorder and interchain coupling are far from being negligible. Furthermore, these effects may be intertwined in a complicated way. As an example we mention the observed smooth optical absorption edge (instead of the expected sharp threshold with a square-root singularity at $2\Delta_0$) which could be attributed to quantum fluctuations, excitons, disorder or interchain coupling or combinations of them. This does not necessarily imply that the

Hückel framework is useless. It is, for instance, conceivable that in certain regions of parameter space (namely within the BOW phase) the Hartree–Fock solution presented in Section 5.1 yields a good approximation both for the electronic gap and the dimerization amplitude (since the gap produced by bond alternation eliminates certain singularities associated with a finite density of states at the Fermi level). In this case it would be possible to describe several properties in terms of a Hückel model with appropriately renormalized parameters. More subtle effects like the binding of electron–hole pairs due to Coulomb attraction could however not be described using an effective single-electron Hamiltonian. There is good hope that by improving the theoretical approach on the one hand, and both sample quality and experimental techniques on the other, more detailed information will become available in order to disentangle the complicated interplay of instabilities, localization effects, correlations and fluctuations. This combined effort will lead to a better understanding of the nature of the semiconductor–metal transition and the contribution of polaronic defects to transport.

Acknowledgements

I would like to thank the Max Planck Institute in Stuttgart for the hospitality in the initial stage of this work and NORDITA for financial support during the period spent in Copenhagen. I have benefited from discussions with M. Aldissi, A. R. Bishop, N. A. Cade, D. K. Campbell, K. Carneiro, T. C. Clarke, R. Friend, G. Harbeke, P. Horsch, M. Horton, H. Kiess, Yu Lu, K. Maki, S. Mazumdar, E. Mulazzi, S. Roth, J. R. Schrieffer, J. C. Scott, G. B. Street, and G. Zerbi.

Appendix A: Local Bond Order and Local Density in the Peierls Distorted State

Using the parametrization of Equations (15) and assuming lattice dimerization, i.e. $r_n - r_0 = (-1)^n \Delta_0/(2\alpha)$, the electronic part of the Hamiltonian (1) takes the following form

$$H_e = -\sum_{ns}[t_0 - \tfrac{1}{2}(-1)^n \Delta_0](c^+_{ns}c_{n+1s} + c^+_{n+1s}c_{ns}). \tag{A1}$$

It is convenient to distinguish between 'right-moving' electrons with $k > 0$ and 'left-moving' electrons with $k < 0$. Therefore, we write the transformation (7) as follows

$$c_{ns} = \frac{1}{\sqrt{N}}\sum_k \mathrm{e}^{ikna}(i^n c_{k_F+ks} + (-i)^n c_{-k_F+ks}), \tag{A2}$$

where k is restricted to the Brillouin zone of the dimerized lattice $-\pi/(2a) < k \leqslant \pi/(2a)$ and $(\pm i)^n = \exp(\pm i k_F na)$. The Hamiltonian (A1) becomes

$$H_e = \sum_{ks}[2t_0 \sin ka(c^+_{k_F+ks}c_{k_F+ks} - c^+_{-k_F+ks}c_{-k_F+ks}) - \\ - i\Delta_0 \cos ka(c^+_{k_F+ks}c_{-k_F+ks} - c^+_{-k_F+ks}c_{k_F+ks})]. \tag{A3}$$

It is diagonalized by the Bogoliubov transformation to fermion operators a_{ks} and b_{ks}

$$\begin{aligned} c_{k_F+ks} &= \cos\theta_k a_{ks} + i\sin\theta_k b_{ks} \\ c_{-k_F+ks} &= \cos\theta_k b_{ks} + i\sin\theta_k a_{ks} \end{aligned} \tag{A4}$$

provided that $\tan 2\theta_k = \delta\cot ka$, where $\delta = \Delta_0/(2t_0)$ and $-\pi/2 < 2\theta_k \leqslant \pi/2$. The Hamiltonian (A3) becomes

$$H_e = \sum_{ks}\operatorname{sign}(k)E_k(a_{ks}^+a_{ks} - b_{ks}^+b_{ks}), \tag{A5}$$

where $E_k = (4t_0^2\sin^2 ka + \Delta_0^2\cos^2 ka)^{1/2}$. As a consequence of the transformation (A4) which shifts the Fermi points $\pm k_F$ to $k = 0$, the gap in the spectrum occurs now at $k = 0$. The expectation value of the bond order (defined by Equation (2)) for a full valence and an empty conduction band is then easily calculated by using Equations (A3) to (A5) together with the gap equation (obtained by minimalizing Equation (23)). We find

$$\langle p_n\rangle = \frac{2}{\pi}[E(1 - \delta^2) - \delta^2/(2\lambda)] - (-1)^n\,\delta/(\pi\lambda). \tag{A6}$$

This has to be compared with the local density which is independent of the amount of dimerization

$$\langle \rho_n\rangle = \sum_s\langle c_{ns}^+c_{ns}\rangle = 1. \tag{A7}$$

Appendix B: Polaron-Type Solutions

We want to show that the 'polaron' Ansatz, Equations (74) and (75), provides an exact solution of the continuum model, i.e. both of the BdG equations (38) and of the self-consistency condition (70). Starting from the second-order differential equation (60), one verifies that Δ satisfies the relation

$$\Delta^2 \mp \hbar v_F\frac{d\Delta}{dx} = \Delta_0^2 - 2(\hbar v_F\kappa_0)^2\operatorname{sech}^2[\kappa_0(x \mp x_0)], \tag{B1}$$

from which one obtains a Schrödinger equation with a short-range attractive potential,

$$(\hbar v_F)^2\left\{-\frac{d^2}{dx^2} - 2\kappa_0^2\operatorname{sech}^2[\kappa_0(x \mp x_0)]\right\}f_\pm = (\varepsilon^2 - \Delta_0^2)f_\pm. \tag{B2}$$

Using the analysis of Landau and Lifshitz (1958b) we find exactly two bound states with eigenvalues in the gap $\varepsilon = \pm\varepsilon_0$, where

$$\varepsilon_0 = [\Delta_0^2 - (\hbar v_F\kappa_0)^2]^{1/2}. \tag{B3}$$

Direct inspection of the BdG equations (59) shows that the corresponding wave functions are

$$f_\pm = (\kappa_0/2)^{1/2}\,e^{\pm i\sigma\pi/4}\operatorname{sech}[\kappa_0(x \mp x_0)], \tag{B4}$$

where $\sigma = \text{sign}\ \varepsilon$. The extended states which satisfy the BdG equations with $\varepsilon = \sigma E_k$ are found to be

$$f_{\pm} = A_k \,\mathrm{e}^{\pm i(\sigma\pi/4 - \phi k/2)} \{k/\kappa_0 + i \tanh[\kappa_0(x \mp x_0)]\}\,\mathrm{e}^{ikx}, \tag{B5}$$

where the normalization constant A_k is given by

$$A_k^2 = \frac{\kappa_0^2}{k^2 + \kappa_0^2} \Bigg/ \left(L - \frac{2\kappa_0}{k^2 + \kappa_0^2}\right). \tag{B6}$$

Assuming periodic boundary conditions $f_{\pm}(-L/2) = f_{\pm}(L/2)$ we find

$$kL = (2n + 1)\pi + 2\theta_k, \tag{B7}$$

where $\theta_k = \tan^{-1}(k/\kappa_0)$. This yields the following density of states

$$\frac{\mathrm{d}n}{\mathrm{d}k} = \frac{1}{2\pi}\left(L - \frac{2\mathrm{d}\theta_k}{\mathrm{d}k}\right) = \frac{1}{2\pi}\left(L - \frac{2\kappa_0}{k^2 + \kappa_0^2}\right) \tag{B7}$$

which implies that one state has been removed from the valence band and one from the conduction band, thus compensating the two levels in the gap. It is now straightforward to find out under which conditions the self-consistency equation is fulfilled. The filled valence band yields a contribution

$$-\pi {\sum_{ks}}' (u_k^* v_k + v_k^* u_k) = \Delta \int_{-\Lambda}^{\Lambda} \mathrm{d}k \frac{1}{E_k} - \\ - \varepsilon_0 \,\mathrm{sech}[\kappa_0(x - x_0)]\,\mathrm{sech}[\kappa_0(x + x_0)] \\ \int_{-\Lambda}^{\Lambda} \mathrm{d}k \frac{\kappa_0^2}{k^2 + \kappa_0^2} \cdot \frac{1}{E_k}. \tag{B9}$$

Adding the states in the gap (with $n_{\pm}$ electrons in the levels $\pm\varepsilon_0$), using the gap equation (71) to get rid of the first integral and performing the second we find for the r.h.s. of the self-consistency condition (70)

$$\Delta + 2\lambda\Delta_0 \sin\theta[\theta - \frac{\pi}{4}(n_+ - n_- + 2) + \\ + \gamma \tan\theta]\,\mathrm{sech}[\kappa_0(x - x_0)]\,\mathrm{sech}[\kappa_0(x + x_0)], \tag{B10}$$

where $\sin\theta = \hbar v_F \kappa_0/\Delta_0$, $0 \leqslant \theta \leqslant \pi/2$. Therefore, Equation (70) is exactly satisfied if Equation (76) holds. The calculation of the defect energy proceeds in a similar way as for the case of a kink. There is a decrease in elastic energy of

$$(2\pi\hbar v_F \lambda)^{-1} \int [(\Delta - \Delta_e)^2 - (\Delta_0 - \Delta_e)^2]\,\mathrm{d}x \\ = \frac{4\Delta_0}{\pi}\left[\gamma \tanh^{-1}(\sin\theta) - \frac{1}{2\lambda}\sin\theta\right]. \tag{B11}$$

For the electronic energy we use again a modified cut-off in order to compare equivalent physical situations. We require that the number of electrons in the valence band (including spin) has decreased by 2, as compared to the ground state. Using the density of states of Equation (B8), this yields a modified cut-off

$$\Lambda' = \Lambda - 2\kappa_0/(L\Lambda). \tag{B12}$$

The electronic contribution to the defect energy is therefore

$$\begin{aligned} &-2\int_{-\Lambda'}^{\Lambda'} \mathrm{d}k \frac{\mathrm{d}n}{\mathrm{d}k} E_k + (n_+ - n_-)\varepsilon_0 + 2\int_{-\Lambda}^{\Lambda} \mathrm{d}k \frac{L}{2\pi} E_k \\ &= \frac{4}{\pi}\Delta_0 \left\{ \left[\frac{\pi}{2} - \theta + \frac{\pi}{4}(n_+ - n_-) \right] \cos\theta + \right. \\ &\quad \left. + \left[1 + \frac{1}{2\lambda} - \gamma \right] \sin\theta \right\}. \end{aligned} \tag{B13}$$

Adding Equations (B11) and (B13) gives the defect energy, Equation (77).

Appendix C: Acoustic Phonons

In order to rewrite the Hückel Hamiltonian (Equations (1) and (15)) in terms of optic and acoustic modes of the dimerized system, we define variables y_n and z_n by the relation

$$r_n - r_0 = (-1)^n y_n + z_n. \tag{C1}$$

In the discrete case only wavevectors $|q| \leqslant \pi/(2a)$ would be involved. In the continuum limit y_n and z_n are replaced by functions $y(x)$ and $z(x)$, respectively, which are assumed to be slowly varying. The Hamiltonian is then given by Equation (35) and additional terms generated by the acoustic displacement field $z(x)$,

$$\begin{aligned} H' &= (2\pi\hbar v_{\mathrm{F}}\lambda)^{-1}\int \mathrm{d}x\,\Gamma^2 - \frac{p_0}{a}\int \mathrm{d}x\,\Gamma + \\ &\quad + ia\sum_s \int \mathrm{d}x\,\phi_s^+ \left(\Gamma\frac{\mathrm{d}}{\mathrm{d}x} + \frac{1}{2}\frac{\mathrm{d}\Gamma}{\mathrm{d}x} \right) \sigma_3 \phi_s, \end{aligned} \tag{C2}$$

where we have introduced the acoustic order parameters $\Gamma(x) = 2\alpha z(x)$. The ground-state energy is now a functional of Δ, Γ and $\mathrm{d}\Gamma/\mathrm{d}x$. The self-consistency equation for Δ is still given by Equation (43), whereas for Γ we find, using the Hellmann–Feynman theorem,

$$\begin{aligned} \Gamma &= \pi\hbar v_{\mathrm{F}}\lambda \left\{ \frac{p_0}{a} - \frac{1}{2} ia \sum_{\nu s}{}' \left(u^* \frac{\mathrm{d}u}{\mathrm{d}x} - \frac{\mathrm{d}u^*}{\mathrm{d}x} u - \right. \right. \\ &\quad \left. \left. - v^* \frac{\mathrm{d}v}{\mathrm{d}x} + \frac{\mathrm{d}v^*}{\mathrm{d}x} v \right)_\nu \right\}. \end{aligned} \tag{C3}$$

Using the wave functions for the homogeneous ground state we find $\Gamma = 0$ provided that p_0 and the cut-off Λ are chosen appropriately (for $p_0 = 2/\pi$ we find $\Lambda \approx \sqrt{2}/a$). For the kink Ansatz, Equation (57), one obtains from Equation (C3)

$$\Gamma = -\frac{\Delta_0^2}{4t_0}\operatorname{sech}^2(x/\xi). \tag{C4}$$

This is not a self-consistent solution but the lowest order distortion which is small in the weak-coupling limit. It corresponds to a slight contraction of the whole system towards the kink centre.

References

1. D. Baeriswyl and J. J. Forney, *J. Phys.* **C13**, 3203 (1980).
2. D. Baeriswyl, G. Harbeke, H. Kiess, and W. Meyer, in J. Mort and G. Pfister (eds), *Electronic Properties of Polymers*, Wiley, New York, Chap. 7 (1982).
3. D. Baeriswyl, *Helv. Phys. Acta* **56**, 639 (1983a).
4. D. Baeriswyl, unpublished results (1983b).
5. D. Baeriswyl, *J. Phys. (Paris)* **44**, C3-381 (1983c).
6. D. Baeriswyl and K. Maki, *Phys. Rev.* **B28**, 2068 (1983).
7. D. Baeriswyl, G. Harbeke, H. Kiess, E. Meier, and W. Meyer, *Physica* **117B** + **118B**, 617 (1983).
8. R. Ball, W. P. Su, and J. R. Schrieffer, *J. Phys. (Paris)* **44**, C3-429 (1983).
9. J. C. J. Bart and C. H. MacGillavry, *Acta Cryst.* **B24**, 1569 (1968).
10. R. H. Baughman and G. Moss, *J. Chem. Phys.* **77**, 6321 (1982).
11. A. R. Bishop, *Solitons and Condensed Matter Physics* (eds A. R. Bishop and T. Schneider), Springer Series in Solid-State Sciences, Berlin, Vol. 8, p. 85 (1978).
12. A. R. Bishop, J. A. Krumhansl, and S. E. Trullinger, *Physica* **1D**, 1 (1980).
13. A. R. Bishop and D. K. Campbell, *Nonlinear Problems: Present and Future* (eds A. R. Bishop, D. K. Campbell, and B. Nicolaenko), North-Holland, Amsterdam, p. 195 (1982).
14. A. R. Bishop, D. K. Campbell, P. S. Lomdahl, B. Horovitz, and S. R. Phillpot, *Phys. Rev. Lett.* **52**, 671 (1984).
15. G. B. Blanchet, C. R. Fincher, T.-C. Chung, and A. J. Heeger, *Phys. Rev. Lett.* **50**, 1938 (1983).
16. T. Bohr and S. A. Brazovskii, *J. Phys.* **C16**, 1189 (1983).
17. D. Boyanowsky, *Phys. Rev.* **B27**, 6763 (1983).
18. J. W. Bray, L. V. Interrante, I. S. Jacobs and J. C. Bonner, *Extended Linear Chain Compounds* (ed J. S. Miller), Plenum, New York, Vol. 3, p. 353 (1983).
19. S. A. Brazovskii, *Pis'ma Zh. Eksp. Teor. Fiz.* **28**, 656 (1978) [*Sov. Phys. JETP Lett.* **28**, 607 (1978)].
20. S. A. Brazovskii and N. N. Kirova, *Pis'ma Zh. Eksp. Teor. Fiz.* **33**, 6 (1981) [*Sov. Phys. JETP Lett.* **33**, 4 (1981)].
21. J. L., Brédas, R. R. Chance, and R. Silbey, *Phys. Rev.* **B26**, 5843 (1982).
22. J. L. Brédas, B. Thémans, and J. M. André, *Phys. Rev.* **B27**, 7827 (1983).
23. J. L., Brédas, J. C. Scott, K. Yakushi, and G. B. Street, *Phys. Rev.* **B30**, 1023 (1984).
24. A. D. Bruce, *Adv. Phys.* **29**, 111 (1980).
25. G. W. Bryant and A. J. Glick, *Phys. Rev.* **B26**, 5855 (1982).
26. D. K. Campbell and A. R. Bishop, *Phys. Rev.* **B24**, 4859 (1981).
27. D. K. Campbell and A. R. Bishop, *Nucl. Phys.* **B200**, 297 (1982).
28. D. K. Campbell, A. R. Bishop, and K. Fesser, *Phys. Rev.* **B26**, 6862 (1982).
29. R. R. Chance, D. S. Boudreaux, J. L. Brédas, and R. Silbey, *Phys. Rev.* **B27**, 1440 (1983).
30. C. K. Chiang, C. R. Fincher, Y. W. Park, A. J. Heeger, H. Shirakawa, E. J. Louis, S. C. Gau, and A. G. MacDiarmid, *Phys. Rev. Lett.* **39**, 1098 (1977).
31. J. A. Chroboczek and S. Summerfield, *J. Phys. (Paris)* **44**, C3-517 (1983).
32. T.-C. Chung, F. Moreas, J. D. Flood, and A. J. Heeger, *Phys. Rev.* **B29**, 2341 (1984).

33. W. G. Clark, K. Glover, G. Mozurkewich, C. T. Murayama, J. Sanny, S. Etemad, and M. Maxfield, *J. Phys. (Paris)* **44**, C3-239 (1983).
34. T. C. Clarke, R. D. Kendrick, and C. S. Yannoni, *J. Phys. (Paris)* **44**, C3-369 (1983).
35. R. Comès, P. Bernier, J. J. André, and J. Rouxel (eds), *Proc. Int. Conf. on the Physics and Chemistry of Conducting Polymers*, *J. Phys. (Paris)* **44**, C3-1 (1983).
36. C. A. Coulson, *Proc. R. Soc. London* **A164**, 383 (1938).
37. P. G. de Gennes, *Superconductivity of Metals and Alloys*, Benjamin, New York, Chap. 5 (1966).
38. V. J. Emery, *Phys. Rev.* **B14**, 2989 (1976).
39. A. J. Epstein, H. Rommelmann, M. Abkowitz, and H. W. Gibson, *Phys. Rev. Lett.* **47**, 1549 (1981).
40. A. J. Epstein and E. M. Conwell (eds), *Proc. Int. Conf. on Low-Dimensional Conductors Mol. Cryst. Liq. Cryst.* **77** (1981), **83** (1982), **85** (1982).
41. S. Etemad, A. J. Heeger, and A. G. MacDiarmid, *Ann. Rev. Phys. Chem.* **33**, 443 (1982).
42. K. Fesser, A. R. Bishop, and D. K. Campbell, *Phys. Rev.* **B27**, 4804 (1983).
43. C. R. Fincher Jr, M. Ozaki, M. Tanaka, D. Peebles, L. Lauchlan, A. J. Heeger, and A. G. MacDiarmid, *Phys. Rev.* **B20**, 1589 (1979).
44. C. R. Fincher Jr, C.-E. Chen, A. J. Heeger, A. G. MacDiarmid, and J. B. Hastings, *Phys. Rev. Lett.* **48**, 100 (1982).
45. J. D. Flood, E. Ehrenfreund, A. J. Heeger, and A. G. MacDiarmid, *Sol. St. Commun.* **44**, 1055 (1982).
46. E. Fradkin and J. E. Hirsch, *Phys. Rev.* **B27**, 1680 (1983).
47. H. L. Frisch and S. P. Lloyd, *Phys. Rev.* **120**, 1175 (1960).
48. H. Fröhlich, *Proc. R. Soc. London* **A215**, 291 (1952).
49. H. Fröhlich, *Proc. R. Soc. London* **A223**, 296 (1954).
50. H. Fukutome and M. Sasai, *Progr. Theor. Phys.* **67**, 41 (1982).
51. H. Fukutome and M. Sasai, *Progr. Theor. Phys.* **69**, 1, 373 (1983).
52. I. B. Goldberg, H. R. Crowe, P. R. Newman, A. J. Heeger, and A. G. MacDiarmid, *J. Chem. Phys.* **70**, 1132 (1979).
53. P. M. Grant and I. P. Batra, *J. Phys. (Paris)* **44**, C3-437 (1983).
54. J. Harada, M. Tasumi, H. Shirakawa, and S. Ikeda, *Chem. Lett.*, p. 1411 (1978).
55. W. Hayes, C. N. Ironside, J. F. Ryan, R. P. Steele, and R. A. Taylor, *J. Phys.* **C16**, L729 (1983).
56. A. J. Heeger and A. G. MacDiarmid, *Mol. Cryst. Liq. Cryst.* **77**, 1 (1981).
57. J. E. Hirsch, *Phys. Rev. Lett.* **51**, 296 (1983).
58. T. Holstein, *Ann. Phys.* **8**, 325 (1959).
59. T. Holstein, *Mol. Cryst. Liq. Cryst.* **77**, 235 (1981).
60. B. Horovitz, *Phys. Rev. Lett.* **46**, 742 (1981a).
61. B. Horovitz, *Mol. Cryst. Liq. Cryst.* **77**, 285 (1981b); *Phys. Rev. Lett.* **47**, 1491 (1981b). In these papers it is argued that the frequencies of the modes are independent of the charge configuration and therefore not sensitive indicators for kinks.
62. P. Horsch, *Phys. Rev.* **B24**, 7351 (1981).
63. E. Hückel, *Z. Phys.* **70**, 204 (1931).
64. E. Hückel, *Z. Phys.* **76**, 628 (1932).
65. S. Ikehata, J. Kaufer, T. Woerner, A. Pron, M. A. Druy, A. Sivak, A. J. Heeger, and A. G. MacDiarmid, *Phys. Rev. Lett.* **45**, 1123 (1980).
66. F. Inagaki, M. Tasumi, and T. Miyazawa, *J. Raman Spectrosc.* **3**, 335 (1975).
67. T. Itô, H. Shirakawa, and S. Ikeda, *J. Polym. Sci. Polym. Chem. Ed.* **12**, 11 (1974).
68. T. Kakitani, *Prog. Theor. Phys.* **51**, 656 (1974).
69. M. Kertész, *Adv. Quantum Chem.* **15**, 161 (1982).
70. H. Kiess, D. Baeriswyl, and G. Harbeke, *Mol. Cryst. Liq. Cryst.* **77**, 147 (1981).
71. H. Kiess, R. Keller, D. Baeriswyl, and G. Harbeke, *Sol. St. Commun.* **44**, 1443 (1982).
72. S. Kivelson, *Phys. Rev.* **B25**, 3798 (1982a).
73. S. Kivelson, *Phys. Rev.* **B26**, 7093 (1982b).
74. S. Kivelson and D. E. Heim, *Phys. Rev.* **B26**, 4278 (1982).
75. S. Kivelson, T.-K. Lee, Y. R. Lin-Liu, I. Peschel, and Lu Yu, *Phys. Rev.* **B25**, 4173 (1982).

76. J. Kondo, *Physica* **98B**, 176 (1980).
77. H. Kuhn, *J. Chem. Phys.* **16**, 840 (1948).
78. H. Kuhn, *J. Chem. Phys.* **17**, 1198 (1949).
79. H. Kuzmany, *Phys. Stat. Sol. (B)* **97**, 521 (1980).
80. G. V. Kventsel and Y. A. Kruglyak, *Theor. Chim. Acta* **12**, 1 (1968).
81. L. D. Landau and E. M. Lifshitz, *Statistical Physics*, Pergamon, London, p. 39 (1958a).
82. L. D. Landau and E. M. Lifshitz, *Quantum Mechanics*, Pergamon, London, p. 69 (1958b).
83. L. Lauchlan, S. Etemad, T.-C. Chung, A. J. Heeger, and A. G. MacDiarmid, *Phys. Rev.* **B24**, 3701 (1981).
84. L. Lauchlan, S. P. Chen, S. Etemad, M. Kletter, A. J. Heeger, and A. G. MacDiarmid, *Phys. Rev.* **B27**, 2301 (1983).
85. R. B. Laughlin, *Phys. Rev. Lett.* **50**, 1395 (1983).
86. P. Y. Le Daëron and S. Aubry, *J. Phys.* **C16**, 4827 (1983).
87. P. A. Lee, T. M. Rice, and P. W. Anderson, *Sol. St. Commun.* **14**, 703 (1974); notice that the factor $\frac{4}{3}$ in the expression for ω_+^2 in this paper has to be replaced by $\frac{1}{3}$.
88. P. A. Lee, T. M. Rice, and R. A. Klemm, *Phys. Rev.* **B15**, 2984 (1977).
89. J. E. Lennard-Jones, *Proc. R. Soc. London* **A158**, 280 (1937).
90. L. S. Lichtmann, D. B. Fitchen, and H. Temkin, *Synth. Metals* **1**, 139 (1979/80).
91. H. C. Longuet-Higgins and L. Salem, *Proc. R. Soc. London* **A251**, 172 (1959).
92. K. Maki, *Phys. Rev.* **B26**, 2181, 2187, 4539 (1982).
93. K. Maki, *Synth. Met.* **9**, 185 (1984).
94. D. C. Mattis and E. H. Lieb, *J. Math. Phys.* **6**, 304 (1965).
95. S. Mazumdar and A. N. Bloch, *Phys. Rev. Lett.* **50**, 207 (1983).
96. S. Mazumdar and S. N. Dixit, *Phys. Rev. Lett.* **51**, 292 (1983).
97. M. Mehring, H. Seidel, W. Müller, and G. Wegner, *Sol. St. Commun.* **45**, 1075 (1983).
98. E. J. Mele and M. J. Rice, *Sol. St. Commun.* **34**, 339 (1980a).
99. E. J. Mele and M. J. Rice, *Phys. Rev. Lett.* **45**, 926 (1980b).
100. E. J. Mele and M. J. Rice, *Phys. Rev.* **B23**, 5397 (1981).
101. E. J. Mele, *Phys. Rev.* **B26**, 6901 (1982).
102. J. W. Mintmire and C. T. White, *Phys. Rev. Lett.* **50**, 101 (1983); *Phys. Rev.* **B28**, 3283 (1983).
103. I. A. Misurkin and A. A. Ovchinnikov, *Optika i Spektroskopiya* **16**, 228 (1964) [*Optics and Spectroscopy* **16**, 125 (1964)].
104. D. Moses, A. Feldblum, E. Ehrenfreund, A. J. Heeger, T.-C. Chung, and A. G. MacDiarmid, *Phys. Rev.* **B26**, 3361 (1982).
105. E. Mulazzi, G. P. Brivio, E. Faulques, and S. Lefrant, *Sol. St. Commun.* **46**, 851 (1983).
106. M. Nakahara and K. Maki, *Phys. Rev.* **B25**, 7789 (1982).
107. T. Nakano and H. Fukuyama, *J. Phys. Soc. Japan* **49**, 1679 (1980).
108. M. Nechtschein, F. Devreux, F. Genoud, M. Guglielmi, and K. Holczer, *Phys. Rev.* **B27**, 61 (1983).
109. K. Nishimoto and N. Mataga, *Z. Phys. Chem.* **12**, 335 (1957).
110. I. Ohmine, M. Karplus, and K. Schulten, *J. Chem. Phys.* **68**, 2298 (1978).
111. Y. Ooshika, *J. Phys. Soc. Japan* **14**, 747 (1959).
112. J. Orenstein and G. L. Baker, *Phys. Rev. Lett.* **49**, 1043 (1982).
113. A. A. Ovchinnikov, I. I. Ukrainskii, and G. V. Kventsel, *Usp. Fiz. Nauk* **108**, 81 (1972) [*Sov. Phys. Usp.* **15**, 575 (1973)].
114. R. E. Peierls, *Quantum Theory of Solids*, Clarendon, Oxford, p. 108 (1955).
115. M. Peo, S. Roth, K. Dransfeld, B. Tieke, J. Hocker, H. Gross, A. Grupp, and H. Sixl, *Sol. St. Commun.* **35**, 119 (1980).
116. M. Peo, Doctoral Thesis, MPI, Stuttgart (unpublished) (1982).
117. L. Pietronero and S. Strässler, *Phys. Rev. Lett.* **47**, 593 (1981a).
118. L. Pietronero and S. Strässler, *Synth. Met.* **3**, 213 (1981b).
119. D. Pines, *Elementary Excitations in Solids*, Benjamin, New York (1963).
120. L. Piseri, R. Tubino, and G. Dellepiane, *Sol. St. Commun.* **44**, 1589 (1982).
121. J. A. Pople and S. H. Walmsley, *Molec. Phys.* **5**, 15 (1962).

122. J. F. Rabolt, T. C. Clarke, and G. B. Street, *J. Chem. Phys.* **71**, 4614 (1979).
123. M. J. Rice, *Phys. Lett.* **71A**, 152 (1979).
124. M. J. Rice, S. Strässler, and P. Fulde, (unpublished) (1980).
125. M. J. Rice, *Phys. Rev. Lett.* **51**, 142 (1983).
126. S. Roth, K. Ehinger, K. Menke, M. Peo, and R. J. Schweizer, *J. Phys.* (*Paris*) **44**, C3-69 (1983).
127. L. Salem, *Molecular Orbital Theory of Conjugated Systems*, Benjamin, London (1966).
128. J. R. Schrieffer, preprint (1983).
129. H. J. Schulz, *Phys. Rev.* **B18**, 5756 (1978).
130. A. C. Scott, F. Y. F. Chu, and D. W. McLaughlin, *Proc. IEEE* **61**, 1443 (1973).
131. J. C. Scott, P. Pfluger, M. T. Krounbi, and G. B. Street, *Phys. Rev.* **B28**, 2140 (1983).
132. J. C. Scott, J. L. Brédas, K. Yakushi, P. Pfluger, and G. B. Street, *Synth. Met.* **9**, 165 (1984).
133. J. C. Scott and T. C. Clarke, *J. Phys.* (*Paris*) **44**, C3-365 (1983).
134. C. V. Shank, R. Yen, R. L. Fork, J. Orenstein, and G. L. Baker, *Phys. Rev. Lett.* **49**, 1660 (1982).
135. B. S. Shastry, *J. Phys.* **A16**, 2049 (1983).
136. N. S. Shiren, Y. Tomkiewicz, T. G. Kazyaka, A. R. Taranko, H. Thomann, L. Dalton, and T. C. Clarke, *Sol. St. Commun.* **44**, 1157 (1982).
137. J. Sólyom, *Adv. Phys.* **28**, 201 (1979).
138. W. P. Su, J. R. Schrieffer, and A. J. Heeger, *Phys. Rev. Lett.* **42**, 1698 (1979).
139. W. P. Su, *Sol. St. Commun.* **35**, 899 (1980).
140. W. P. Su and J. R. Schrieffer, *Proc. Natl. Acad. Sci. USA* **77**, 5626 (1980).
141. W. P. Su, J. R. Schrieffer, and A. J. Heeger, *Phys. Rev.* **B22**, 2099 (1980); Erratum *Phys. Rev.* **B28**, 1138 (1983).
142. W. P. Su, S. Kivelson, and J. R. Schrieffer, 1981, *Physics in One Dimension*, (eds J. Bernasconi and T. Schneider), Springer Series in Solid State Sciences, Berlin, Vol. 23, p. 201 (1981).
143. W. P. Su, *Sol. St. Commun.* **42**, 497 (1982).
144. Z. Su and Lu Yu, *Phys. Rev.* **B27**, 5199 (1983); Erratum *Phys. Rev.* **B29**, 2309 (1984); *Commun. in Theor. Phys.* **2**, 1203, 1323, 1341 (1983).
145. K. R. Subbaswamy and M. Grabowski, *Phys. Rev.* **B24**, 2168 (1981).
146. N. Suzuki, M. Ozaki, S. Etemad, A. J. Heeger, and A. G. MacDiarmid, *Phys. Rev. Lett.* **45**, 1209, 1463 (1980).
147. H. Takayama, Y. R. Lin-Liu, and K. Maki, *Phys. Rev.* **B21**, 2388 (1980).
148. H. Thomann, L. R. Dalton, Y. Tomkiewicz, N. S. Shiren, and T. C. Clarke, *Phys. Rev. Lett.* **50**, 533 (1983).
149. Y. Tomkiewicz, T. D. Schultz, H. B. Brom, A. R. Taranko, T. C. Clarke, and G. B. Street, *Phys. Rev.* **B24**, 4348 (1981).
150. I. I. Ukrainskii, *Zh. Eksp. Teor. Fiz.* **76**, 760 (1979) [*Sov. Phys. JETP* **49**, 381 (1979)].
151. I. I. Ukrainskii, *Phys. Stat. Sol.* (*B*) **106**, 55 (1981).
152. D. Vanderbilt and E. J. Mele, *Phys. Rev.* **B22**, 3939 (1980).
153. Z. Vardeny, J. Orenstein, and G. L. Baker, *J. Phys.* (*Paris*) **44**, C3-325 (1983); *Phys. Rev. Lett.* **50**, 2032 (1983).
154. C. T. White, M. L. Elert, and J. W. Mintmire, *J. Phys.* (*Paris*) **44**, C3-481 (1983).
155. E. G. Wilson Jr, J. C. Decius, and P. C. Cross, *Molecular Vibrations*, McGraw-Hill, New York (1955).
156. Y. Yacoby, S. Roth, K. Menke, F. Keilmann, and J. Kuhl, *Sol. St. Commun.* **47**, 869 (1983).
157. G. Zannoni and G. Zerbi, *J. Molec. Struct.* **100**, 485 (1983a).
158. G. Zannoni and G. Zerbi, *J. Molec. Struct.* **100**, 505 (1983b); *Sol. St. Commun.* **47**, 213 (1983b).

Notes added in proof

Since this article has been completed the theory of conducting polymers has been reviewed by several authors; two review papers have already been published (S. A. Brazovskii, *Sov. Sci. Rev.* **A5**, 99 (1984); H. W. Streitwolf, *Phys. Stat. Sol.* (**B**) **127**, 1 (1985)). Additional material can be found in recent conference proceedings (*Proc. Int. Conf. on the Physics and Chemistry of Low-Dimensional Synthetic Metals*, Abana Terme 1984, to be published in *Mol. Cryst. Liq. Cryst.*; *Proc. Int. Winterschool on Electronic Properties of Polymers and Related Compounds*, Kirchberg 1985, to appear in Springer Series of Solid State Sciences).

Some progress has been made in the understanding of the nonlinear dynamics of the coupled electron-phonon system. Numerical simulations show that energetic kinks emit long-lived breather modes which may play a role in photogeneration experiments (A. R. Bishop, D. K. Campbell, P. S. Lomdahl, B. Horovitz, and S. R. Phillpot, *Synth. Met.* **9**, 223 (1984); F. Guinea, *Phys. Rev.* **B30**, 1884 (1984)). Analytical studies for various scattering processes show that kinks and polarons pass freely through each other (S. Okuno and Y. Onodera, *J. Phys. Soc. Japan* **52**, 3495 (1983)); the same is true for polarons and bipolarons (Y. Onodera, *Phys. Rev.* **B30**, 775 (1984)), whereas two free polarons decay into a pair of charged kinks (Y. Onodera and S. Okuno, *J. Phys. Soc. Japan* **52**, 2478 (1983)). The interchain tunnelling of polarons has been investigated by S. Jeyadev and J. R. Schrieffer (*Phys. Rev.* **B30**, 3620 (1984)) and found to be strongly reduced relative to the corresponding rate for band electrons. The possible relevance of polarons and bipolarons for systems with a non-degenerate ground state like polypyrrole or polythiophene has been reviewed by J. L. Brédas (*Mol. Cryst. Liq. Cryst.*, to be published).

The discussion of electronic correlation has recently been stimulated by optical absorption experiments (B. R. Weinberger, C. B. Roxlo, S. Etemad, G. L. Baker, and J. Orenstein, *Phys. Rev. Lett.* **53**, 86 (1984)). Unfortunately, reliable calculations for excited electronic states including electron–electron interactions are only available for short chains (S. Ramasesha and Z. G. Soos, *Synth. Met.* **9**, 283 (1984)). Nevertheless, correlation effects can also be extracted from ground state properties, in particular if an "unbiased" value for the electron–phonon coupling constant is used (D. Baeriswyl and K. Maki, *Phys. Rev.* **B31**, 6633 (1985)). One has to conclude that in conjugate polymers the Hubbard parameter U is of the order of the π bandwidth. The effects of U on energetics and internal structure of intrinsic defects has been investigated using Quantum Monte Carlo methods (J. E. Hirsch and M. Grabowski, *Phys. Rev. Lett.* **52**, 1713 (1984); D. K. Campbell, T. A. DeGrand, and S. Mazumdar, *Phys. Rev. Lett.* **52**, 1717 (1984)).

The combined effects of disorder and lattice relaxation have been considered by several authors. W. P. Su (*Solid St. Commun.* **47**, 947 (1983)) has studied a disordered array of charged kinks and antikinks, in particular in view of the semiconductor-metal transition. The addition of diagonal disorder (which breaks the electron-hole symmetry) can lead to trapping of kinks (S. Phillpot, D. Baeriswyl, A. R. Bishop, and P. S. Lomdahl, preprint), an effect which appears to play an important role in photogeneration experiments (J. Orenstein, Z. Vardeny, G. L. Baker, G. Eagle, and S. Etemand, *Phys. Rev.* **B30**, 786 (1984)). A new scheme for interpreting Raman lineshapes involving a distribution of parameters λ has been used for discriminating between "intrinsic" and "extrinsic" disorder (Z. Vardeny, E. Ehrenfreund, O. Brafman, and B. Horovitz, *Phys. Rev. Lett.* **54**, 75 (1985)).

EQUILIBRIUM PROPERTIES OF TTF-TCNQ

S. BARIŠIĆ

Department of Physics, Faculty of Science,
P.O.B. 162, 41001 Zagreb, Croatia, Yugoslavia

and

A. BJELIŠ

Institute of Physics of the University,
P.O.B. 304, 41001 Zagreb, Croatia, Yugoslavia

1. Introduction

The purpose of this text is to discuss the thermodynamical properties of TTF-TCNQ, this being a prototype of organic conductors with two families of conducting chains. It belongs to an even wider class of linear chain compounds which are currently attracting a lot of attention by their unusual properties. A recent paper [1] (henceforth Paper I) reviewed the thermodynamical properties of conducting tri-chalcogenides and gave a short historical introduction to the physics of quasi-one-dimensional (quasi-1d) materials. In particular, its introduction places TTF-TCNQ among other quasi-1d materials. Therefore, we shall not repeat the same discussion here but rather emphasize that the main idea of the two joint papers is an attempt at synthesizing experiment and theory, which is focused on a given class of materials. Such approach has the advantage that it rationalizes the experimental situation in a specific way but the obvious price to pay is that some definite, though more or less justified options are then required. While we shall try to point out carefully the remaining alternatives, it should be kept constantly in mind that some definite choices have been made here. Like an earlier similar attempt [2], this paper thus represents rather a view than a review of TTF-TCNQ, intention being to give a broad basis for further discussions and improvements in the understanding of this interesting class of materials. Such approach has been chosen because TTF-TCNQ and related materials have been thoroughly reviewed lately [3, 4]. With this in mind we shall not be exhaustive in recounting all experimental facts, but rather limit ourselves to those that are directly relevant to the understanding of the equilibrium properties of TTF-TCNQ. However, some additional care will be devoted to the most recent experimental and theoretical results which have not been reviewed earlier.

It has been emphasized many times, and particularly in Paper I, that the quantities which determine the nature of a given type of materials are coupling constants and the associated characteristic (cuf-off) energies. These coupling constants describe the electron–electron interactions mediated by acoustic phonons, optic phonons as well as by the Coulomb interactions. The corresponding cut-off energies

H. Kamimura (ed.), Theoretical Aspects of Band Structures and Electronic Properties of Pseudo-One-Dimensional Solids, 49–122.

are the Debye energy, the optic phonon frequencies, and the intraband plasma frequency. In addition, it is important to take into account the interchain electron hopping. By considering the interrelation of these quanties it proved possible in Paper I to classify the conducting trichalcogenides (at ambient pressure) as materials dominated by a strongly retarded (low Debye energy) interaction via acoustic phonons. Since the interchain electron hopping is not large they appear thus as Peierls quasi-1d conductors.

Many aspects of theoretical reasoning which lead to such conclusions do not pertain exclusively to conducting trichalcogenides and can be directly applied to TTF-TCNQ. Nevertheless, we shall repeat here most of this discussion because the general point of view regarding TTF-TCNQ is somewhat different than in conducting trichalcogenides. However, this will be done briefly, just in order to keep the paper self-contained. Those developments which are common to both papers will be marked by giving the corresponding expressions two equation numbers, one for the present paper and the second for Paper I. This should enable the reader to find in Paper I a more detailed explanation of considered facts, also in different wording.

TTF-TCNQ and similar materials differ from conducting trichalcogenides in two important respects. First, the Coulomb rather than the phonon interactions seem to dominate in these materials [2]. In contrast to phonon frequencies the characteristic cut-off frequency of the Coulomb interaction, namely the plasma frequency, is large with respect to the temperatures of interest, i.e. the dominant interactions are unretarded. Second, the electron interchain hopping matrix element can be thought of as being small compared with the temperatures of interest. Accordingly TTF-TCNQ represents the first physical example where the many-body effects peculiar to one-dimensional systems with unretarded interactions play undoubtedly an important role (KCP like Pt salts are also quite one-dimensional but they are presumably dominated by the phonon mediated retarded interactions, similarly to conducting trichalcogenides). From this point of view TTF-TCNQ salts are perhaps even more interesting than the single-chain materials of the $(TMTSF)_2PF_6$ type. The latter are certainly less one-dimensional than TTF-TCNQ, but their peculiar superconducting properties have brought them into the focus of current investigations [5].

Instead of superconductivity or magnetism (spin density waves, SDW) the TTF-TCNQ-like materials exhibit the charge density wave (CDW) instabilities. As the electrons are (linearly) coupled to the lattice, the CDW instabilities manifest themselves as lattice instabilities. The lattice properties play here only a secondary role. It is nevertheless interesting to analyze the lattice instabilities by themselves, especially in view of the fact that they involve the sequence of commensurate–incommensurate–commensurate transitions, which include the first-order transition preceded and accompanied by hysteresis effects. In this respect TTF-TCNQ is among the first solid-state systems described in terms of nonintegrable differential equations. The prerequisite for such an approach is a careful symmetry analysis of the second-order transitions which precede the first-order transition. This includes the explanation of the unprecedentedly large temperature dependence of the positions of superlattice Bragg spots in the Brillouin zone.

2. Interaction Scheme

In this section we wish to define a set of elementary excitations and their interactions. In subsequent sections the interactions among these excitations will lead to the collective properties of the system, the main empirical features of which are described in Section 2.4.

2.1. Electron Bands

TTF-TCNQ is a two-chain material. TTF and TCNQ molecules form segregated stacks, as shown [6] in Figure 1(a) and (b). The main conducting axis is the b-axis along the stacks. Indeed, the overlaps of the molecular wave functions are by far the most pronounced along this axis [7]. Each chain contributes a band at the Fermi level, as shown in Figure 2(a). There is a charge transfer ρ between the inequivalent chains [6, 9] so that the number of electrons in the TCNQ band equals the number of holes in the TTF band. The factors which might control this charge transfer will be discussed in Section 3.1. Let us only mention at this point that the Fermi vector is

$$k_F = \frac{\pi}{b}\frac{\rho}{2} \tag{1}$$

with $\rho \approx 0.6$. This value of ρ follows from the structural measurements [10] of the $2k_F$ superlattices (see also Section 2.4).

The width of the bands, i.e. the values of the overlap integrals along the chain can be determined either from tight-binding calculations [7] or from appropriate experiments. The calculations [7, 11] lead to the overlaps t of the order of 0.1 eV. The best experiments to determine the bare bandwidth are probably the optical ones, because they involve high frequencies, where various renormalizations are expected to be nonimportant. Indeed, the plasma frequency ω_0 in TTF-TCNQ was found [12, 13] to be 1.4 eV. ω_0 is related to the long-range Coulomb matrix element by the RPA expression ($\hbar = 1$) [14–16]

$$n_F \frac{e^2}{b} = \frac{1}{4\pi}\left(\frac{\omega_0}{4t}\frac{1}{k_F b}\right)^2 \frac{ac}{4b^2}, \qquad (2);\ (I.5)$$

where

$$n_F = \frac{1}{4\pi t k_F b} \tag{3}$$

is the electron density of states per molecule per spin, averaged over the two chains. Using Equations (1)–(3), together with the structural data [6] for lattice parameters a, b, c, we find the average bandwidth of the order of $4t \approx 0.5$ eV. Such values of t are consistent with the thermopower data [3, 17, 18].

In the next step we proceed to distinguish the bandwidths $4t_Q$ and $4t_F$ on TCNQ and TTF chains respectively (cf. Figure 2). In addition to calculations [7] which

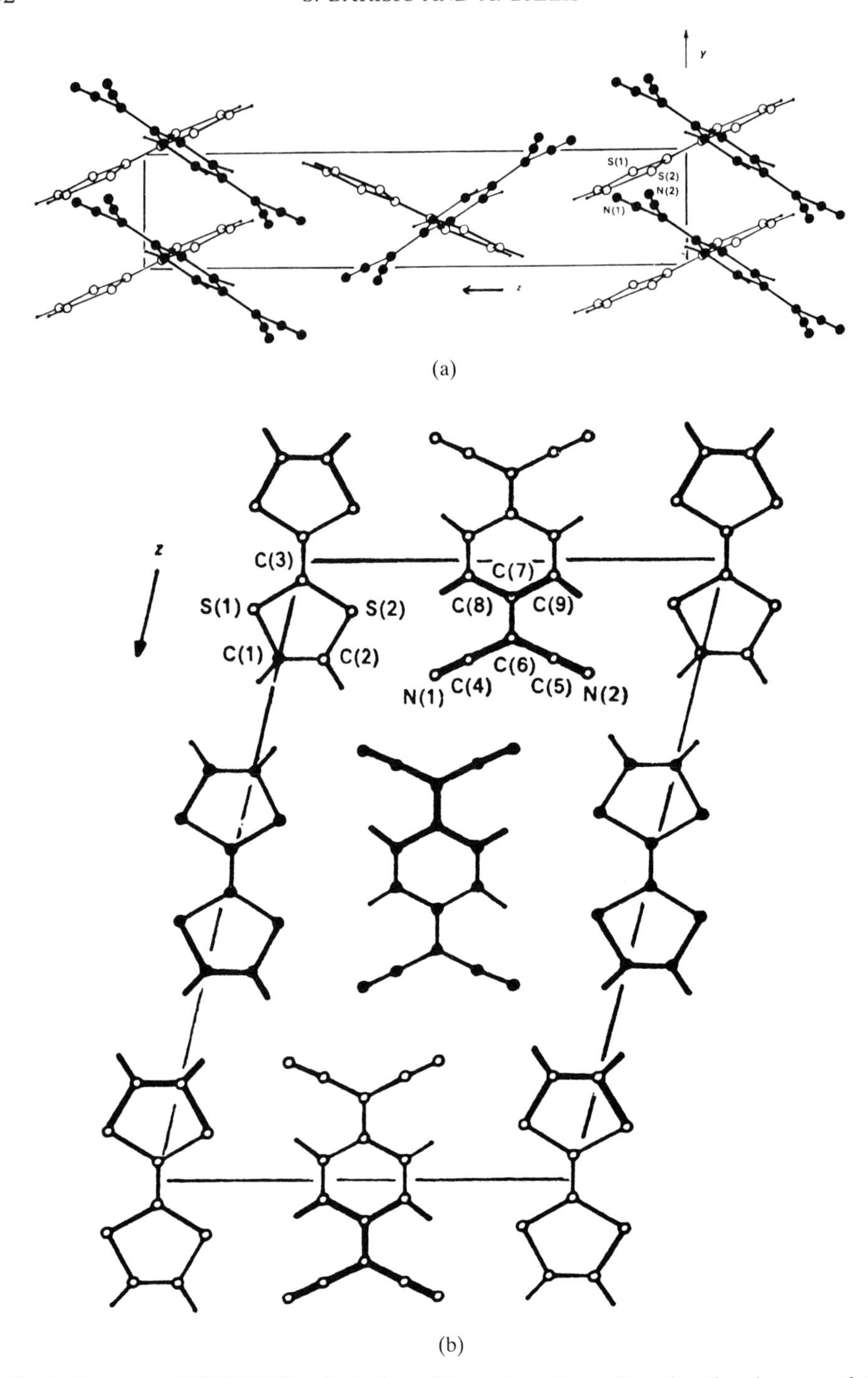

Fig. 1. Structure of TTF-TCNQ: (a) *ab* plane; (b) *ac* plane, heavy lines denoting the parts of tilted molecules above this plane (after Reference [6]). (© International Union of Crystallography.)

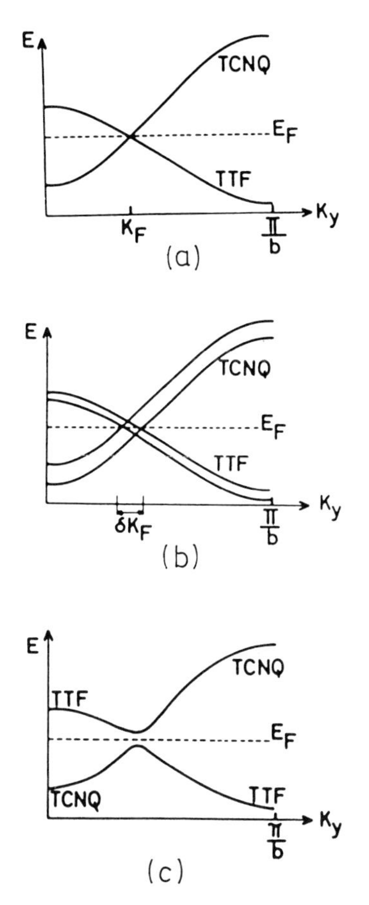

Fig. 2. Electron band structure of TTF-TCNQ (a) $t_\perp$ neglected; (b) with $t_\perp$ in c-direction; (c) with $t_\perp$ in a-direction (after Reference [8]). (© Plenum Publ. Corp., N.Y.)

suggest $t_Q \gtrsim |t_F|$ it can be observed that those overlaps are separately involved in local properties such as the NMR on different chains. The recent C^{13} experiments on TTF [19] combined with previous data [21–23], allowed a rather accurate separation of the uniform magnetic susceptibilities χ_Q and χ_F, shown in Figure 3. The temperature dependence of χ_Q and χ_F is to be associated with interactions between electrons, as will be discussed in Section 3.2. It will be argued there that the low-temperature extrapolation of metallic susceptibilities should lead to the Pauli values $\chi_P \sim 1/t$. In such picture χ_F at 100 K should be close to the Pauli susceptibility of the TTF chain, whereas the low-temperature extrapolation of χ_Q is more ambiguous, because of the structural transition and its precursor effects. Nevertheless, the behavior of χ_Q is not inconsistent with the inequality $t_Q \gtrsim |t_F|$, obtained in band calculations mentioned above.

In addition to t_Q and t_F there are of course weak overlaps $t_\perp$ between the chains, in both the a and c directions (cf. Figure 2). The inclusion of these overlaps intro-

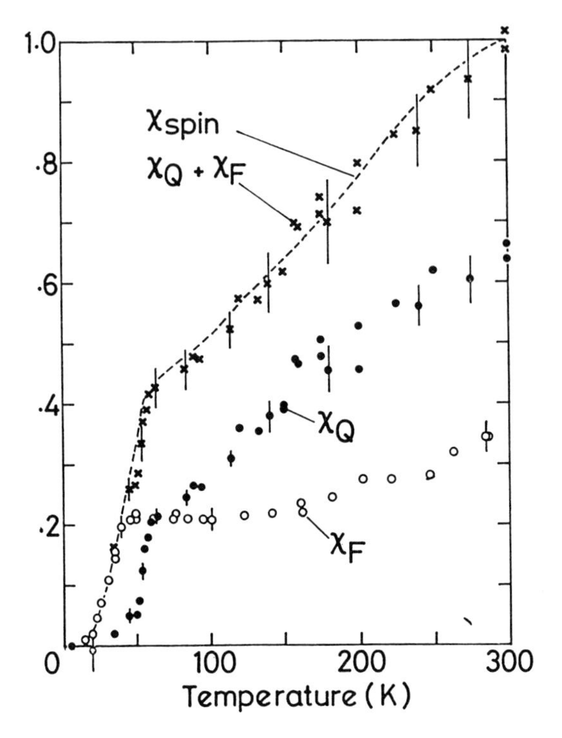

Fig. 3. Uniform magnetic susceptibilities χ_F and χ_Q of TTF and TCNQ chains, respectively, as functions of temperature (after Reference [19]). (© The Institute of Physics.)

duces the full symmetry $P2_1/c$ into the band calculation [8]. As discussed in Section 5.4, this space group has only one-dimensional small representations, i.e. $t_\perp$ (in the *a*-direction [8]) opens the (anticrossing) gap at the Fermi level. This gap should, however, be quite small,

$$T > t_\perp \qquad (4); \ (\text{I}.24)$$

because it is unobservable at temperatures of interest (50 K). Another piece of evidence that $t_\perp$ is small in both the *a* and *c* directions comes from the magnetic field dependence [20] of the NMR relaxation time T_1 which describes the relaxation rate of the spin created at an initial moment on a given site. The spin can either undergo flips or linger away to adjacent sites. The latter process is presumably dominant [20] in TTF-TCNQ. It leads to quite different behaviors in 1d and 3d regimes and the observed crossover [20] is consistent with the inequality (4). The inequality (4) also seems to hold in other salts of this family, with the exception of HMTSF-TCNQ. In the latter material the Hall effect [24] and magnetoresistance [25] measurements have demonstrated the existence of small electron and hole pockets around 100 K, which arise from the $t_\perp$ hybridization of inequivalent chains.

As will be seen in Section 3, the inequality (4) plays an important role in introducing nontrivial many-body effects in the theory of TTF-TCNQ.

2.2. Coulomb interactions

The many-body theory of TTF-TCNQ involves the interacting electron gas. On the other hand, as was stated in the preceding section, the electron bands in TTF-TCNQ are quite narrow. In such cases the Coulomb forces are usually important [26] and therefore will be investigated here in some detail.

Equation (2) is already written in a way which emphasizes the proportionality between the Coulomb matrix element and the ratio of the plasma frequency ω_0 to the bandwidth $4t$. Using the previously found $4t \approx 0.5\,\text{eV}$, we find from Equation (2) that

$$n_F \frac{e^2}{b} \approx 1. \tag{5}$$

This value is obtained from the small wavevector behavior of plasma and represents therefore the extrapolation of the long-range Coulomb matrix element to the first neighbor distance b. Although valid at long distances the point charge approximation used in Equation (2) overestimates the first neighbor matrix element, which is thus actually somewhat lower than the estimate (5).

Let us next turn our attention to the intramolecular Coulomb interaction U. The temperature dependences in Figure 3 suggest that Coulomb interactions are larger on TCNQ chains than on TTF ones,

$$n_F^Q U_Q > n_F^F U_F \tag{6}$$

in spite of $n_F^Q \lesssim n_F^F$ found in Section 2.1. In fact it will turn out in Section 3.2 that the temperature behavior of χ is determined by U rather than by the off-site interactions (5), and therefore the stronger temperature dependence is associated here with larger $n_F U$. The probable reason for the difference between U_Q and U_F resides in the difference of the intramolecular polarizabilities of the two chains. Such approach, currently used in reducing the local Coulomb interaction in transition metals [27] prior to introduction of band effects, was also invoked early [28, 29] for the TTF-TCNQ salts. The recent determination [30] of charge charts in $(TMTSF)_2$ AsF_6 corroborates to some extent this point of view. As can be seen in Figure 4, the counterion AsF_6 attracts the conducting charge on the molecule TMTSF. This molecule is a parent of TTF. Hence similar effect (of opposite sign) is expected to occur when an extra electron arrives on TTF or in its neighborhood. Unfortunately, the analogous results are not available for TCNQ. Being thus unable to compare directly the two molecules, we will take the inequality (6) as indicating that TCNQ is less polarizable than TTF, $\alpha_F > \alpha_Q$ where $\alpha_{Q,F}$ denote the molecular polarizabilities.

Let us finally consider the Coulomb interchain interactions. This will be done very briefly, because the discussion does not differ in any important respect from that given in Section I.3.2 for conducting trichalcogenides. The most important interchain parameter is related to the interaction of $2k_F$ CDWs on different chains. The leading idea in evaluating [31] the corresponding coupling constant $g_1^\perp$ is that the potential created by CDW charge redistribution along the chain is small at distances $R_\perp$ from the chain which, in turn, are large with respect to the CDW

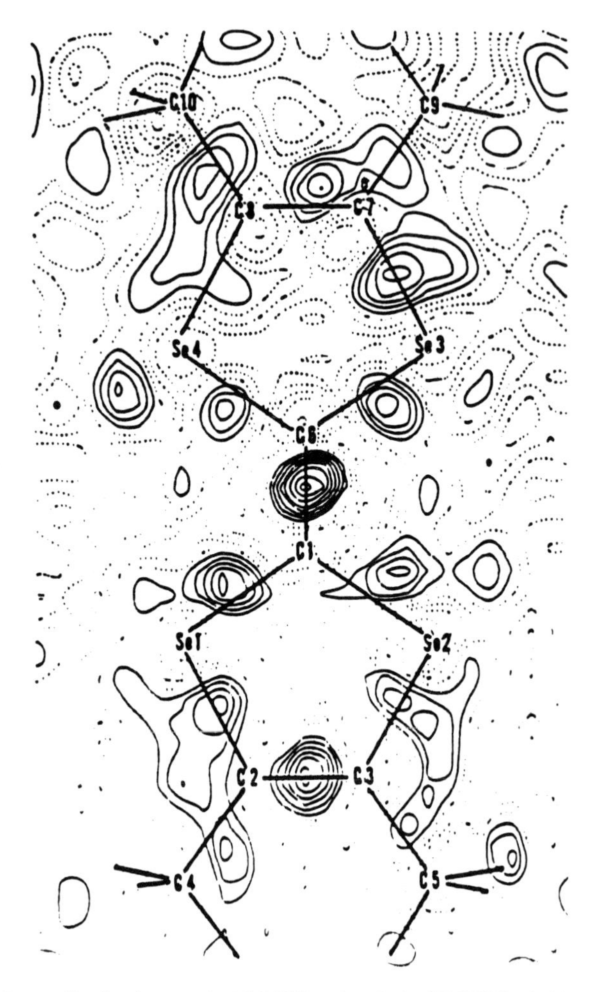

Fig. 4. Electron charge distribution on the TMTSF molecule in $(TMTSF)_2$ AsF_6 conducting salt (after Reference [30]). (© American Association for the Advancement of Science.)

wavelength $2\pi/2k_F$. This is easily understood from Figure 5, where we see that the CDW can be visualized as a linear sequence of compensated dipoles. The interchain distances $a/2$, $c/2$ in TTF are sufficiently large on the π/k_F scale to make the interchain interaction small [32]. Apparently, however, the point charge approximation implicit here is not particularly accurate for short wavelengths in TTF-TCNQ, and the finite size of molecules may be taken into account. This is usually done [33] by splitting the CDW on each chain into (four) components along the

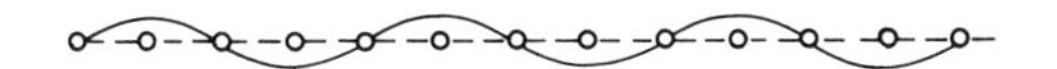

Fig. 5. Schematic view of the $2k_F$ CDW.

strings formed by overlapping wave functions, which are involved in intermolecular tight binding. The (four) component CDWs on the chain are assumed to be in phase (i.e. intramolecular polarizabilities α are ignored). The interaction of the CDWs on the two stacks is thus decomposed into the interaction of component CDWs on the strings (Equation (I.10)). As a result, the overall interaction of CDWs on two stacks is characterized by the unique relative phase $\phi_\perp$ and by the amplitude $g_1^\perp$, composed of interstring interactions. Although all interstring distances are not particularly large with respect to $2\pi/2k_F$, $g_1^\perp$ turns out to be comparatively small [33]

$$g_1^\perp < 0.1. \tag{7}$$

The complete interaction is $g_1^\perp \cos\phi_\perp$, which shows further that the CDWs on neighboring chains prefer to be of opposite phase, $\phi_\perp = \pi$, as is usual in Coulomb–Madelung problems [32].

2.3. Phonons and Electron–Phonon Interactions

The vibrational motions of molecules in molecular crystals are usually classified [34] into translations, librations (rigid rotations of the molecule), and internal distortions of the molecule. In the TTF-TCNQ case translations involve both accoustic and optic modes due to the complex structure of the unit cell. Setting aside the minor differences between such acoustic and optic modes we shall characterize them by a single (cut-off, Debye) frequency $\omega_{ac} \approx 50$ to 100 K, as obtained from neutron [35, 36] and infrared [37–39] investigations. This corresponds to the elastic constants of the order of

$$C \sim 5 \times 10^3\,\mathrm{K\AA}^{-2} \tag{8}$$

obtained also from structural data under pressure [40], which show furthermore that TTF-TCNQ is not quasi-1d from the elastic point of view. The transverse elastic constants $C_\perp$ are even larger than the longitudinal one [3, 40]

$$C_\perp \gtrapprox C. \tag{9}$$

The librations and the internal distortions of the molecules can be observed by infrared [37–39] and Raman [41] spectroscopy. In particular, a number of such modes becomes optically active [42–44] below the structural phase transition, as explained in more detail at the end of this section. Figure 6 illustrates the remarkable resolution, which has been achieved in observing those modes by using the bolometric method [37].

According to the assignments of the bolometric infrared analysis [37], the librations fall into the low-frequency region, between 50 and 100 K. This is related to the fact that the moments of inertia of large molecules are also large themselves [34]. On the other hand, the internal molecular vibrations have frequencies in the range between 100 and 3000 K. The modes mentioned are not quite pure. The mixing is especially important for modes with similar frequencies. This applies in particular to the translational and librational modes, which in addition can mix [37] with twisting and bending about the central bond of TCNQ. This is mentioned here only to

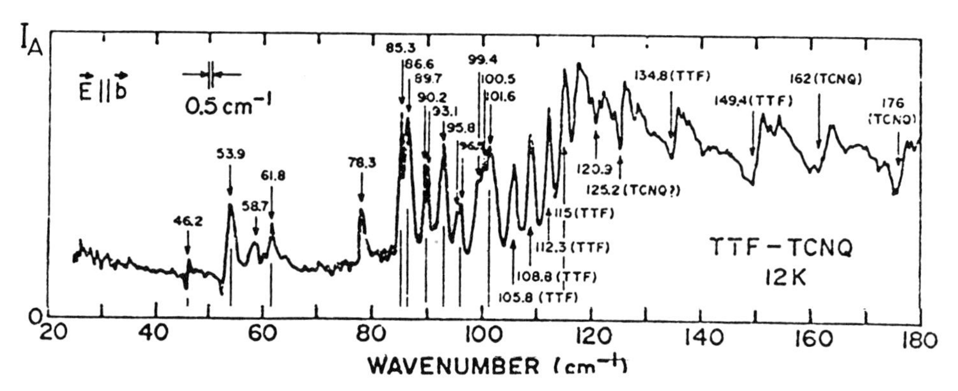

Fig. 6. High-resolution portion of the bolometric spectrum of TTF-TCNQ for $\mathbf{E}\|b$ at 12 K. The molecular assignment is given for the identifiable intramolecular modes (after Reference [37]). (© Canadian Journal of Physics.)

illustrate how detailed the mode analysis in TTF-TCNQ can be, but for our present purpose it is sufficient to note that the frequencies ω_{ac} of translations (and librations) satisfy the condition

$$2\pi T > \omega_{ac} \qquad (10); \ (I.16)$$

for temperatures of interest, i.e. around 50 K. In contrast to that, most of the optical modes fall in the opposite category

$$\omega_{opt} > 2\pi T. \qquad (11)$$

The importance of the distinction between (10) and (11) will be further discussed in Section 4.1.

Here we shall proceed by investigating the coupling of phonons to the conducting electrons. In the case of translational modes we can use the usual estimate of the coupling constant [45]

$$\lambda_{ac} \approx \frac{t}{2\pi C q_0^{-2}} \qquad (12); \ (I.9)$$

which characterizes the effective electron–electron interaction mediated by phonons. Here q_0 denotes the Slater coefficient which describes the spatial extension of bonding wave functions and thereby [45] the rate of change of overlap integrals with deformation. With $q_0 \approx 1\,\text{Å}^{-1}$ we find $\lambda_{ac} \approx 0.1$. Such λ_{ac} certainly satisfies

$$\lambda_{ac} < n_F \frac{e^2}{b}, \ n_F U \qquad (13); \ (I.14)$$

of Equations (5) and (6). As already pointed out in Section I.3.2, the Coulomb coupling constants scale as t^{-1}, whereas the acoustic phonon mediated interactions go as t. Therefore, the Coulomb interactions dominate in TTF-TCNQ, in contrast to the case of conducting trichalcogenides, where the situation is just the opposite.

Comparing further $\lambda_{ac} \approx 0.1$ with the result (7) for $g_1^{\perp}$ we find

$$\lambda_{ac} \gtrsim g_1^{\perp}. \tag{14}$$

This probably is not a strong inequality, but even an equality would roughly do for our later purposes (Section 4.1).

The effect of possibly important [34] librations on electrons in an isolated chain of TTF or TCNQ is of the second order in displacements [34]. The first-order effect may thus be associated with important librations only by their mixing with other modes.

Turning now to electron–electron coupling through internal modes, we can use the estimate based on the appropriately modified Equation (12). The intramolecular overlaps are measured by the splitting ΔE of the intramolecular electron states. Such $\Delta E \approx 1\,\text{eV}$ replaces t in Equation (12), the Slater coefficient remains essentially unchanged, whereas C describes now the molecular rigidity, which is considerably larger than the crystal rigidity (8). This indicates that it is reasonable to assume

$$\lambda_{opt} < n_F U,\ n_F \frac{e^2}{b}. \tag{15}$$

However, in a more careful approach λ_{opt} should be evaluated by summing over the separate contributions of various intramolecular modes, which are linearly coupled to electrons through the variation of intramolecular overlaps. All such modes, seven on TTF and ten on TCNQ [37] belong to the fully symmetric (a_g) class of deformations. Further analysis is usually based on semi-empirical arguments [44] rather than on refinements [46] of Equation (12). The totally symmetric (a_g) modes are infrared inactive at high temperatures. However, at low temperatures TTF-TCNQ develops [10, 36] a static CDW, illustrated for example in Figure 5. As already mentioned in connection with this figure, the total dipole moment of such CDW vanishes. However, the long-wavelength deformations of the CDW structure (phasons) produce a dipole moment and thereby the CDW infrared activity [47–50], more commonly known as Fröhlich conductivity [51, 52]. CDW and, consequently, its deformations (phasons) are in turn linearly coupled to the a_g modes. Infrared resonances are thus obtained for frequencies of a_g modes, which so become infrared [44] active at low temperatures. The strength of these resonances is determined by the values of individual electron–(a_g) phonon couplings [44]. Although the detailed analysis, carried out for TEA $(TCNQ)_2$ yields [44] only an indication about λ_{opt} in TTF-TCNQ, it appears reasonable to accept the assumption (15). The consistency of this assumption with other experimental findings will be further examined in Section 3.4.

2.4. Instabilities

Until now we have defined electron bands, phonon branches, electron–phonon, and electron–electron Coulomb interactions. This involved some simple theoretical

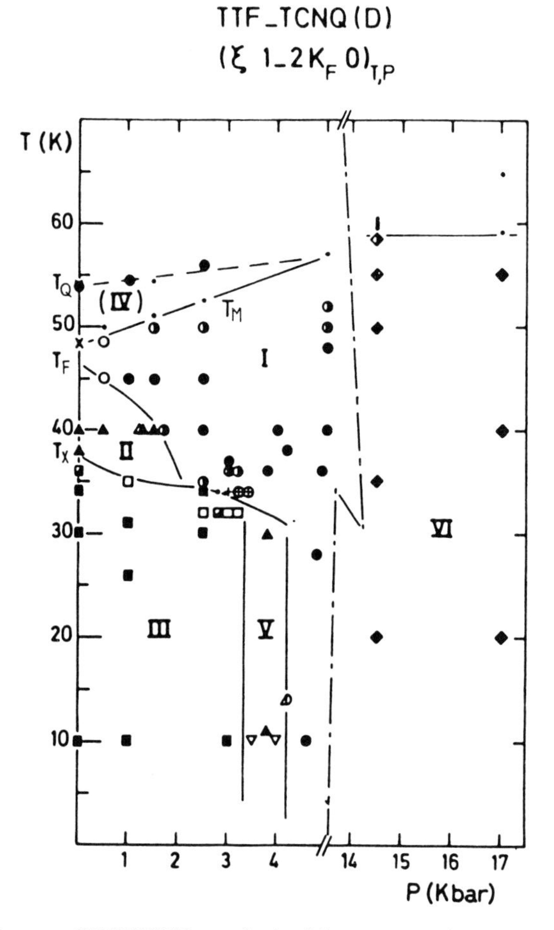

Fig. 7. p–T phase diagram of TTF-TCNQ, as obtained from structural measurements. Transition lines are associated with transverse (interchain) orderings of displacements discussed in Section 5 (after S. Megtert, These, Orsay, 1984, and Reference [3]). (© Taylor & Francis Ltd.)

considerations combined as much as possible, with the analysis of high-temperature and high-frequency data. The result is contained in a set of inequalities which will lead in forthcoming chapters to the appropriate theoretical description of the low-temperature collective properties of the system. However, before going to the theory it seems useful to review the main experimental results concerning the low-temperature phases, in order to make the preceding and the future theoretical choices more obvious.

At low temperatures (~50 K) and ambient pressure TTF-TCNQ undergoes a sequence of structural [10, 36, 53, 54] phase transitions (cf. Figure 7). At first sight this may seem to contradict the inequalities (13) and (15) which are relegating the lattice to the secondary role. However, it should be noted that all transitions occur on a very narrow temperature scale, within some ten degrees (38–54 K). As will be argued in Section 4, different lattice modes and different chains are involved in those transitions. In spite of the fact that their bare frequencies differ noticeably, as shown in Figure 8, the transition temperatures are nearly the same. This leads us to believe that a strong Coulomb singularity in the electron system only reflects itself in the lattice properties of the crystal.

Such point of view is corroborated by pretransitional behaviour of TTF-TCNQ. The temperature dependences of the $2k_F$ lines differ appreciably from the slow logarithmic behavior predicted in the Peierls theory (Equation (I.19)). Instead, only a very weak $2k_F$ anomaly is present in the phonon spectrum down to temperatures as low as 100 K, as shown in Figures 9, 10, and 11. In fact, the neutron data [35, 36] of Figure 9 alone are insufficient for such conclusion, because their poor resolution in the low-frequency range can possibly hide the formation of the central (quasi-elastic) peak. This is why Figures 10 and 11 present the X-ray measurements [36, 53, 54], which give the scattering intensity integrated over all frequencies. Comparison of those data with neutron scattering results gives a clear although indirect evidence

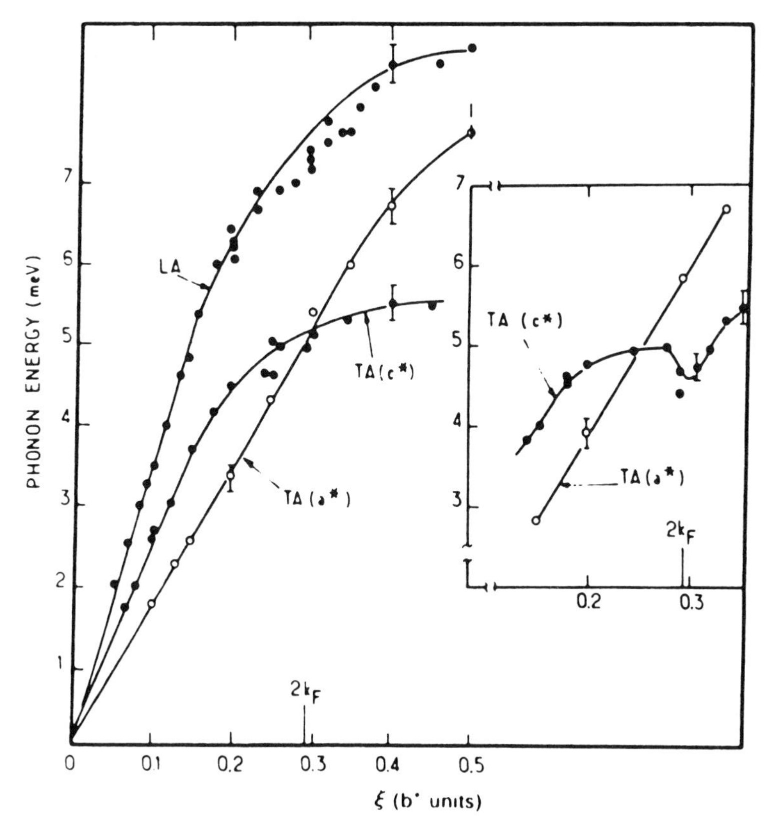

Fig. 8. Dispersion curves from TTF-TCNQ(D) for acoustic modes propagating along the chain at 295 K and 84 K (insert) (after Reference [35]). (© The American Physical Society.)

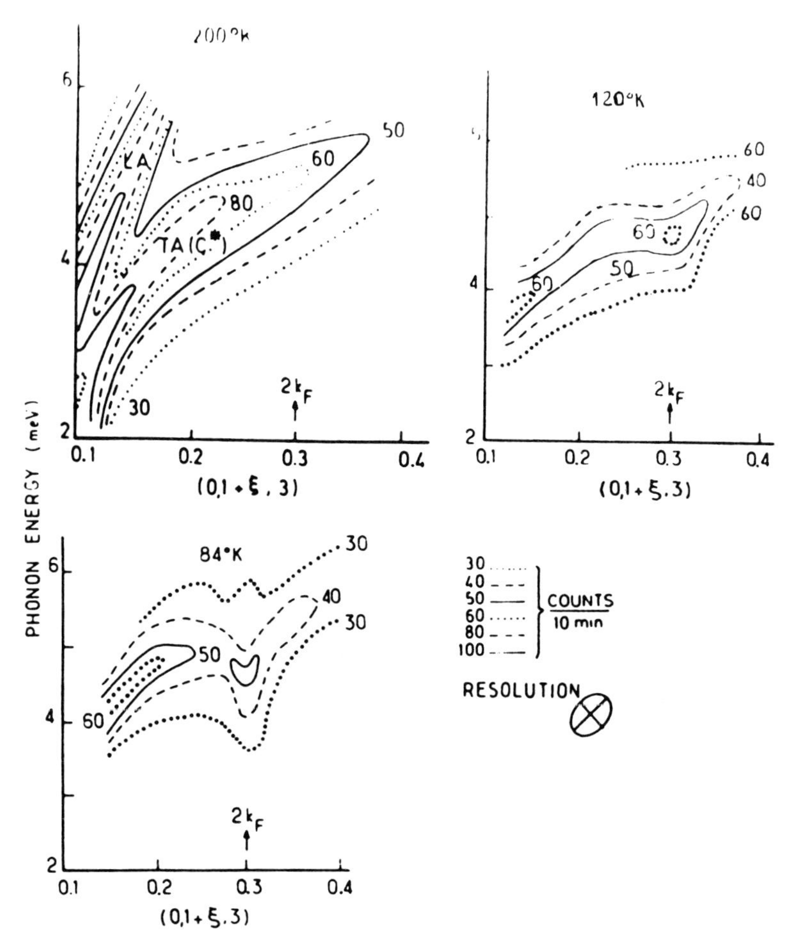

Fig. 9. Normalized ω, q intensity contours of the neutron scattering of TTF-TCNQ from the acoustic excitations mainly polarized in the c^*-direction and at 200, 120, and 84 K (after Reference [35]). (© The American Physical Society.)

that the central peak develops somewhere below 100 K, on the very narrow energy scale.

Further evidence for the non-Peierls behavior of TTF-TCNQ comes from the $4k_F$ scattering. The weak diffuse $4k_F$ lines are observed in X-ray data [53, 54] of Figures 10 and 11 already at temperatures as high as 300 K. Figure 11 shows clearly that this $4k_F$ scattering cannot be interpreted as a harmonic of the $2k_F$ one because the temperature behaviors of the two scattering intensities are quite different. This difference will find its natural explanation within the Coulomb interaction model of the next section.

Let us finally turn our attention to temperatures well below the structural phase transitions. In this temperature range the behavior of TTF-TCNQ is characterized by the set of gaps. The gap determined from the electrical d.c. conductivity mea-

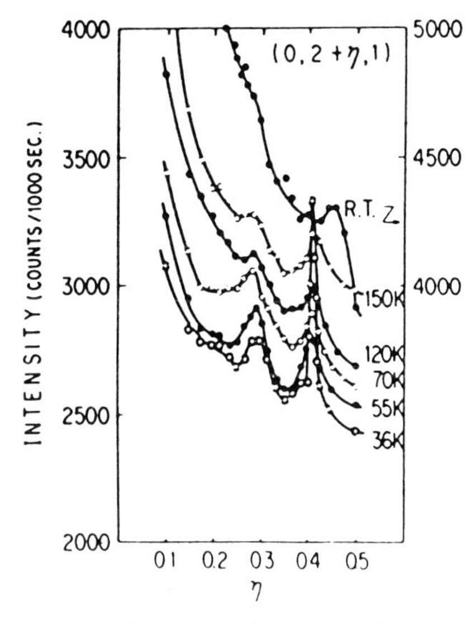

Fig. 10. X-ray counter measurements of the $2k_F$ and $4k_F$ scattering as a function of temperature (after Reference [54]). (© Physical Society of Japan.)

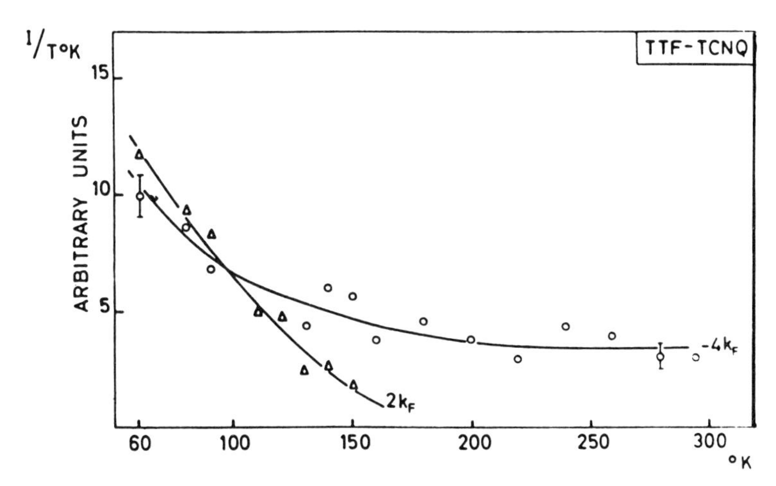

Fig. 11. The temperature dependence of the peak intensity I of the $2k_F$ and $4k_F$ scatterings. I is divided by T to eliminate the phonon population factor (after Reference [53]). (© The American Physical Society.)

surements [55] is $\Delta_\sigma \approx 200$ K. The gaps Δ_χ found by NMR [19], EPR [23], and static magnetic susceptibility [56] measurements do not differ substantially from this figure, especially if we think in terms of the two-chain model. For example, the static magnetic susceptibility at low temperatures was successfully fitted [56] with $\Delta_\chi^Q \approx 250$ K and $\Delta_\chi^F \approx 140$ K, the latter value being not too far from the recent [19] C^{13} NMR estimate. All those gaps are undoubtedly related to the lattice instability, and we shall symbolize them by Δ_L. It is customary to compare the $T = 0$ values of gaps Δ_L^0 to the corresponding transition temperatures T_L, because the Peierls theory indicates that $\Delta_L^0 \approx 2T_L^0$, (Equation (I.38)), where T_L^0 denotes the mean-field (MF) transition temperature. In Section 4.2 it will turn out that a similar relation also holds for the strongly correlated electron gas, coupled to the lattice. Therefore, in contrast to the temperature behavior, this relation or the departures from it, cannot be used to distinguish between the Peierls and Coulomb cases. Rather, the fact that the actual transition temperatures T_L ($\approx$50 K) are lower than $T_L^0 \approx \Delta_L^0/2$ will be used in Section 4.2 to introduce the departure from the mean-field theory, whatever its microscopic background happens to be.

Before closing this section it should be mentioned that besides the gaps Δ_L, which show up only below T_L, there are structures which persist up to room temperature. One such structure is the already mentioned plasma frequency ω_0 of Equation (2), which is actually the mid-zone gap in the CDW spectrum. In addition to that, Figure 12 shows that the low-temperature gap Δ_L evolves at room temperature in the structure at $\omega_R \approx 500$ K, sometimes named gap or pseudogap [57]. As will be discussed below – and in particular in Section 4.3 – so far this feature has not received a definitive interpretation.

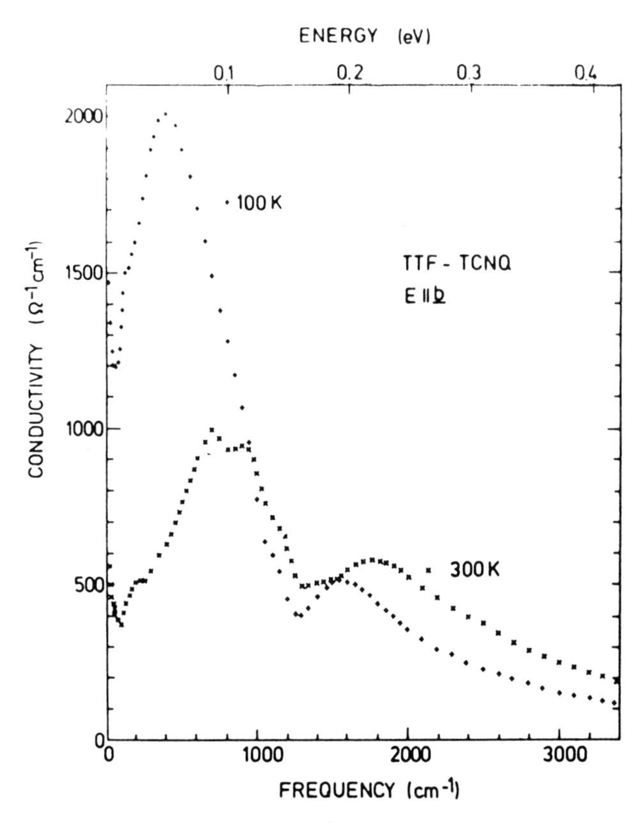

Fig. 12. Frequency dependent conductivity for $\mathbf{E}\|b$ at 300 and 100 K (after Reference [57]). (© Springer-Verlag, Heidelberg.)

3. Coulomb Correlated Electron Gas

In Sections 2.2 and 2.3 it appeared that due to the small bandwidth of TTF-TCNQ the Coulomb interactions are larger than the acoustic phonon mediated coupling λ_{ac}. Therefore we shall start by ignoring the CDW coupling to the translational modes of the lattice. It is also likely (cf. Equation (15)) that Coulomb interactions dominate over the attractive couplings mediated by optic phonons. However, in contrast to acoustic phonons most of the optic phonons have frequencies ω_{opt} comparable to the plasma frequency ω_0 (and $\omega_0 \approx \varepsilon_F = v_F k_F$). Therefore, λ_{opt} act additively [58, 59] with Coulomb interactions, i.e. they can be investigated within the same theoretical scheme as the Coulomb interactions. Actually we shall discuss here in some detail the situation where λ_{opt} is dominant, mainly in order to argue that the theoretical consequences of such assumption as opposed to Equation (15), appear to be at variance with most experimental results in TTF-TCNQ.

The effect of the acoustic phonons will be discussed in Section 4, where the present results will be incorporated in the appropriate theory of lattice instabilities. Such order is inverse to that chosen in Paper I, Section IV, where the emphasis was on acoustic phonon interaction λ_{ac}, important in conducting trichalcogenides. However, most results derived in I are actually independent of the relative values of λ_{ac} and the Coulomb couplings, i.e. of the order of presentation, and can be directly transferred to TTF-TCNQ. In this paper we shall thus only briefly rephrase the results of Paper I and supplement them to emphasize the importance of Coulomb forces in TTF-TCNQ.

The Coulomb forces in TTF-TCNQ are not only larger than λ_{ac} but also they are quite large in absolute value. Indeed, $n_F(e^2/b)$ is of the order of unity, according to the estimate (5). The on-site values $n_F U$ can hardly be smaller. In other words, TTF-TCNQ is an intermediate coupling material, and therefore, strictly speaking, requires a numerical approach. Until now such numerical calculations were performed only for some simple purely 1d cases [60–63]. On the other hand the results obtained in the weak coupling limit [16, 59, 64, 65], besides their elegance, include the analysis of the respective roles of the interchain hopping and the interchain electron–electron interactions [59, 65, 67–71], which are quite important for the proper understanding of real materials. Our general attitude here will thus be to discuss the weak-coupling theory, attempting to extrapolate its conclusions to intermediate couplings in a physically reasonable manner. This approach is supported by the fact that TTF-TCNQ is a good metal above 50 K. Moreover, in all available cases the theory gives no qualitative difference between the weak- and intermediate-coupling results.

3.1. Cohesion of TTF-TCNQ

Let us start by stating a few additional facts which support the weak-coupling (metallic) approach from the cohesive point of view: The condensation of $2k_F/4k_F$ superlattices occurs at temperatures T_L and with gaps Δ_L which are much smaller than the bandwidth. Moreover, this condensation does not affect critically either the charge transfer ρ [10, 36], the lattice parameters a, b, c [3, 72–74], or the elastic constants C, $C_\perp$ [3, 40], although a, b, c and especially C do show a significant but smooth temperature dependence. This tends to rule out the $2k_F/4k_F$ superlattice contribution as the main source [75, 76] of cohesion in TTF-TCNQ. At this stage we are facing two possibilities. The first is that there are large 'metallic' terms in the cohesion, unrelated to the $2k_F/4k_F$ superstructure, which determine the parameters ρ, a, b, c, C, $C_\perp$ almost independently of the superlattice effects. The second alternative is that the electron band formation is simply irrelevant to the stability of the main lattice, which is thus again independent of the superlattice.

Let us consider this latter possibility in some more detail assuming that, like aromatic insulators, such as anthracene and naphthalene, TTF-TCNQ is bound together by the intermolecular van der Waals (vdW) interactions [3, 40, 77, 78]. It can be noted in this respect that the geometry of the TTF-TCNQ lattice is consistent with such a model. Figure 1, schematized in Figure 13(a), shows that the equivalent chains are aggregated in planes along the c-direction. The propensity of equivalent molecules to segregate in chains and planes can tentatively be attributed to the vdW forces, because the Madelung energy of charged molecules favors on the contrary the Q, F alternation in all three directions. This can be seen from the London approximation for the vdW energy which expresses it through the product of polarizabilities α of the molecules involved in the particular bond [77, 79]. The vdW energy change on going from the configuration 13(b) to 13(a) is thus [3, 40, 77, 79]

$$\Delta E_{\mathrm{vdW}} \sim -(\alpha_Q - \alpha_F)^2. \tag{16}$$

As required, the TTF-TCNQ configuration has a lower vdW energy. The intramolecular polarizabilities $\alpha_{Q,F}$ have already been discussed in connection with Equation (6) and Figure 4, with the conclusion that they differ considerably. It can be thus speculated that ΔE_{vdW} is sufficient to compensate for the loss ΔE_M of the Madelung energy. In fact the simultaneous existence of both configurations in nature (e.g TTF-TCNQ [6] and HMTTF-TCNQ [80] lattices) indicates that ΔE_{vdW} and ΔE_M should be of the same order of magnitude. Unfortunately the available theoretical estimates give [75] α's too small to make for this energy balance, or,

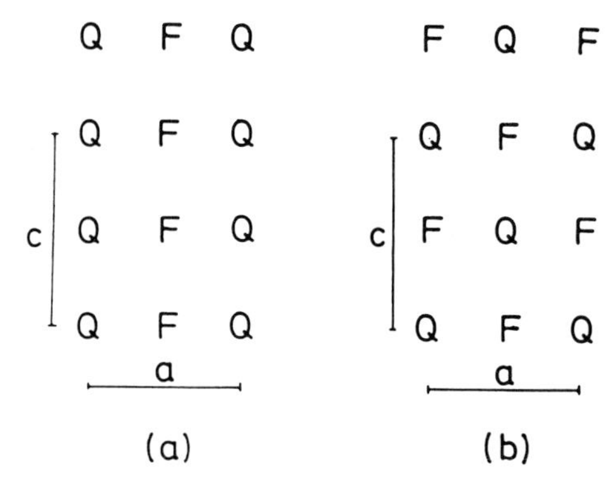

Fig. 13. Schematic view of the ac plane in (a) TTF-TCNQ; (b) HMTTF-TCNQ.

what is similar, for the stability of each lattice separately. This discrepancy becomes even more significant on realizing [3] that organic conductors are already quite stiff at room temperature, twice as stiff as aromatic insulators [78, 81], so that large polarizabilities are in fact required. An additional difficulty is inherent to the molecular bonding model. Empirically, the charge transfer ρ is remarkably stable, not only at T_L in TTF-TCNQ, but also from one salt to another. This requires [82] an E_{vdW} strongly dependent on ρ, which together with $E_{\mathrm{M}} \sim \rho^2$ would create a pronounced maximum of the cohesive energy at $\rho \approx 0.6$. Although such $E_{\mathrm{vdW}}(\rho)$ (i.e. $\alpha(\rho)$) cannot be completely ruled out in the molecular coupling model, this leads us definitely to turn our attention to the model of metallic cohesion.

In the metallic model [83] the Madelung energy coincides with the Hartree term in the weak-coupling expansion, Figure 14(a). The other low-order terms are shown in Figures 14(b) and (c). Figure 14(b) represents the Fock term. This is an on-chain

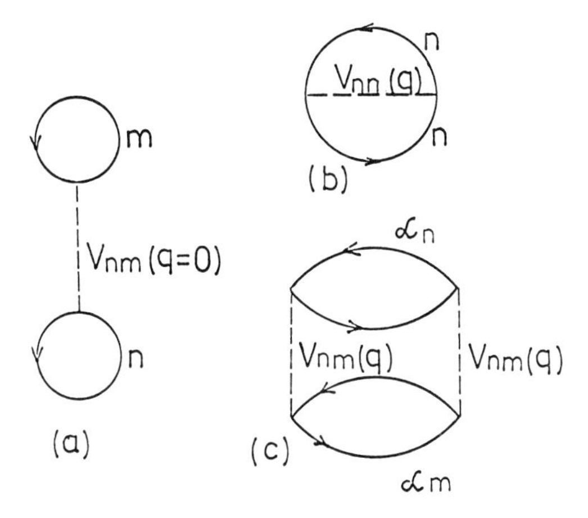

Fig. 14. Low-order conducting electron contributions to the cohesive energy: (a) Hartree–Madelung term; (b) Fock intrachain term; (c) van der Waals interaction between chains.

term which is dominated by the large momentum ($q \approx 2k_{\mathrm{F}}$) transfer part of the (on-chain) Coulomb interaction. All on-chain terms cancel out from the difference ΔE_c of the cohesion energies corresponding to the configurations 13(a) and (b), and to this extent the Fock term can be ignored. The next term, that of Figure 14(c), has an interesting meaning in chain materials. It represents [83] the energy of correlated charge fluctuations (dynamic CDWs) on chains n and m, i.e. it is the metallic counterpart of the molecular vdW terms discussed above. The difference is that the charge fluctuations occur here along the chains rather than within the molecules. Following the idea of separation of the main lattice and the superlattice cohesion, we shall consider here only the long waves $q \ll 2k_{\mathrm{F}}$. Their contribution to cohesion is comparatively large due to large values of the corresponding Coulomb forces, and in particular of their interchain components. Indeed, $n_{\mathrm{F}}(e^2/R_{\perp})$ is larger for many interchain distances $R_{\perp}$ than $g_1^{\perp}$ of Equation (7). With this in mind it is easy to understand the metallic vdW energy of the pair of chains (n, m) at distance $R_\perp = R_{nm}$ [83],

$$E_{\mathrm{vdW}}^{(n,m)} = -\frac{e^4}{\pi^2 R_{nm}^2}\frac{n_{\mathrm{F}}^{(n)} n_{\mathrm{F}}^{(m)}}{n_{\mathrm{F}}}, \tag{17}$$

remembering that the long wavelength metallic chain polarizabilities are proportional to the corresponding electron densities of states $n_{\mathrm{F}}^{(n,m)}$ (n_{F} is the average value).

In contrast to the usual vdW forces their metallic counterparts (17) are quite long range. However, it should be kept in mind that the charge fluctuation on the nth chain cannot excite the fluctuation on the remote mth chain, because its potential is screened out by the fluctuation induced on intermediate chains. In consequence, the summation over R_{nm} has to be cut off at distances of the order the Thomas–Fermi screening distance k_{TF}^{-1}, related to the plasma frequency (2) in the usual way $\omega_0 \approx v_{\mathrm{F}} k_{\mathrm{TF}}$. Omitting for convenience the intrachain term in the n, m summation we get in this way [83]

$$E'_{\mathrm{vdW}} \approx \frac{n_{\mathrm{F}} e^4}{ac} \log(k_{\mathrm{TF}}\sqrt{ac}). \tag{18}$$

According to Equations (2) and (5) the logarithm in Equation (18) is of the order of unity in TTF-TCNQ, signifying that the intermediate couplings are attained. As for the magnitude of E'_{vdW}, the intrachain overlaps in $n_{\mathrm{F}}^{-1} \sim t$ are somewhat smaller than the intramolecular overlaps ΔE in the molecular polarizabilities [75] $\alpha \sim (\Delta E)^{-1}$, as already discussed in connection with λ_{opt} of Equation (15). The metallic E'_{vdW} thus tends to be somewhat larger than its molecular counterpart. Turning now to the dependence of E'_{vdW} on ρ, we realize that the main effect comes from $n_{\mathrm{F}} \sim \rho^{-1}$. Using essentially this result some reasonable values have been obtained [82] for the optimal charge transfer in TTF-TCNQ salts.

The long-range effects are absent in ΔE_{vdW}, whatever k_{TF}, and from Equation (17) we find

$$\Delta E_{\mathrm{vdW}} \approx -\frac{e^4}{c^2}\frac{(n_{\mathrm{F}}^{\mathrm{Q}} - n_{\mathrm{F}}^{\mathrm{F}})^2}{n_{\mathrm{F}}^{\mathrm{Q}} + n_{\mathrm{F}}^{\mathrm{F}}}. \tag{19}$$

Not surprisingly such ΔE_{vdW} has the same structure as that of Equation (16), i.e. the metallic vdW forces also favor the TTF-TCNQ configuration 13(a). Comparing, further, Equation (19) to ΔE_{M} we see that

$$\left|\frac{\Delta E_{\mathrm{vdW}}}{\Delta E_{\mathrm{M}}}\right| \approx n_{\mathrm{F}} \frac{e^2}{b}\frac{b^2}{c^2\rho^2} \tag{20}$$

is of the order of unity. This is in satisfactory agreement with the simultaneous appearance of lattices 13(a) and (b) in nature.

It should be kept in mind, however, that the above results are obtained by using the weak-coupling expressions for intermediate couplings. It is therefore interesting to test this procedure on an exactly solvable model. Such is the Tomonaga model in which the interactions with large momentum transfer ($\sim 2k_F$) are completely neglected, and not only selectively as was done here. The exact Tomonaga results, obtained by bosonization [84, 85], agree with the perturbative results in the range of $n_F(e^2/b)$ which depends on the choice of $n_F U$. For $n_F U \approx 0$ this goes up to only $n_F(e^2/b) \lesssim 0.1$ [85]. For large $n_F(e^2/b)$ the perturbative expressions (18) and (19) overestimate E'_{vdW} and ΔE_{vdW}, especially the latter. However, the large local terms are not treated consistently within the Tomonaga model. It might thus be that the proper treatment of those terms would extend the agreement to larger values of $n_F e^2/b$. Alternatively, we can think that the metallic vdW terms determine ρ [82], whereas the molecular [3, 40, 77] vdW terms make the TTF-TCNQ structure relatively more stable than the HMTTF-TCNQ one. Apparently, these and other alternatives [61, 75, 86, 87] require further investigation, using more realistic models and better approximations.

3.2. Parquet Approximation – Coupling Constants

Having discussed the cohesion of the main TTF-TCNQ lattice, we turn to the formation of electron superlattices (CDWs, Wigner lattices [88, 89]). This discussion is usually [59, 90, 91] carried out in terms of correlation functions rather than of condensation energies used in the previous section. Here we shall follow the usual approach.

When $T > t_\perp$ [65] (Equation (4)), the simplest approximation which can be employed in the investigation of electron gas with Coulomb interactions is the parquet approximation. As was explained in detail in Paper I, two regimes [92] are further encountered according to the relative values of Coulomb interaction and the retarded [45, 65, 92–94] ($2\pi T > \omega_{ac}$) acoustic–phonon mediated interaction λ_{ac}. If λ_{ac} is large the Coulomb corrections to the unrenormalized electrons are small for all temperatures above $T_P^0 \approx \varepsilon_F e^{-1/2\lambda_{ac}}$ [92], where $\varepsilon_F = v_F k_F$. In T_P^0 we recognize the mean-field Peierls transition temperature (Equation (I.20)), which thus describes the lattice instability induced by the nearly free electron gas. On the other hand, if λ_{ac} is smaller than the Coulomb matrix elements, as is the case in TTF-TCNQ according to Equation (13), the parquet renormalizations of electrons are important and the parquet series has to be summed to all orders [2, 65].

In dealing with parquet summation it is customary [59, 64, 91] to use the Fourier transforms of the Coulomb interactions, defined by Equations (2) or (5) and (6). As in the preceding section the small momentum transfer ($q < k_F$, forward) scattering is distinguished in this way from the large momentum transfer (backward, $2k_F$ scattering).

The bare Coulomb interaction is long range, as already stated in connection with Equation (17). Its Fourier transform thus contains the 3d $|\mathbf{q}|^{-2}$ singularity for wave-vectors smaller than $d_\perp^{-2}$, where $d_\perp$ symbolizes $a/2$ and $c/2$. This singularity falls in the forward region if $k_F \approx b^{-1}$ satisfies

$$k_F d_\perp > 1. \tag{21}$$

Such a strong singularity requires, first of all to be screened. This is achieved within

RPA [14–16]. Roughly speaking, RPA cuts off all the interactions at distances beyond k_{TF}^{-1} or times beyond ω_0^{-1}. According to Equation (2) ($\omega_0 \approx v_F k_{TF}$)

$$k_{TF} d_\perp \lesssim 1 \tag{22}$$

in the weak-coupling limit $n_F(e^2/b) < 1$. Combining Equations (21) and (22) we see that $k_F > k_{TF}$ or $\varepsilon_F > \omega_0$: The RPA screening is confined to the 3d regime (Equation (22)), which in turn falls in the forward scattering range.

The result of the RPA screening is particularly simple within the parquet approximation if $g_1^\perp$ of Equation (7) is disregarded [16]. Indeed, this $g_1^\perp$ is small and can certainly be neglected in the 1d regime ($T > 80\,\mathrm{K}$ in TTF-TCNQ according to Section 2.4). In this case the only important forward interaction in parquet diagrams is that between two electrons on the same chain. It is obtained by integrating the RPA screened 3d Coulomb interaction over the transverse momentum, i.e. the screening medium for the on-chain interaction involves the neighboring chains. The resulting forward interaction is [16]

$$g_2 = n_F\left(U + \frac{e^2}{b}\log\frac{\varepsilon_F}{\omega_0}\right). \tag{23); (I.27}$$

In analogy with Equation (18), the long-range nature of the Coulomb forces increases g_2 over the first neighbor value, the cut-off being ensured by k_{TF} or ω_0. The appropriate weak-coupling requirement is thus

$$n_F\frac{e^2}{b}\log\frac{\varepsilon_F}{\omega_0} < 1 \tag{24}$$

rather than $n_F(e^2/b) < 1$. However, in materials like TTF-TCNQ where $b \approx d_\perp$, $n_F(e^2/b) < 1$ implies (24), according to Equation (2). In fact in TTF-TCNQ, $n_F(e^2/b) \approx 1$ and the logarithmic term in Equation (23) should be taken of the order of unity, similarly to what was assumed in connection with Equation (18).

With this in mind let us proceed with the weak-coupling theory. Under the assumption (21), the backward scattering g_1 is a regular function of the momentum. For usual band fillings, Equation (1), it can be taken as [16, 31]

$$g_1 \approx n_F U. \tag{25}$$

The important combination of the coupling constants turns out to be $2g_2 - g_1$. In Equation (25) we are therefore omitting the terms of the order of $n_F(e^2/b)$ with respect to $n_F(e^2/b)\log(\varepsilon_F/\omega_0)$ of Equation (23), just as in this equation itself [16].

3.3. Parquet Sums

The next step consists in introducing the nonsingular (effectively short-range) forces (23) and (25) in the singular processes [16]. The relevant low-order vertex corrections to g_1 and g_2 are shown in Figure 15. It is the particularity of the 1d electron gas with short-range interactions, that the e–h diagrams of Figure 15 are logarithmically singular, besides the Cooper (e–e) processes, which are the same irrespective of the

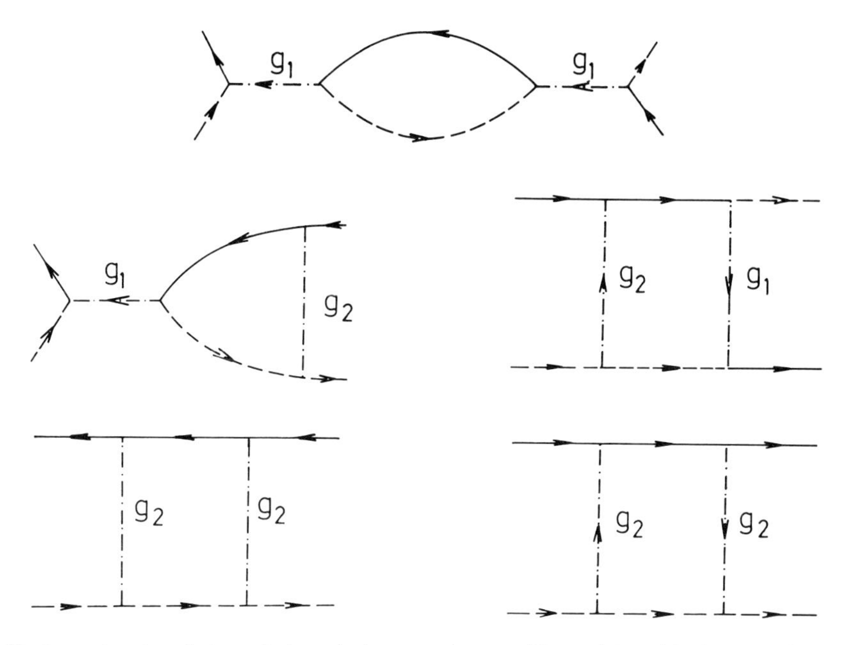

Fig. 15. Lowest order electron–hole and electron–electron (Cooper) logarithmic corrections to the Coulomb interactions. Full line: (k_F); dashed line: $(-k_F)$ electron or hole; dotted-dashed line: Coulomb interaction.

dimensionality [64]. The vertex corrections of Figure 15 iterated [16] within the parquet scheme [64] result in the logarithmic screening of the 'bare' interactions (23) and (25). It should be noted in this respect that the logarithmic screening involves the RPA bubble of Figure 15 for wavevectors around $2k_F$, whereas the screening (22), (23) of the forward Coulomb interaction involved this bubble only for wavevectors smaller than k_{TF} ($k_{TF} < k_F$). This shows that the RPA diagrams are not double counted in the described scheme. The logarithmically screened vertices turn out to be [64, 66]

$$g_1(T) = \frac{g_1}{1 + g_1 \log(\varepsilon_F/T)} \qquad (26); \ (I.58)$$

and

$$g_2(T) \approx g_2 - \frac{g_1}{2} + \frac{g_1(T)}{2} \qquad (27); \ (I.59)$$

with g_2 and g_1 given by Equations (23) and (25), respectively.

To conclude this part of the discussion let us emphasize that the treatment [16] of screening in two separate steps (pre-logarithmic and logarithmic) is possible due to the weak-coupling assumption (22). This separability justifies also the RPA approach to the cohesive energy in the preceding section: the logarithmic effects are the superstructure on the basic, metallic, RPA scheme.

Coming back to the results (26) and (27), it is clear that the sign of g_1 plays an important role. We shall close this section by considering the $g_1 > 0$ case, leaving $g_1 < 0$ for the next section. $g_1 > 0$ makes $g_1(T)$ small, i.e. the parquet approximation valid at all temperatures [59, 66]. The first parquet result we wish to quote is that for the uniform magnetic susceptibility [70, 90, 95, 96]

$$\chi = \chi_P \frac{1 + g_1(T)}{1 - g_1} \tag{28}$$

For $g_1 > 0$ such χ decreases on lowering the temperature, in qualitative agreement with experimental results shown in Figure 3 and discussed in Section 2.2. The Pauli value χ_P enhanced by $(1 - g_1)^{-1}$ is reached for temperatures well below

$$T_g \approx \varepsilon_F \, e^{-1/|g_1|} \qquad (29);\ (I.60)$$

In fact Equation (28) should be compared to experiments in the temperature range well above the structural transition in order to avoid the pretransitional effects. It should be noted in this respect that the temperature dependence (26) is however too weak to account for the behavior of $\chi_Q(T)$ observed in Figure 3. It is likely that the intermediate coupling corrections [61] would improve the fit, but careful comparison has not been carried out so far. Accepting thus tentatively $g_1 > 0$ (i.e. the assumption (15)), we can quote further the corresponding behavior of the $2k_F$ CDW and SDW correlation functions. The latter follows from the parquet equations [90] shown in Figure 16: the shaded triangular (3-leg) vertices are intimately related [90] to the logarithmically screened square (4-leg) vertices (26) and (27) and this results in [59, 66]

$$\chi_{1d}^{\mathrm{CDW/SDW}} \approx \frac{1}{2g_2^*} \left(\frac{\varepsilon_F^2}{\varepsilon^2} \right)^{g_2^*} \tag{30}$$

with

$$\varepsilon^2 \approx \omega^2 + v_F^2 (q \pm 2k_F)^2 + T^2. \tag{31}$$

The result (30) is given in a somewhat simplified form valid at $\varepsilon < T_g$ when

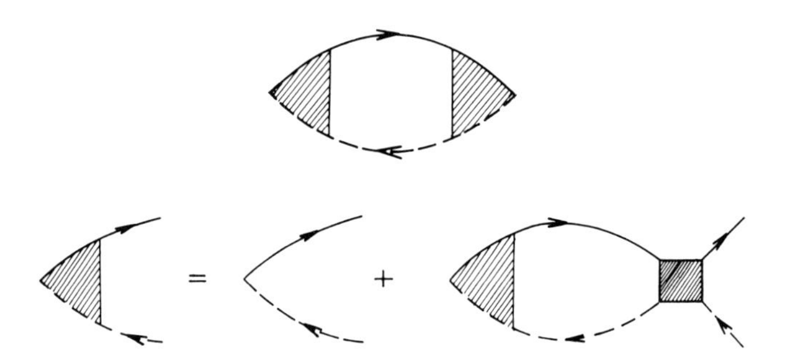

Fig. 16. CDW correlation function χ^{CDW}. Shaded triangular vertices follow from the logarithmically screened Coulomb interactions (shaded square).

$$g_2(T) \approx g_2^* = g_2 - \frac{g_1}{2}$$

in Equation (27). $\chi_{1d}^{CDW/SDW}$ is singular for $g_2^* > 0$. The $2k_F$ CDW/SDW instability is thus due to large g_2, or according to Equation (23), to imperfect screening of the forward Coulomb interaction in the parquet correlation function (30) [16]. Electron one-dimensionality (i.e. parquet) plays here an important role because in the opposite ($t_\perp > T$, RPA) limit, strong repulsive interactions inhibit the CDW instability [45], even when the nesting properties of the Fermi surface allow for it. In the RPA case only the SDW instability remains for sufficiently large g_2 [67].

3.4. Exact Theory

In this section we proceed to discuss the case $g_1 < 0$. As already mentioned several times, and in particular at the beginning of Section 3, an effective $g_1 < 0$ can arise from optic phonons if the condition (15) is not obeyed. Although this seems unlikely, we shall discuss briefly such a possibility in order to appreciate better the consequences of the assumption $g_1 < 0$.

In this case $g_1(T)$ of Equation (26) and $\chi(T)$ of Equation (28) increase on lowering the temperature and become singular at T_g of Equation (29) [64, 90]. The 'exact' theory [59, 91, 97, 101], which must replace the parquet approximation when the effective interactions are large, rounds up this singularity and shows that, at well below T_g, χ has an activated behavior [97]

$$\frac{\chi}{\chi_P} \approx \left(\frac{\Delta_{SDW}}{T}\right)^{1/2} e^{-\Delta_{SDW}/T} \tag{32}$$

with [59, 101, 102] $\Delta_{SDW} \approx T_g$. Δ_{SDW} appears to be the gap in the spin-density-wave spectrum at small q.

The measured magnetic susceptibility of Figure 3 does not behave in an activated way. Moreover, its behavior above the structural transition extrapolates to a finite value, in agreement with Equation (28) for $g_1 > 0$, but contrary to Equation (32). Nevertheless, there are some advantages in introducing the gap Δ_{SDW} into the TTF-TCNQ picture [50]. This gap is unrelated to structural phase transitions and it is therefore tempting to identify it with the infrared 'gap' ω_R, which has the same property, as discussed at the end of Section 2.4. Although not directly involved, the SDW gap Δ_{SDW} can show up in optical properties in the presence of magneto-elastic coupling [50]. This possibility cannot be entirely ruled out and we may therefore quote the corresponding $2k_F$ CDW correlation function ($g_2 - g_1/2 > 0$) [101]

$$\chi_{1d}^{CDW} \sim \left(\frac{\varepsilon_F}{\Delta_{CDW}}\right)^{2g_2^*} \left(\frac{\Delta_{SDW}^2}{\varepsilon^2}\right)^{1/2-g_2^*}, \tag{33}$$

where ε is the same as in Equation (31). Again we are encountering a power law, somewhat faster than in Equation (30). It is thus hard to distinguish between $g_1 \gtrless 0$ considering CDWs, but unlike in Equation (30) SDW is absent in Equation (33), which is a major difference. In fact it will turn out that the distinction between regimes (30) and (33) is not very important for the forthcoming discussion of $2k_F$ lattice instabilities, although we favor $g_1 > 0$ on the basis of the magnetic susceptibility data.

3.5. Coulomb Interchain Coupling

In this section we examine the role of the interchain Coulomb coupling $g_1^\perp$ of Equation (7). In Section 2.2 it was argued that

$$g_1^\perp \ll g_{1,2}$$

using the notations (23) and (25) for the on-chain Coulomb matrix elements. The effect of $g_1^\perp$ can thus appear only at low temperatures in agreement with the observation of the 1d regimes in TTF-TCNQ.

The small $g_1^\perp$ is usually taken into account through the mean-field (MF) approximation based on the accurate solution of the 1d problem [69]. The corresponding diagrams which involve χ_{1d}^{CDW} (but not χ_{1d}^{SDW}) are shown in Figure 17 and yield

$$\chi^{CDW} = \frac{\chi_{1d}^{CDW}}{1 + g_1^\perp \cos\phi_\perp \chi_{1d}^{CDW}} \tag{34}$$

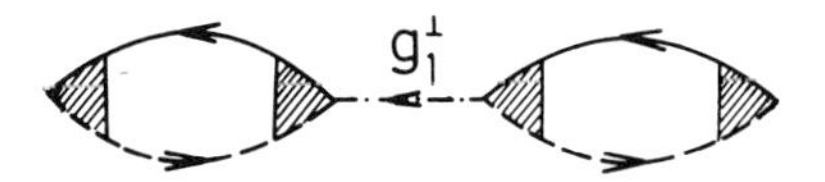

Fig. 17. Leading interchain contribution to $\omega_{2k_F}^{CDW}$.

where $\phi_\perp$ denotes the relative phase of the $2k_F$ CDWs, assumed the same for all pairs of neighboring (e.g. TCNQ) chains. Similar approximation was used [103–109] successfully in Landau quasi-1d problems, where it leads to practically exact [107, 108] results. In the case considered here, the MF approximation holds well only above [71] the MF transition temperature $T_\perp$, associated with the $\phi_\perp = \pi$ zero of the denominator in Equation (34). For example, for $g_1 > 0$ this temperature follows from Equation (30) as [69, 71]

$$T_\perp \approx \varepsilon_F \left(\frac{g_1^\perp}{2g_2^*} \right)^{1/(2g_2^*)} \tag{35}$$

In the vicinity of $T_\perp$ it is necessary to employ at least the full parquet scheme, which treats the interchain couplings on the same footing as the intrachain ones [65, 68–70]. This, the so-called fast parquet regime, is quite intricate and not yet fully understood although some qualitative features of the solutions have been found [65]. In particular there are cross-effects between forward and backward interchain scatterings [65, 70]. In some simple cases involving two chains only [70] the cross-effects enhance the tendency towards the CDW instability. However, the parquet results with RPA screened forward interaction and $g_1^\perp \neq 0$, which would extend

Equation (23) [16], are not yet available. Fortunately, only the values of χ^{CDW} above $T_\perp$ are needed for electrons coupled to the lattice, as will be seen in Section 4, and for these we shall rely on Equation (34).

3.6. $4k_F$ CORRELATIONS

The presence of the $4k_F$ scattering is usually considered as undeniable proof that (Coulomb) interactions are substantial in TTF-TCNQ. Not surprisingly then the early explanations of this effect were formulated in the strong-coupling limit [110–114]. However, in order to be consistent with our general approach we shall show here how the $4k_F$ correlations are built up, starting from the weak-coupling side treated within the parquet approximation [70, 101, 114].

The leading logarithmic correction to the $4k_F$ correlation function is shown in Figure 18. In this figure [70] the incoming $4k_F$ momentum is split into two $2k_F$ channels, and the two $2k_F$ bubbles are convoluted one with the other. As shown in Figure 18 the higher order diagrams are obtained by inserting the interactions g_1 and g_2 so as to generate further logarithmic terms [70]. It is clear that all such diagrams must have one $3k_F$ propagator and one $2k_F$ interaction (together roughly equal to g_1) next to each $4k_F$ external vertex. g_1^2 can thus be factorized out, and the summation gives [70]

$$\chi_{1d}^{4k_F} \approx g_1^2 \left(\frac{\varepsilon_F^2}{\varepsilon^2} \right)^{4g_2^*-1} \tag{36}$$

for $T < T_g$ of Equation (29). ε is given by Equation (31) and $g_2^* = g_2 - g_1/2$ by Equation (27).

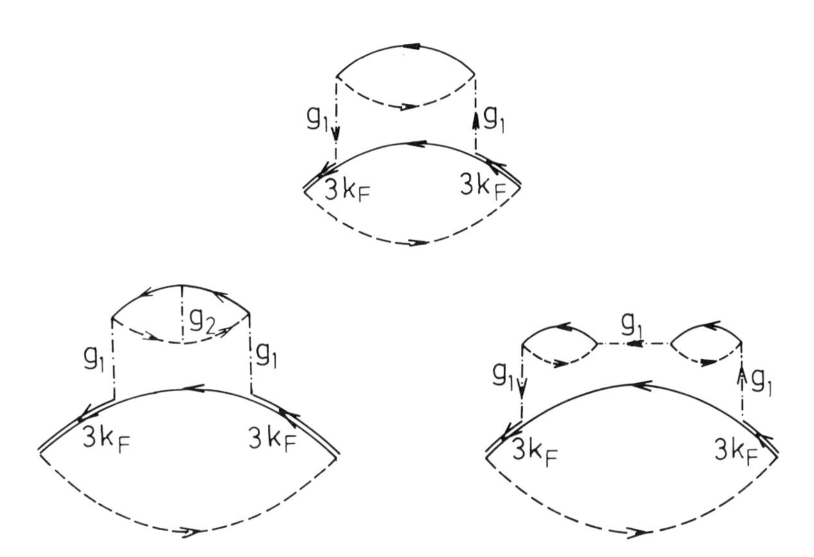

Fig. 18. Lowest order logarithmic correction to the $4k_F$ correlation function $\chi_{4k_F}^{CDW}$ and next order logarithmic corrections to $\chi_{4k_F}^{CDW}$.

A few remarks are in order concerning the result (36). First, it is obvious that the interaction g_2 has to be of the order of unity and larger than g_1 to produce the $4k_F$ correlations at low temperatures. The resulting $4k_F$ singularity is, however, weaker than the $2k_F$ one (cf. Equation (30)), and is expected to have a longer high-temperature tail. This result is consistent with observations in TTF-TCNQ shown in Figures 9 and 10 and discussed in the preceding sections. In particular, g_2 is large in TTF-TCNQ according to Equations (5) and (23). The inequality (6) together with Equation (25) situates through Equation (36) the $4k_F$ scattering on the TTF chain rather than on the TCNQ chain: this latter has a larger g_1 and therefore the $4k_F$ anomaly is absent, but the uniform magnetic susceptibility is more temperature dependent. The intramolecular and metallic screening properties again underly this result.

Admittedly, these conclusions are drawn here for the values of coupling constants which do not fall precisely into the range of the weak-coupling theory. This leads us to a brief discussion of larger couplings. One way to improve the result is to use better approximation in terms of coupling constants, keeping the electron spectrum linear in $k \pm k_F$, as in the parquet approximation. Such model can be suitably bosonized [113] and the parquet result generalized correspondingly. In particular, the exponent in Equation (36) is replaced by [91, 113] $1 - 2(1 - 2g_2^*)^{1/2}(1 + 2g_2^*)^{-1/2}$, which shows that the parquet result is qualitatively correct for $g_2^* < \frac{1}{2}$.

For larger values of coupling it is usual to use the full tight-binding electron band structure and seek the expansions in terms of t/U. In this limit the on-site and off-site interactions play somewhat less symmetric roles than suggested by Equation (36). This can be best seen in the $U = \infty$ limit of the Hubbard model. Then the system of ρ electrons with spin decomposes [91, 112, 115] into a set of ρ free spins and ρ free spinless fermions. The latter produce the weak logarithmic singularity at $2\tilde{k}_F$, where the Fermi momentum $\tilde{k}_F$ of spinless fermions equals $2k_F$ due to the removal of the spin degeneracy. This weak $4k_F$ singularity is enhanced by introducing [115] the off-site interactions. The interacting spinless fermions have been investigated by various methods [91]. For sufficiently small off-site interactions g_2 the $4k_F$ enhancement can be obtained from the parquet results (30) for the $2k_F$ correlation function [115], because we are dealing with the $2k_F$ singularity of spinless fermions. In parallel with that, the Curie magnetic susceptibility of free spins is cut off [61] by the departure of U from infinity and by off-site interactions. The situation is particularly clear [91, 101] for the half-filled band of the extended Hubbard model, with repulsive U when the spin subsystem is governed by the antiferromagnetic Heisenberg Hamiltonian with the effective exchange matrix element proportional to t^2/U (independent of V for positive U). Otherwise one has to rely on numerical results [60–63], mostly obtained by the Monte Carlo simulation [60]. Those results agree essentially with the qualitative picture described above.

It can finally be mentioned that the interchain coupling of the $4k_F$ modes is expected to be comparable to $g_1^\perp$ of Equations (7) because, by Umklapp, $2k_F$ and $4k_F$ correspond to similar charge distributions [1, 31]. Equation (34) applied to the $4k_F$ CDW suggests then that the 1d nature of the $4k_F$ CDW is to be attributed to the weakness of the 1d $4k_F$ singularity, Equation (36), relative to the $2k_F$ one, Equation (30). This question will be taken up again in Section 5.3.

4. CDW Coupling to the Lattice

In Section 3 we found that the electron gas interacting through Coulomb forces tends to be unstable with respect to the formation of the $2k_F$ and $4k_F$ CDWs. On the other hand, it was seen in Section 2.3 that CDWs are coupled to the lattice through electron–phonon coupling. It can be expected, therefore, that CDWs induce the lattice displacements. This intuitive picture underlines the developments of the present section. TTF and TCNQ chains will here be distinguished only occasionally; the corresponding refinements will be introduced in the forthcoming sections.

4.1. Harmonic Theory

The condition which ensures the applicability of the picture in which the lattice is simply driven by the CDW is given by

$$2\pi T > \omega_{ac} \qquad (10); \; (I.16)$$

for all temperatures above the structural transition. The condition (10) allows us to neglect [65, 92] the virtual exchange of low-frequency acoustic (translation, libration) phonons provided that [1]

$$T > T_L^0 \qquad (37); \; (I.17)$$

is also satisfied. As already mentioned in Section 2.4, T_L^0 is the mean-field temperature of the lattice instability, to be defined more precisely below. The condition (10) eliminates the quantum fluctuations of the phonon field making it classic, whereas Equation (37) renders the theory harmonic in terms of this field [1]. Indeed, with the conditions (10) and (37) satisfied, the soft phonon propagator is given by Figure 19. The acoustic phonon lines are retained only in real processes, which are conserving energy $\tilde{\omega} \approx 0$ [65, 92]. This is the leading acoustic phonon contribution because the acoustic propagators involved in virtual processes give rise to much smaller contributions due to the conditions (10) and (37). Consequently, the shaded bubbles in Figure 17 contain only the Coulomb and the optic phonon renormalizations [65]. The renormalized phonon frequency is hence given by [65, 92]

$$\tilde{\omega}^2 = \omega_{ac}^2 - \Pi(\tilde{\omega}) \qquad (38)$$

with

$$\omega_{ac}^{-2}\Pi = \lambda_{ac}\chi^{CDW}, \qquad (39)$$

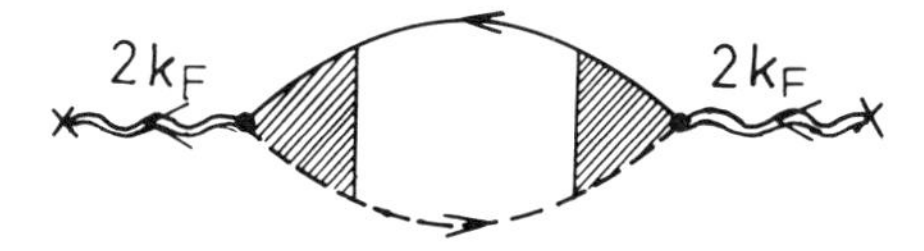

Fig. 19. Coupling of χ^{CDW} to the lattice through terms linear in displacements. Wavy lines are acoustic phonon propagators and the dot is the linear electron–phonon vertex.

i.e. the Coulomb CDW correlation function $\chi^{\rm CDW}$ (calculated approximately in the preceding section) plays the role of the acoustic phonon self-energy [65]. A word of caution should be inserted here. In fact it is not the CDW correlation function that couples to the lattice but rather [60, 92] the charge transfer one in which $c_{i+1}^{+}c_i$ replaces the CDW site density $\rho_i = c_i^{+}c_i$. As pointed out earlier [92], the difference between the two is unessential in the weak-coupling limit, except when the band is half-filled. We shall thus proceed with Equation (39), but it should be mentioned that the strong-coupling limit does require the distinction between the two correlation functions to be made even for arbitrary band fillings [60].

Multiplying Equation (39) by $\frac{1}{2}M|u|^2$, where $|u|^2$ is the square of the displacement and M the corresponding (weighted) mass in motion, gives the effective harmonic energy [45] associated with the acoustic phonon

$$E^{(2)} = \lambda_{\rm ac} C\chi^{\rm CDW}\left[\frac{1}{2\lambda_{\rm ac}\chi^{\rm CDW}} - 1\right]|u|^2. \tag{40}$$

Note that the linear electron–phonon coupling enters Equation (40) to the same order as u, because the virtual phonon processes are neglected in $\chi^{\rm CDW}$. It is thus legitimate to call $E^{(2)}$ the harmonic energy.

Apparently Equations (39) and (40) agree with the physical picture described at the beginning of Section 4. The mean-field transition temperature for the lattice instability is defined as corresponding to vanishing of the bracket in Equation (40). It is noteworthy that this occurs above the temperature at which $\chi^{\rm CDW}$ itself is singular. The lattice instability induced by a CDW precedes the CDW instability of the electron gas with Coulomb interactions [65].

As pointed out in Paper I, it is important to distinguish at this stage between $\lambda_{\rm ac} \gtrless g_{1,2}$. In the case $\lambda_{\rm ac} > g_{1,2}$ the bracket in Equation (40) vanishes at $2k_{\rm F}$ for a temperature sufficiently high that $\chi^{\rm CDW}$ can be approximated by the free electron gas expression $\log \varepsilon_{\rm F}/T$ [92]. Indeed, the lattice transition temperature $T_{\rm P}^0 \approx \varepsilon_{\rm F}\,{\rm e}^{-1/2\lambda_{\rm ac}}$ determined in this way, lies above the Coulomb correlation temperature T_g of Equation (29). This, Peierls, limit seems appropriate for trichalcogenides as was thoroughly discussed in Paper I. On the other hand, for $\lambda_{\rm ac} < g_{1,2}$ the structural transition is triggered by the correlated electron gas: T_L^0 lies below T_g [65].

A brief digression concerning the $4k_{\rm F}$ instability is in order here. The preceding discussion applies to both $2k_{\rm F}$ and $4k_{\rm F}$ provided that this latter is coupled to the lattice via the linear electron–phonon coupling (i.e. $\lambda_{\rm ac}$), but this is not the only possibility. In fact, the Coulomb matrix elements themselves depend on the intermolecular distance. This produces the phonon coupling to a term bilinear in the electron density [91, 113]. The corresponding vertex is shown in Figure 20(a) and can [70] replace that of Figure 20(b) used in Equations (39) and (40). This amounts to replacing $\lambda_{\rm ac} g_1^2$ (g_1^2 from Equation (36) or Figure 18) in Equations (39) and (40) by a quantity of the order of $n_{\rm F}(e^4/Cb^4)$. Both quantities are of the same order of magnitude according to Equations (12) and (25) and it is hard to assess which one is dominant in TTF-TCNQ. In any case there is no essential difference between the $2k_{\rm F}$ and $4k_{\rm F}$ expressions for $E^{(2)}$ and from now we shall consider the $2k_{\rm F}$ instability

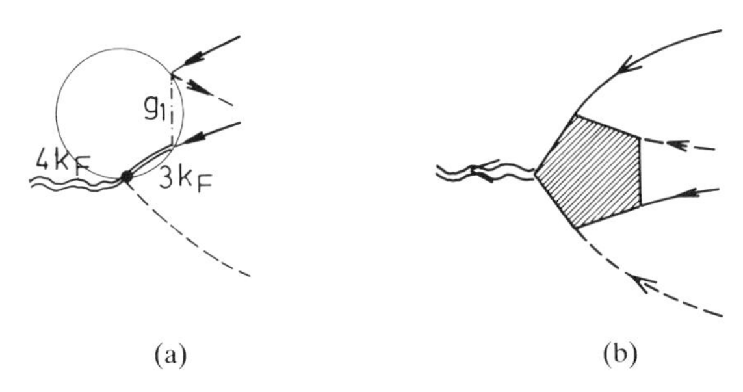

Fig. 20. Two alternative ways of coupling of $\chi^{CDW}_{4k_F}$ to the lattice: (a) through the usual linear electron–phonon vertex, as in Figure 19; (b) through linear variation of the Coulomb matrix elements with displacements.

only. If necessary, the resulting $2k_F$ expressions can be easily adapted to the $4k_F$ case.

Inserting Equation (34) in Equation (40) leads to

$$E^{(2)} \approx (g_1^{\perp} + \lambda_{ac})\chi^{CDW}\left[\frac{T - T_L^0}{T_L^0} + \xi_0^2(q \pm 2k_F)^2 + \right.$$

$$\left. + \xi_{0\perp}^2 d_{\perp}^2(\phi_{\perp} - \phi^0)^2\right] C\,|u|^2 \tag{41}$$

with the longitudinal characteristic length ξ_0 equal to

$$\xi_0 \approx \frac{b}{2\pi n_F T_l^0}. \tag{42}$$

The interchain Coulomb coupling results in the characteristic transverse distance

$$\xi_{0\perp} \approx d_{\perp}\sqrt{\frac{g_1^{\perp}}{\lambda_{ac} + g_1^{\perp}}} \tag{43a}$$

appearing in the expansion around $\phi_0 = \pi$ favored by Coulomb interactions. The elastic interchain coupling will also favor some value of ϕ_0, expectedly 0 or π. The corresponding contribution to the transverse distance can be roughly estimated from the uniform (long wavelength) elastic constants (9) as

$$\xi_{0\perp} \approx d_{\perp}\sqrt{\frac{C_{\perp}}{C}} \tag{43b}$$

lacking the more precise information about the transverse dispersion of the $2k_F\omega_{ac}^2$ in Equation (38).

It is interesting to compare Equations (41)–(45) to the corresponding Peierls expressions (I.30a, c). The difference arises from the fact that χ^{CDW} is given by the power law in the former and by a logarithmic law in the latter case. Unlike Equation (I.31), Equation (41) has χ^{CDW} in the prefactor. Consequently, the λ_{ac} dependence of $\xi_{0\perp}$'s differs in the two cases. Small λ_{ac} enhances $\xi_{0\perp}$'s more in the Peierls case

(Equations (I.30a, c)) than in Equations (43a, b) (there is one power of λ_{ac} more in the denominators of Peierls $\xi_{0\perp}^2$'s). This fact, valid for both $g_1 \gtrless 0$, was first pointed out [50] in the $g_1 < 0$ case for the elastic coupling, corresponding to Equation (43b).

We have started the discussion of Equation (41) with $\xi_{0\perp}$'s because they do not depend explicitly on the sign of g_1. By contrast, the value of T_L^0 does depend on it, according to Equations (30) and (33). For $g_1 > 0$, Equations (30), (34), and (40) give

$$T_L^0 \approx \varepsilon_F \left(\frac{\lambda_{ac} + g_1^\perp}{2g_2^*} \right)^{1/(2g_2^*)}. \tag{44}$$

This expression differs only in detail from the corresponding $g_1 < 0$ expression which follows from Equation (33). However, both expressions for T_L^0 differ essentially in their λ_{ac} dependence from the Peierls $T_P^0 \approx \varepsilon_F \exp(-1/2\lambda_{ac})$, and that for the same reason as for $\xi_{0\perp}$'s.

The differences similar to those in $\xi_{0\perp}$'s are also found in the time behaviors. Expanding Π of Equation (38) in terms of $\tilde{\omega}$ we find

$$\frac{\tilde{\omega}^2}{\omega_{ac}^2} \approx \frac{T - T_L^0}{T_L^0} + \frac{\tilde{\omega}^2}{T_L^{0^2}} \tag{45}$$

irrespective of the sign of g_1 and assuming for simplicity the 1d regimes (30) or (33). The characteristic frequency scale of $\omega_{ac}^{-2}\Pi \approx 2\lambda_{ac}\chi^{CDW}$ is thus

$$\omega_C^L \approx T_L^0 \tag{46}$$

as shown in Reference [50] for $g_1 < 0$, whereas the same expansion in the Peierls case gives [116]

$$\omega_C^P \approx \frac{T_P^0}{\sqrt{\lambda_{ac}}} \tag{47}$$

for small λ_{ac}. It should be remembered now that we are assuming $2\pi T_L^0 > \omega_{ac}$, which eliminates the quantum fluctuations of the lattice. According to Equations (46) and (47) this condition not only makes the phonon field classic but also the electron response to it adiabatic, i.e. the second term of Equation (45) negligible with respect to its left-hand side. The adiabatic approximation is, however, better founded [50] in the Peierls case, due to the presumably small λ_{ac} in the denominator of Equation (47). This is why we are showing in Figures 21(a) and (b) the two limits $\omega_C^L \gtrless \omega_{ac}$ of Equations (38), (39), and (45), although, strictly speaking, Equation (39) was derived only for $\omega_C^L > \omega_{ac}$. It is obvious from Figure 21(a) that $\omega_C^L > \omega_{ac}$ means the simple [116] softening of the $2k_F$ phonon, $\tilde{\omega}^2 \approx \omega_{ac}^2(T_L^0 - T)/T_L^0$. In contrast to that, $\omega_C^L < \omega_{ac}$ of Figure 21(b) is associated with two solutions [116–118] of Equation (38). The first is roughly equal to the bare ω_{ac}, because ω_{ac} is too large for electrons to respond to it. The second solution appears only for T close to T_L^0 when $\tilde{\omega} \approx 0$. The appearance of this latter solution can obviously be associated with the growth of the central peak [116] in the ω-resolved scattering cross-section at $2k_F$.

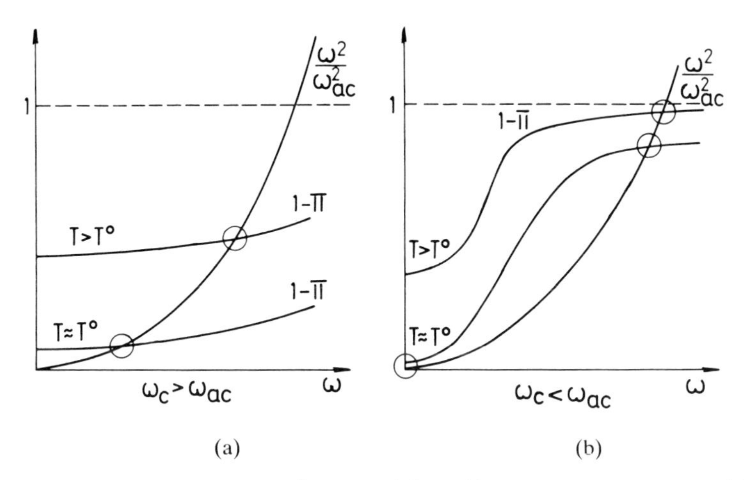

Fig. 21. Graphical solutions of equation $\bar{\omega}^2 = 1 - \Pi(\bar{\omega})$ for (a) adiabatic $\omega_{ac} < \omega_C$ case; (b) $\omega_{ac} > \omega_C$ case when the central peak grows close to T_L^0.

The central peak is presumably present in TTF-TCNQ (Section 2.4). However, in the next section an alternative interpretation of the central peak will also prove possible.

Let us thus conclude this section with a few remarks concerning the validity of Equation (41). This equation has the appearance of the straightforward harmonic term in the Landau expansion around the transition temperature T_L^0. The corresponding order parameter is the $2k_F$ displacement rather than the $2k_F$ CDW. It is important, however, to realize that the (exact) susceptibility χ^{CDW} involved in this term, Equation (40), has the singularity at (or more precisely close below) $T_\perp$ of Equation (35). $T_L^0 > T_\perp$ (see Equation (44)), as it should be, but if $\lambda_{ac} < g_1^\perp$ the range of validity $T_L^0 - T_\perp$ of the Landau expansion (41) is very limited. Physically this means that on decreasing $\lambda_{ac}/g_1^\perp$ the lattice is less and less important and χ^{CDW} coupled to it has to be calculated with better and better accuracy in the interchain couplings. According to Section 2, in TTF-TCNQ the situation seems to be intermediate, $\lambda_{ac} \gtrsim g_1^\perp$. Lacking a better approximation we shall continue to use Equation (34) for χ^{CDW} in the vicinity of T_L^0.

4.2. Quartic Interaction of Phonons

The harmonic theory of the preceding section ceases to hold close to T_L^0 because the $2k_F$ soft phonons start to interact. The corresponding quartic interaction is depicted in Figure 22. This intrachain interaction term is large for two reasons. First, the internal electron remains attached to the Fermi surface after each backward scattering it undergoes on its way around the vertex. Second, the triangular vertices at the extremity of each phonon 'half'-line are renormalized by Coulomb interactions, according to the parquet prescription [119]. The triangular vertices in Figure 22 are hence the same as those involved in χ_{1d}^{CDW} of Figure 16. It follows that

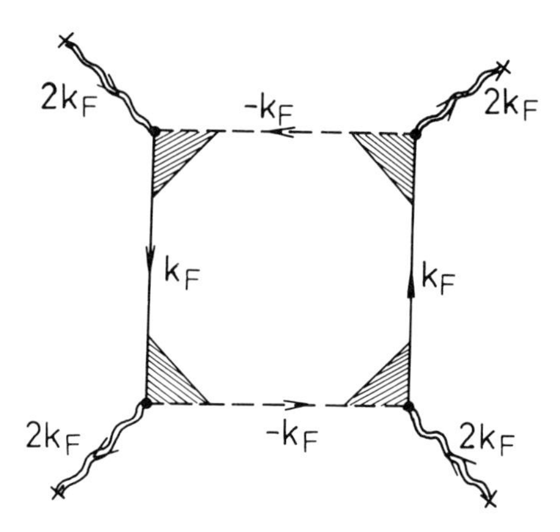

Fig. 22. Anharmonic interaction of $2k_F$ soft phonons. Shaded vertices are the same as in Figure 17, leading to the results proportional to $(\chi^{CDW})^2$.

$$E_{1d}^{(4)} \sim \frac{\lambda_{ac}^2 \chi_{1d}^2}{n_F T_P^{0^2}} C^2 |u|^4 \tag{48}$$

up to a numerical coefficient. A similar expression was obtained from the bosonization approach for $T < \Delta_{SDW}$ in the $g_1 < 0$ case [50]: χ_{1d}^{CDW} of Equation (33) replaces that of Equation (30).

The vertex of Figure 22 was originally evaluated [119] as the quartic interaction of 'external' fields in the mean-field approach [119–121] to the $g_1^\perp$ interchain coupling of CDWs beyond Equation (35). In this way it describes the interaction of CDWs on neighboring chains by exciting a given chain. In our case the external fields are the displacements on this same chain, and the problems [122] related to the mean-field approach are therefore absent.

The nature of corrections related to the term (48) can be well appreciated on joining together two phonon lines in Figure 22. This yields a correction to the soft phonon propagator, which contains an electron at the Fermi surface, renormalized by the $2k_F$ soft phonon. The corresponding electron self-energy insertion describes the opening of the pseudo-gap in the electron spectrum, [123, 124], provided that the adiabatic condition $\omega_C^L > \omega_{ac}$ (Equation (46)) is fulfilled. This is the case for λ_{ac} small (Equation (47)) in the Peierls case discussed in Section I.4.3, but, as we know, may fail for the correlated electron gas. If so, there are departures from the pseudo-gap picture, which were discussed earlier on the basis of the semi-phenomenological model [125].

However, the perturbative approach to the pseudo-gap just described is of limited validity. It is justified only [49, 50] in the vicinity of T_L. This is sufficient provided that the true transition temperature T_L is not much depressed with respect to T_L^0 by the low dimensionality of the system. In the opposite case better approximations, to be discussed now, are required. In this regard a distinction should be made between $\xi_{0\perp} \gtrless d_\perp$ in the harmonic part of the problem, Equations (43a, b). The respective behaviors of the soft phonon frequency $\tilde{\omega}^2$ (or more precisely, of the inverse elastic scattering cross-section) are shown in Figures 23(a) and (b). Let us consider separately these two cases.

4.2(a) $\xi_{0\perp} > d_\perp$

In Figure 23(a) the anomaly is localized in the Brillouin zone. By rescaling the wavevectors it is possible to symmetrize such problem entirely [92, 126]. In contrast to that there is no affinity transformation which would symmetrize the case $\xi_{0\perp} < d_\perp$ of Figure 23(b).

Proceeding to the anharmonic interactions we note that the term (48) is independent of wavevectors (local in direct space). Therefore the scale transformation, which symmetrizes the harmonic energy in the $\xi_{0\perp} > d_\perp$ case, leaves it invariant, i.e. renders 3d the entire anharmonic Landau problem [92, 126]. The 3d Ginzburg criterion can thus be used to estimate the width ΔT of the critical region. Taking into

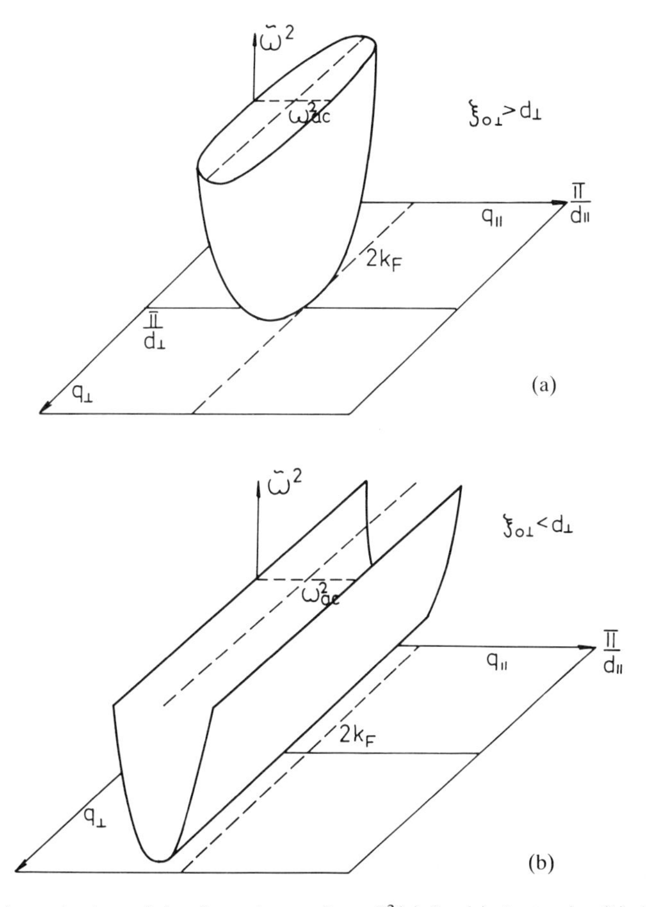

Fig. 23. Schematic view of the dispersion surfaces $\tilde{\omega}^2(\mathbf{q})$ for (a) $\xi_{0\perp} > d_\perp$; (b) $\xi_{0\perp} < d_\perp$.

account that the scale transformation leads to the multiplicative renormalization of the coefficients in the integrated Landau energy, we obtain [2, 65, 92, 126]

$$\frac{\Delta T}{T_L^0} \approx \frac{T_L^{0^2} B^2}{\xi_0^6 A'^4} \frac{d_\perp^4}{b^4} \frac{\xi_0^4}{\xi_{0\perp}^4}. \qquad (49); \ (I.57)$$

The result (49) is expressed here in terms of the original coefficients, i.e. A' and B in Equation (49) are the short-hand notations for the prefactors of u's in Equations (41) and (48). This result generalizes the Ginzburg criterion to the anisotropic $\xi_0 > \xi_{0\perp} > d_\perp$ systems. When the particular values of A', B, and ξ_0 of Equation (41) are inserted in Equation (49), retaining only the leading $g_1^\perp$ contribution, it follows that

$$\frac{\Delta T}{T_L^0} \approx \frac{d_\perp^4}{\xi_{0\perp}^4} \qquad (50)$$

Due to the special relation between A', B, and ξ_0, ΔT in Equation (50) could be expressed solely in terms of the harmonic coefficients. It is interesting to note that Equation (50) is also valid in the corresponding Peierls case, although the structure of the coefficients is somewhat different there, as discussed in the preceding section. In both cases the 'harmonic' condition $\xi_{0\perp} > d_\perp$ implies the 'anharmonic' relation $\Delta T < T_L^0$. However, as was pointed out in connection with Equations (43a, b), with small λ_{ac} it is easier to reach the regime $\xi_{0\perp} > d_\perp$ in the Peierls than in the strongly correlated case [50]. Indeed, the values of $\xi_{0\perp}/d_\perp$, estimated for TTF-TCNQ from the parameters found in Section 2 and Euations (43a, b) turn out to bc only of the order of unity.

Note that our approach is valid if

$$\frac{\Delta T}{T_L^0} < \frac{T_L^0 - T_\perp}{T_L^0}, \qquad (51)$$

where $T_\perp$ is given by Equation (35). This condition follows simply from the arguments given at the end of the preceding section and together with (10) ensures the validity of the Landau expansion. Hopefully it holds in TTF-TCNQ.

ΔT represents not only the width of the critical region but also [92] the shift of the transition temperature $\Delta T \approx T_L^0 - T_L$. This shows that two energy scales tend to appear in the strongly anisotropic systems. One is related to the mean-field transition temperature T_L^0 and to pseudo-gap, and the other to the real transition temperature T_L.

4.2(b) $\xi_{0\perp} < d_\perp$

The use of the picture with two energy scales introduced here on approaching intermediate couplings $\xi_{0\perp} \approx d_\perp$ from the 3d side is, however, fully justified only if $\xi_{0\perp} < d_\perp$. Let us thus describe how the two scales appear in the language of this limit. The order parameter of the Landau expansion (41), (48) is the $2k_F$ deformation. Above the transition temperature there is no long-range order but rather the order parameter undergoes slow variations in space. As is well known (cf. also Section I.4.3), for $2k_F \neq \pi/b$ this can be described through variations of the amplitude and phase of the order parameter. For $T < T_L^0$ the amplitude becomes fixed at

$$|u_{2k_F}|^2 \approx \frac{n_F T^{0^2}}{C} \qquad \frac{T_L^0 - T}{T_L^0}.$$

This follows by minimizing Equations (41) and (48), assuming for simplicity $\chi^{CDW} = \chi_{1d}^{CDW}$. Concerning now the phases, they beome correlated by interchain interaction $\xi_{0\perp}$ only at low temperatures [104, 105, 108], which leads to the transition at $T_L \ll T_L^0$. In this respect there is no difference between the Peierls and the correlated electron case and we can borrow the Peierls result (I.47)

$$T_L \approx n_F T_L^{0^2} \frac{\xi_{0\perp}}{d_\perp} \tag{53}$$

replacing only T_P^0 by T_L^0 and using $\xi_{0\perp}$ from Equations (43) rather than from Equations (I.30).

Below the transition at T_L the phase fluctuations remain gapless if the transition involves the same type of chains [47, 49, 50]. This might explain, in a way alternative to that of the preceding section, the growth of the central peak in the scattering cross-section. If, on the contrary, the oppositely charged chains were involved in the transition with equal amplitudes there would be a gap ω_L of the order of T_L in the phase fluctuation spectrum [47, 50]. This is important for dielectric and conducting properties of the model, because phase fluctuations are responsible for the Fröhlich contribution to the conductivity [48–52, 127]. In phason model the central peak in the scattering cross-section at $2k_F$ is identical with the low-temperature and low-frequency features observed in infrared measurements [57, 127].

Let us focus now on the electron subsystem. In the Peierls limit it is well known (see, e.g., Section I.4.3) that the amplitude of the order parameter $|u_{2k_F}|$ is intimately related to the gap Δ_P in the electron spectrum, which in turn is of the order of the mean-field transition temperature T_P^0, $\Delta_P^0 \sim T_P^0$ (Equation (I.38)). If this latter relation is also to hold in the correlated electron case, namely if

$$\Delta_L^0 \sim T_L^0, \tag{54}$$

the gap Δ_L is to be related to $|u_{2k_F}|$ by

$$\Delta_L^2 \approx n_F^{-1} C |u_{2k_F}|^2 \approx T_L^{0^2} \frac{T_L^0 - T}{T_L^0} \tag{55}$$

according to Equation (52). Unlike in the Peierls case, there is no λ_{ac} in the factor relating Δ_L^2 and $|u_{2k_F}|^2$. This difference is similar to those encountered before for $\xi_{0\perp}$, and ω_C.

The result (55), conjectured here on the basis of the fourth-order expansion, is confirmed by the infinite order parquet or bosonization ($g_1 < 0$) [50] calculations.

It appears therefore that the fourth-order Landau theory with appropriately truncated coefficients ($T = T_L^0$ in A', B, ξ_0) leads to essentially correct results even for $T \ll T_L^0$, i.e. in the 1d anharmonic regime $\xi_{0\perp} < d_\perp$. As in the Peierls case (cf. Section I.4.3), the reason is that this truncation procedure treats the phase behavior in an essentially exact way, fixing the amplitude (52) roughly [49, 50] correctly. Since the phase fluctuations are more important at low temperatures, the resulting accuracy is satisfactory, and with such accuracy the Landau theory can be used not only for $\Delta T < T_L^0$, as stated [49, 50] usually in the literature, but also at $T_L \ll T_L^0$, as was done in Equation (53).

4.3. Microscopic picture – summary

The purpose of this section is to summarize the interpretations of various experimental results which were scattered over the preceding text with the intention to substantiate the theoretical discussions. To some extent this section can also be viewed as an iteration of Section 2.4 on a different level of theoretical understanding.

In Sections 2.2 and 3.6 it has been shown that plasma edge measurements and $4k_F$ scattering both imply large g_2. The uniform magnetic susceptibility behavior discussed in Sections 2.2 and 3.3 indicates that g_1 is large and positive, at least on the TCNQ chains. This tends to rule out optic phonons as determinant in TTF-TCNQ, although they do contribute to the coupling constants, as observed in infrared measurements described in Section 2.3. The acoustic phonons are also assumed to play a secondary role owing to the narrowness of the observed critical energy scales. Theoretically this is illustrated by relatively weak dependence of T_L on lattice properties, when $g_1^\perp \approx \lambda_{ac}$ in Equation (44).

According to Section 2.4 there is a propensity for two energy scales to apepar in TTF-TCNQ at low temepratures, one related to Δ_L^0 and the other to the real transition temperature T_L. Through the law of corresponding states (54) $\Delta_L^0 \sim T_L^0$, the mean-field transition temperature T_L^0 at which the phase-amplitude formulation of the previous section becomes useful, is of the order of 100 K. This is the temperature at which the central peak in the scattering starts (presumably) to develop according to Section 2.4. However, the phase transition occurs only at $T_L \approx 50$ K. $T_L^0 > T_L$ makes it tempting to attribute the central peak to anharmonic coupling (i.e. to phasons) rather than to the breakdown of the adiabatic approximation in the harmonic term (45). This is corroborated by the measurements of the TTF-TCNQ conductivity under pressure, which show that an additional conductivity mechanism is activated below about 100 K [3, 128]. This additional conductivity is usually attributed to phasons [50, 127, 128], whereas the precursor effects in the range above 100 K can be understood as phonon drag [129–131]. The uniform magnetic susceptibility of the TCNQ chain also shows an enhanced temperature dependence below 100 K. In addition, there is probably a perturbative effect of the lattice up to 150 K, parallel to those observed in structural ($2k_F$) and conductivity measurements.

It should be noted in this respect that there is some evidence of isotope effect in the magnetic susceptibility at intermediate temperatures [132]. This is an indication that the nonadiabatic effects due either to acoustic or to optic phonons, but neglected in the present adiabatic, acoustic phonon–Coulomb description, might play some role in TTF-TCNQ.

It is thus appropriate to mention again that the present theoretical description of TTF-TCNQ is based on analytic results which are available only in certain limits, whereas the actual situation in TTF-TCNQ is intermediate in many respect. First of all this concerns the weak-coupling limit in Coulomb couplings, which are in fact of the order of unity. Second, the assumption $T_L^0 \gtrsim \omega_{ac}$, which renders the lattice field classic and the electron response adiabatic, is also closer to equality than to inequality. Further investigation is also required to determine how well is satisfied the inequality $T_L^0 > T_\perp$, which justifies the use in a wide temperature range of the Landau theory in terms of the lattice order parameter. This condition is implicit in deriving the law of corresponding states $\Delta_L^0 \approx T_L^0$, which seems to hold well, although the closeness of various transition temperatures indicates that $T_L^0 > T_\perp$ is not a strong inequality in TTF-TCNQ. Once the Landau expansions are accepted another intermediate situation is again encountered, namely $\xi_{0\perp} \approx d_\perp$. This, however, does not present serious difficulties because it is possible to approach both limits $\xi_{0\perp} > d_\perp$ and $\xi_{0\perp} < d_\perp$ with practically the same degree of accuracy.

It is appropriate to mention now that no satisfactory explanation was offered here for the optic structure at ω_R which persists up to room temperature, as described in Section 2.4. This feature cannot be associated with the $2k_F$ pseudo-gap because the $2k_F$ anomaly entirely disappears at about 200 K. The attempt to introduce an additional gap Δ_{SDW} into the theory by assuming $g_1 < 0$, and to associated it with ω_R, proved also inadequate here, being inconsistent with the temperature dependence of the magnetic susceptibility. Alternative explanations have been invoked [57], however: the disorder due to defects and thermal vibrations can produce a kind of mobility gap [133]. Also, optical phonons can be emitted by electrons, assisted by acoustic phonons [134], leading to Fano interferences [135] in the infrared frequency range. Most probably, however this indentation is related to the phonon mode at that frequency.

Another open question is that of $2k_F$ SDW instabilities, which should be present for $g_1 > 0$ according to Equation (30). They should appear at last in the fluctuative 1d form (30), since the interchain coupling of SDWs, direct or through the lattice [114], is certainly weaker than that discussed in Section 4.1. One way to observe the $2k_F$ SDW fluctuations is through the NMR relaxation time T_1, which involves the imaginary part of the magnetic susceptibility (integrated over all wavevectors, $2k_F$ in particular [19–21]). The $2k_F$-contribution to the imaginary part of the susceptibility has been calculated in Ref. [20b]. However experimental data for T_1 in TTF-TCNQ are dominated by $q \approx 0$ wavevectors, so that $2k_F$-contributions cannot be easily separated. Nevertheless, it is noteworthy that the T_1 measurements [19–21, 136] do indicate an enhanced $2k_F$ SDW contribution, in qualitative agreement with the idea [114] of SDW fluctuations. Even more, the recent C^{13} Knight shift data indicate [19] a change in the magnetic state of TTF-TCNQ below 15 K. Unfortunately, neutron experiments for small spins distributed over large molecules are extremely difficult [36], and the direct confirmation of this results is not yet available.

In spite of those open questions it is felt that the microscopic picture, which resulted in the Landau expansion of Section 4, is basically correct. Only in 1d conductors do the strong Coulomb forces lead to the CDW instabilities, reflected in the lattice displacements. This picture deserves therefore further investigation on including the two chain nature of TTF-TCNQ and the vector character of the molecular displacements. This discussion is carried out in Section 5.

5. Three-Dimensional Ordering

In the previous section it was shown how the $2k_F$ and $4k_F$ CDWs of strongly correlated electrons couple to the lattice and the conditions were determined for the structural instability to be described within the Landau theory in terms of the structural order parameters. That discussion was pursued without the details of the TTF-TCNQ structure and even the two-chain nature was invoked only occasionally.

In this section we shall take into account the particular crystal structure of TTF-TCNQ and start with a simplified discussion in which each family of chains is described by a displacive mode of a given polarization, i.e. the CDW amplitudes ρ_Q and ρ_F are assumed to induce only one type of displacements, u_Q and u_F respectively. The vector nature of the displacement field is thus ignored and there is one-to-one correspondence between CDWs ρ and the induced displacements field $u_{Q,F} \sim \sqrt{\lambda_{ac}^{Q,F}}\, \rho_{Q,F}$. Due to this proportionality we may think either in terms of displacement or of CDW ordering, whichever is more convenient. When the CDW language is chosen, however, it must be borne in mind that the CDW–lattice coupling is quite important in making the Landau theory valid, as was explained in Section 4.

This simplified approach led to the first explanations [31, 32, 137–139] of the cascade of phase transitions occurring at 54, 49, and 38 K. After reviewing these results we shall gradually extend the Landau model in order to take into account more subtle features. Section 5.2 is devoted to the unusual hysteresis which exists in the temperature range between 49 and 38 K. The three-dimensional ordering of $4k_F$ displacements and its connection to $2k_F$ ordering is treated in Section 5.3. In Section 5.4 we consider the (p, T) phase diagram. Particular attention is devoted to the splitting of the 49 K transition which becomes strongly enhanced under pressure. This splitting makes it necessary to distinguish among different displacive modes belonging to the same chains, i.e. to the same CDWs. Through this we invoke the full symmetry treatment of the TTF-TCNQ lattice and identify the lattice displacements which participate in the (p, T) range of the three-dimensional ordering.

5.1. Phase transitions at 54, 49, and 38 K

Early transport data indicated that the ordered state in TTF-TCNQ is formed gradually, through at least two phase transitions occurring at 54 K [140, 141] and 38 K [142, 143]. A more detailed insight into this state has been obtained from X-ray

[53, 54, 144, 145] and neutron scattering [10, 146, 147] studies. Early structural measurements had already shown the existence of two modes with different longitudinal wavenumbers which become ordered below 54 K. Their values, $0.295b^*$ and $0.59b^* \equiv 0.41b^*$, are attributed to the magnitude of $2k_F$ and $4k_F$, respectively, in accordance with the microscopic picture reviewed in Section 3. We consider first the properties of the $2k_F$ instability and postpone the inclusion of the $4k_F$ terms for Section 5.3.

5.1(a) $2k_F$ *Ordering.*

The transition to the three-dimensionally ordered phase manifests itself through the condensation of diffuse $2k_F$ lines into the satellite spots. Their positions are defined by the wavevector $(q_a(T), q_b = 2k_F, q_c = 0)$. The interchain coherence is determined by the transverse components q_a and q_c. While q_c shows no visible temperature change, q_a has a peculiar behavior shown in Figure 24(a) [146, 147]. In the range $49\,K > T > 38\,K$, q_a varies with temperature in a continuous way, within the experimental resolution. This is one of the first and best-known examples of the sliding incommensurate superstructure.

The sliding incommensurate ordering is preceded by the temperature range $49\,K < T < 54\,K$ in which the a-direction period is equal to $2a(q_a = a^*/2)$. The significance of the temperature at which the sliding starts becomes more apparent from polarization analysis of the displacements involved [10, 53, 148, 149]. The first measurements revealed [10] two kinds of spots, as shown in Figure 24(b). One of them belongs to the displacement which is polarized in the c-direction and is conventionally called the 'transverse' mode. These spots are formed at 54 K with $q_a = a^*/2$. The second type of spots corresponds to the 'longitudinal' displacements in the b-direction. This mode becomes ordered at just about 49 K, as is clearly seen from the extrapolated curve in Figure 24, and its condensation apparently triggers the sliding of q_a.

The sliding wavenumber q_a from Figure 24(a) as well as the intensity of both modes depicted in Figure 24(b) have a finite jump at 38 K. Below this first-order phase transition q_a remains pinned at the fourth-order commensurate value, $a^*/4$, with no additional changes observed at still lower temperatures.

5.1(b) *Landau Model for the Sliding Regime* ($T > 38\,K$)

The interpretation of the behavior described above follows already from the simplest version of the Landau theory for the system of coupled CDWs. We start by considering the temperature range above 38 K. It is well known that in a system with only one set of chains (like KCP) the CDWs on the neighboring chains prefer to be in opposition of phases ($q_a = a^*/2$) in order to minimize their mutual Coulomb repulsion [31, 47]. This suggests a picture in which only one family of chains (say TCNQ) has ordered CDWs (transverse displacements) between 54 and 49 K, while the sliding of q_a below 49 K starts because of CDWs ordering (longitudinal displacements) on the remaining (TTF) family of chains.

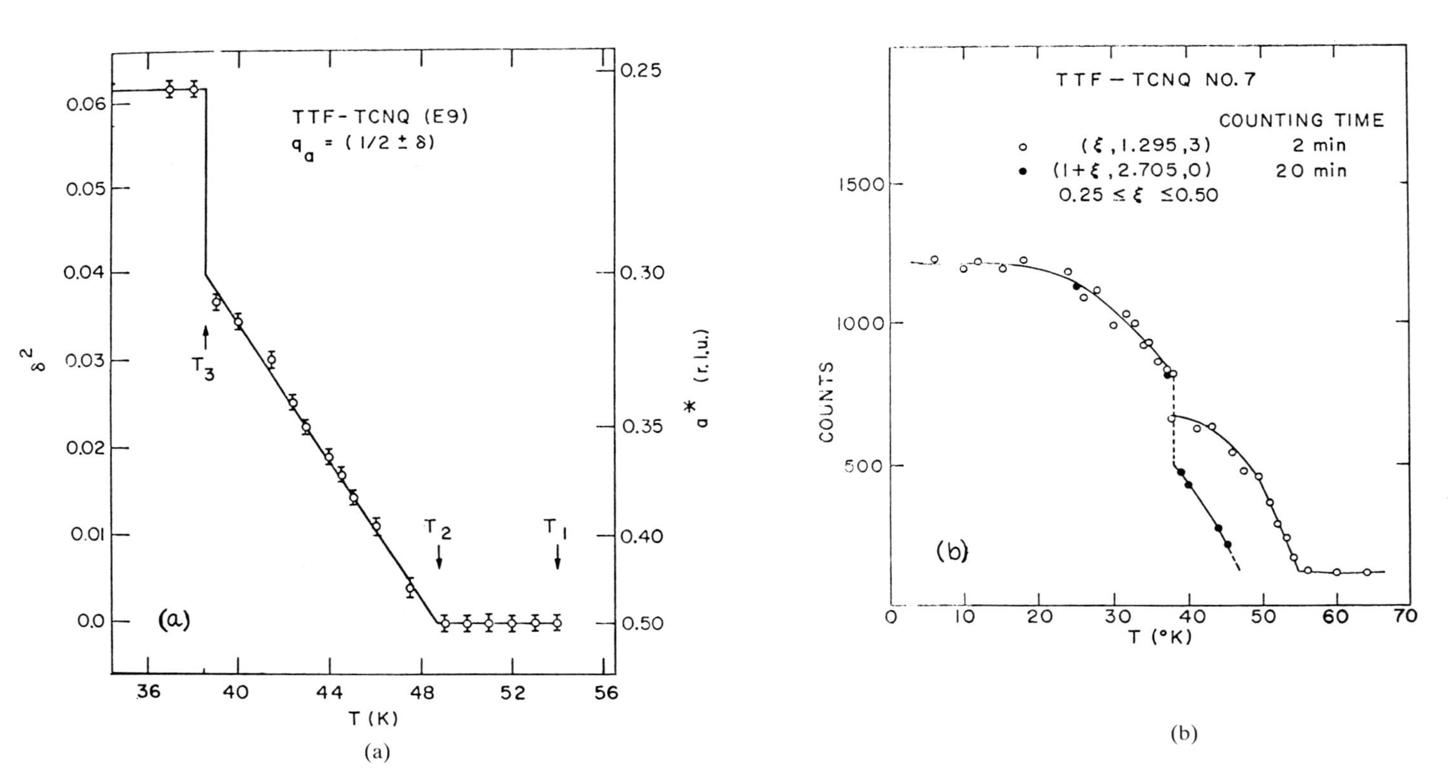

(a)

(b)

Fig. 24. (a) The traverse wavenumber q_a as a function of temperature (after Reference [146]); (b) the intensity of the transverse (○ circles) and longitudinal (● circles) modes as a function of temperature. Note different intensity scales for two modes (after Reference [10]). (© The American Physical Society.)

The corresponding Landau model [32] is quite simple. Let us denote by ρ_Q and ρ_F the amplitudes of the sinusoidal CDWs on the TCNQ and TTF chains, respectively. After assuming that the shifts between neighboring CDWs are constant all over the crystal, the quadratic part of the free energy is expressed as

$$\begin{aligned} F^{(2)} = {} & a_Q \rho_Q^2 + a_F \rho_F^2 + 2\lambda_{c/2,Q} \cos(q_c c/2) \cos(\theta_{Q,c}) \rho_Q^2 + \\ & + 2\lambda_{c/2,F} \cos(q_c c/2) \cos(\theta_{F,c}) \rho_F^2 + \\ & + 2\lambda_{a,Q} \cos(q_a a) \rho_Q^2 + 2\lambda_{a,F} \cos(q_a a) \rho_F^2 + \\ & + 4\lambda_{a/2} \cos(q_a a/2) \cos \theta_{Q,F} \rho_Q \rho_F. \end{aligned} \tag{56}$$

Here a_Q and a_F are intrachain coefficients, the meaning of which will be discussed in detail in Section 5.1(c). The coefficients λ are the bilinear interchain coupling constants which correspond to characteristic transverse distances $\xi_{0\perp}^2$ of Equations (43a, b). These constants are expected to decay fast with the interchain distances for each of the two mechanisms of interchain coupling invoked earlier. For example, in the Coulomb $g_1^\perp$ model of Section 3.5, the coefficients λ decay exponentially [31]. It is therefore sufficient to include in Equation (56) only the most significant neighboring pairs. The relative phases between neighboring unit cells in the a and c directions are denoted by $q_a a$ and $q_c c$, respectively, while $\theta_{Q,F}$, $\theta_{Q,c}$ and $\theta_{F,c}$ are relative phases inside a unit cell. Obviously, the equilibrium values of $q_c c/2$ and θ phases in Equation (56) are either 0 or π, depending on the sign of the corresponding coefficients λ. Thus $q_c = 0$, in accordance with the experiments, and the expression (56) reduces to

$$\begin{aligned} F^{(2)} = {} & [a_Q - 2|\lambda_{c/2,Q}| + 2\lambda_{a,Q} \cos(q_a a)] \rho_Q^2 + \\ & + [a_F + 2|\lambda_{c/2,F}| + 2\lambda_{a,F} \cos(q_a a)] \rho_F^2 - \\ & - 4|\lambda_{a/2}| \cos(q_a a/2) \rho_Q \rho_F. \end{aligned} \tag{57}$$

It remains to determine q_a. There are two possibilities, depending on the values of the amplitudes ρ_Q and ρ_F. If one of them is zero, and with repulsive interaction between next-neighboring chains, $\lambda_a > 0$,

$$q_a = \frac{a^*}{2}, \tag{58}$$

i.e. the neighboring ordered CDWs are in opposition of phases. However, if both families of chains are ordered, the full energy (57) is in equilibrium either for

$$\cos\left(\frac{q_a a}{2}\right) = \frac{|\lambda_{a/2}|}{2\lambda_a} \frac{\rho_Q \rho_F}{\rho_Q^2 + \rho_F^2} \tag{59}$$

or for

$$q_a = 0. \tag{60}$$

Further analysis [32] of the stability of particular solutions [58–60] shows that the range of transverse ordering with $q_a = a^*/2$ (and $\rho_F = 0$) is defined by

$$\tilde{a}_Q \equiv (a_Q - 2|\lambda_{c/2}| - 2\lambda_a) < 0, \tag{61a}$$

$$\tilde{a}_F \equiv (a_F - 2|\lambda_{c/2}| - 2\lambda_a - \lambda_{a/2}^2/\lambda_a) > 0, \tag{62b}$$

i.e. for

$$T_Q > T > T_F, \tag{62}$$

where the left-hand sides in Equations (61a, b) change sign at T_Q and T_F, respectively. T_Q and T_F are just the critical temperatures for the three-dimensional ordering on TCNQ and TTF chains, respectively. At T_F the solution $q_a = a^*/2$ becomes unstable, and the new minimum of free energy (57) corresponds now to the solution (59) in which q_a depends on temperature and decreases from $q_a = a^*/2$ towards $q_a = 0$.

The three-dimensional ordering would proceed in a different way when the temperature range (62) does not exist, i.e. when $T_F > T_Q$ where T_Q and T_F are again defined by Equations (61a, b). Now both families of chains order at a unique critical temperature positioned between T_F and T_Q, and with the wavenumber q_a given either by Equation (59) or by Equation (60) depending on the parameters involved in the expansion (57) [32].

TTF-TCNQ corresponds to the first alternative when there is an interval (62) with the temperatures T_Q and T_F associated with the transition temperatures at 54 and 49 K, respectively. In fact, the phase transition at T_F was first predicted theoretically in Ref. [137]. This completes the interpretation proposed in early two-chain models [31, 32, 137, 138], on which the considerations to follow will be based. Note that the initial free energy of Reference [137] is slightly more general than the form (57). Based primarily on the symmetry considerations, it is expanded in terms of the amplitude ρ_F and the difference $q_a - a^*/2$ (and contains explicitly the corresponding Lifshitz invariant), while the dependence on ρ_Q is kept in the unexpanded form. The form (57) is based on the assumption that the Landau expansion in both ρ_Q and ρ_F up to the fourth order can be extrapolated down to below 38 K. The wavenumber q_a, which varies appreciably in the considered temperature range, enters however into Equation (57) in the unexpanded way. Both approaches lead to the $q_a(T)$ laws which fit well the experimental curve in Figure 24(a). The data on the temperature behavior of the amplitudes ρ_Q and ρ_F are unfortunately masked by the $q(T)$ dependence of the form factors which enter into the scattering cross-sections [53].

The Landau model (56) idealizes to a great extent the actual situation in the chain compounds like TTF-TCNQ and a few additional remarks in this respect are in order. The finite extension of the TTF and TCNQ chains (and of the corresponding CDWs) in the transverse directions influences considerably the parameters in the expansion (56), in particular the values of the interchain coupling constants $\lambda_{a/2}$ and $\lambda_{c/2}$. The detailed analysis in References [33, 34] showed that these constants are reduced in comparison to the results of the point-charge approximation [31, 32]. The same authors also extended the Landau expansion by introducing the libron displacive modes, invoked before in Reference [150]. No trace of the libronic long-range order was found in the structural measurements [53, 151]. However, on the basis of the far infrared data [37], it was claimed that the librations, as well as the torsional distorsions of TTF molecules around the longest molecular axis are present in the static distorsion (see also Section 2.3).

The expansion (56) may also be extended by taking into account the bilinear terms which couple the chains positioned diagonally, e.g. at relative positions **a**/2 ± **c**/2 [152, 153]. However, the possible effect of these terms, namely the temperature variation of the transverse wavenumber q_c, was not observed experimentally. It may be easily seen that due to such terms the phase transition at 54 K also involves a weak ordering of TTF chains. Indeed, the full symmetry analysis [154] showed that the order parameters which enter in the Landau expansion are in principle linear combinations of displacements (or CDW amplitudes) from *both* TTF and TCNQ chains. This point will be discussed in detail in Section 5.4. The NMR [21], EPR [23], and thermopower [17] data indicate that one family of chains, actually the TCNQ, is predominant in the onset of long-range order at 54 K. The conductivity measurements on the samples with the isotopically substituted TCNQ and/or TTF chains [155] lead to the same conclusion.

The situation is less clear concerning the transition at 49 K. The latter was observed in structural [146] and specific heat [157] measurements but no strong anomaly was found in e.g. the transport [140–143] or magnetic susceptibility [157] data. As for the nature of the T_F-transition, the Knight shift measurements [21] on TCNQ (C^{13}) samples gave some evidence that the magnetic susceptibility of TTF chains starts decreasing below 49 K. However, such behavior was not confirmed by the recent measurements [19] on TTF (C^{13}) samples.

5.1(c) *The Commensurate Lock-in at $a^*/4$*

In order to understand the phase transition at 38 K we shall here extend the model (57) and treat it in a more detailed way. The experiments show that the same $2k_F$ modes, i.e. those from Figure 24(b) are involved above and below 38 K, and that no additional $2k_F$ deformations are visible below 38 K. It is therefore tempting to proceed by introducing further contributions in the Landau expansions in terms of

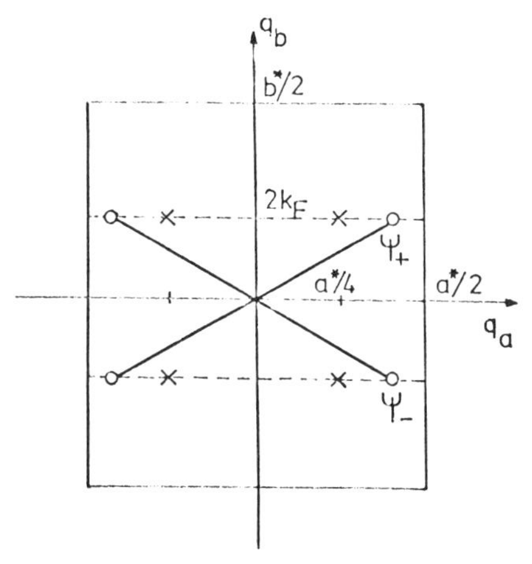

Fig. 25. The star of wavevectors for the ordering in the range 49 K > T > 38 K ($a^*/2 > q_a > a^*/4$) and below 38 K ($q_a = a^*/4$).

the modes already introduced in Equation (57). In order to do that, it is useful to consider briefly the symmetry properties of the ordered phase below 49 K.

The ordering below 49 K is characterized by the wavevector components $q_a(T)$, $q_b = 2k_F$, $q_c = 0$. Evidently, the expression (57) remains unchanged after reversing the signs of these components ($q_a \to -q_a$ and/or $q_b \to -q_b$). This invariance has its origin in the lattice symmetry of TTF-TCNQ. Indeed, the wavevectors ($\pm q_a$, $\pm q_b$, 0) are connected by the operations of the TTF-TCNQ space group $P2_1/c$. For $0 < q_a < a^*/2$, $0 < q_b < b^*/2$, $q_c = 0$ this symmetry group has only one four-dimensional irreducible representation, which is therefore completely defined by the above mentioned star of wavevectors [139, 154], shown also in Figure 25. This means that the real lattice deformation (or the corresponding CDW amplitude) of, for example, TCNQ chains, can be expressed as

$$u(na, jb, pc) = \Psi_+ e^{i(q_a na + q_b jb)} + \\ + \Psi_- e^{i(q_a na - q_b jb)} + \text{c.c.}, \tag{63}$$

with

$$\Psi_\pm = \rho_\pm e^{i\theta_\pm}. \tag{64}$$

Note that the components Ψ_+ and Ψ_- which belong to different legs of the star in Figure 25 cannot be bilinearly coupled. This means that the full harmonic part of the Landau free energy consists of two equivalent contributions of the form (57) which depend only on ρ_+ and ρ_-, respectively.

Having defined the representation (63), we are able to construct the higher order terms in the Landau expansion for the free energy. We note that the transition at 38 K is only partially incommensurate-commensurate, i.e. $q_a(T)$ tends to commensurate $a^*/4$, but the longitudinal component $q_b = 2k_F$ remains incommensurate. It is therefore important to see whether among the symmetry-allowed terms there are Umklapp contributions which favor the commensurate wavenumber $q_a = a^*/4$, and therefore could account for such a transition.

Evidently, with $q_b = 2k_F$ incommensurate, no Umklapp term of the third order is possible. The first Umklapp term appears in the fourth order, and is given by [139]

$$\Psi_+^2 \Psi_-^2 e^{i \cdot 4 q_a na} + \text{c.c.} = 2\rho_+^2 \rho_-^2 \cos[4 q_a na + 2(\theta_+ + \theta_-)]. \tag{65}$$

The cosine factor in Equation (65) resembles the standard incommensurate–commensurate term for the phase $\theta_+ + \theta_-$. After the continuation $na \to x$ and the transformation

$$4\left(q_a - \frac{a^*}{4}\right)x + 2(\theta_+ + \theta_-) - \pi \equiv \beta \tag{66}$$

we arrive at the sine-Gordon model for β with the Lifshitz invariant in the second-order part of the free energy functional, as will appear later (cf. Equation (73)).

It is well known that in the standard sine-Gordon model the transition from the incommensurate to the commensurate state is continuous [158, 159], which contradicts the experimental situation in TTF-TCNQ (Figures 24(a), (b)). However, the particular property of the term (65) is that it is biquadratic in the amplitudes ρ_+ and ρ_- which belong to different legs of the star in Figure 25. Thus, the necessary condition for the lock-in of q_a at $a^*/4$ via the Umklapp term (65) is that *both* amplitudes, ρ_+ and ρ_-, are finite below 38 K and the discontinuity can arise on this basis. Indeed, we shall soon see that for $q_a \neq a^*/4$ it is energetically more favorable to have one leg activated.

The situation with one leg activated will be called phase modulation since then the CDWs on the neighboring TCNQ chains have the same amplitude and are shifted in phase by either $q_a a$ or $-q_a a$. In the same language the ordering with ρ_+ and ρ_- is amplitude modulated. Then the CDWs are not relatively shifted, but their amplitudes vary like $\sin(q_a x)$ by going in the x-direction.

We proceed by proving that the incommensurate ordering is phase modulated. For q_a not close to $a^*/4$ the quartic part of the free energy reduces to the normal contributions given by

$$B_1(\rho_+^4 + \rho_-^4) + B_2\rho_+^2\rho_-^2. \tag{67}$$

This expression is to be added to the harmonic part. It is easy to see that the phase modulation is then stable provided that [139, 160]

$$B_2 - 2B_1 > 0. \tag{68}$$

On the other hand, the dominant fourth-order contributions to the free energy are expected to be of local or, better, intrachain nature as discussed in Section 4.2. The coefficients in Equation (67) are then related by $B_2 = 4B_1 > 0$, so that the phase modulation is strongly preferred.

The situation is different for $q_a = a^*/4$, when the condition (68) is replaced by [139, 160]

$$B_2 - 2B_1 - 2B_3 > 0, \tag{69}$$

where B_3 is the coefficient in front of the Umklapp term (65). For intrachain contributions only $B_3 = B_1$, so that the left-hand side of the condition (69) just vanishes. Thus, even small interchain anharmonic contributions can reverse the sign of this expression, stabilizing the amplitude modulation. It will be therefore necessary to analyse such contributions more carefully. But, even without these details, we can already conclude that at 38 K both the phase β and the modulation should behave discontinuously [139, 161].

The notion of the modulation is most conveniently described by introducing new variables [162]

$$\rho_+^2 + \rho_-^2 \equiv R^2, \tag{70a}$$

$$\frac{\rho_-}{\rho_+} \equiv \tan\frac{\Phi}{4}, \tag{70b}$$

with $0 \leq \Phi \leq 2\pi$. For the amplitude modulation $\Phi = \pi$, while for two degenerate phase modulations $\Phi = 0$ and $\Phi = 2\pi$. The quartic part of the free energy (expressions (65) and (67)) is then given by

$$B_1 R^4 + B_3 R^4 \sin^2\frac{\Phi}{2} \cdot \left(\sin^2\frac{\beta}{2} - b\right), \tag{71}$$

where $b \equiv -(B_2 - 2B_1 - 2B_3)/4B_3$. The modulation Φ enters in the Landau functional as an effective sine-Gordon field too. We are dealing with two nonlinear (sine-Gordon-like) fields which are coupled. Having this in mind, let us proceed with further refinements concerning the ordering above 38 K. On one hand, it can be expected that the regions in the crystal which are locally ordered with q_a and $-q_a$ are connected by the domain walls in which Φ gradually changes from 0 to 2π. On the other hand, one could expect discommensuration effects above 38 K, i.e. a soliton latice in β. However, it follows from Equation (71) that the domain walls in Φ go together with flat regions in the phase β [$\beta = (2n + 1)\pi$], and vice versa, i.e. the discommensurations in β occur across the domains in Φ. This interplay of two nonlinear features is realized within a peculiar mathematical scope to which we shall return after completing the free energy functional.

Being interested in the domain structures above 38 K, in which the components $\Psi_\pm$ vary slowly in the x-direction, we can make the gradient expansion of the free energy around the star of Figure 25 which, by itself, defines the fast sinusoidal variations of CDWs in the x and y directions. The temperature dependence of this star is determined by the ratio $\eta \equiv \rho_F/\rho_Q$ in Equation (59) (to be distinguished from ρ_-/ρ_+). In fact the factor η connects the two CDWs in the TTF-TCNQ unit cell

$$u_F(n) = -\eta \cdot u_Q\left(n - \frac{1}{2}\right). \tag{72}$$

As will be seen later, it is reasonable to assume that this intracell parameter does not vary from one cell to another.

Expressed in the variables introduced by Equations (66) and (70), the second-order part of the free energy reads (primes denote the derivatives with respects to x)

$$F_2 = \frac{1}{a}\int dx\left\{AR^2 + C_\perp\left[R'^2 + R^2\frac{\Phi'^2}{16} + \right.\right.$$
$$+ R^2\left(\frac{1}{4}(\theta_+ - \theta_-)'^2 + \left(\frac{\beta'}{4} - \delta\right)^2\right) +$$
$$\left.\left. + R^2\cos\frac{\Phi}{2}(\theta_+ - \theta_-)' \cdot \left(\frac{\beta'}{4} - \delta\right)\right]\right\} \tag{73}$$

with

$$A = \tilde{a}_Q + \tilde{a}_F\eta^2 + \frac{\lambda_{a/2}^2}{\lambda_a}\frac{\eta^4}{1 + \eta^2} \tag{74a}$$

$$C_\perp = a^2\lambda_a(1 + \eta^2)\left[1 - \frac{1}{4}\frac{\lambda_{a/2}^2}{\lambda_a^2}\frac{\eta^2}{(1 + \eta^2)^2}\right] =$$
$$= a^2\lambda_a(1 + \eta^2)\cdot\sin^2(qa/2) \tag{74b}$$

and

$$\delta \equiv q_a - \frac{a^*}{4}. \tag{74c}$$

It remains to add the quartic terms. In particular, we must take into account the interchain contributions since they have an important role in the transition at 38 K. We therefore include, beside the intrachain terms

$$\frac{1}{6} b_Q \sum_{n,j} u^4\left(n + \frac{1}{2}, j\right), \tag{75a}$$

$$\frac{1}{6} b_F \sum_{n,j} u^4(n, j), \tag{75b}$$

also the leading interchain term which affects the condition (69), and is of the lowest order in η. (Note that Equation (59) suggests that $\eta \ll 1$.) It is of the form

$$\frac{1}{6} b_{1\perp} \sum_n u_Q^2\left(n - \frac{1}{2}\right) u_F(n) u_Q\left(n + \frac{1}{2}\right). \tag{76}$$

The further interchain term, not shown here explicitly, comes from the biquadratic coupling of the neighboring TCNQ and TTF chains, and so on. We also do not include the contributions which involve only TCNQ chains, since these chains are rather far from each other, so that the corresponding coupling constants are expected to be much smaller than $\eta b_{1\perp}$ in Equation (76). The contributions (75) and (76) lead to the following coefficients which define the fourth-order free energy density (71):

$$B_1 \simeq b_Q + b_F \eta^4 - \frac{\sqrt{2}}{3} b_{1\perp} \eta \tag{77a}$$

$$B_3 \simeq b_Q - b_F \eta^4 + \sqrt{2} b_{1\perp} \eta \tag{77b}$$

$$b \simeq \frac{1}{b_Q}\left(\frac{2\sqrt{2}}{3} b_{1\perp} \eta - b_F \eta^4\right). \tag{77c}$$

The complete free energy functional is the sum of the expressions (73) and (71). Beside Φ and β it contains the total amplitude R and the phase difference $\theta_+ - \theta_-$. The amplitude R can be taken as constant, since its weak space variation does not affect significantly the Φ and β dependent terms. The same conclusion can be drawn for the CDW amplitude on TTF chains, ρ_F, and consequently for the parameter η in Equation (72). This is a plausible simplification insofar as the anharmonic interchain coupling is weaker than the bilinear one, so that the CDW amplitudes are essentially determined by the free energy (56) completed with the intrachain part of the fourth-order terms (71).

Regarding $\theta_+ - \theta_-$, the minimization of F_2 gives

$$(\theta_+ - \theta_-)' = -2 \cos\frac{\Phi}{2}\left(\frac{\beta'}{4} - \delta\right). \tag{78}$$

After inserting this result into F_2, and passing from x to the temperature-dependent scale

$$y = 2R\sqrt{\frac{B_3}{C_\perp}}x, \tag{79}$$

the total free energy finally reads

$$F = \frac{1}{a}\int \frac{1}{2R}\sqrt{\frac{C_\perp}{B_3}}\mathrm{d}y\Big\{AR^2 + B_1R^4 + \\ + B_3R^4\Big[\frac{1}{4}\Big(\frac{\mathrm{d}\Phi}{\mathrm{d}y}\Big)^2 + \frac{1}{4}\sin^2\frac{\Phi}{2}\Big(\frac{\mathrm{d}\beta}{\mathrm{d}y} - \lambda\Big)^2 + \\ + \sin^2\frac{\Phi}{2}\Big(\sin^2\frac{\beta}{2} - b\Big)\Big]\Big\}, \tag{80}$$

with

$$\lambda \equiv \frac{2}{R}\sqrt{\frac{C_\perp}{B_3}}\delta. \tag{81}$$

We remark that the star of wavevectors in the expansion (80) is defined by the parameter λ, which depends on the ratio η through the Equations (59) and (74).

The functional (80) is the starting point for the consideration of the domain structures above 38 K. Before doing this, let us shortly analyze some quantitative characteristics of the transition at 38 K, still assuming that the ordering above 38 K consists of a single domain. Then for $T > 38$ K

$$\Phi = 0 \qquad \text{or} \qquad \Phi = 2\pi, \tag{82}$$

and β is arbitrary, while for $T < 38$ K

$$\Phi = \pi \qquad \text{and} \qquad \beta = 0. \tag{83}$$

After inserting these solutions into the free energy (80), it remains to minimize this energy with respect to R and η (or the corresponding parameter δ defined by Equations (59) and (74c)), and find their equilibrium values above and below 38 K, R_{inc}, δ_{inc} and R_{com}, δ_{com}, respectively. The temperature of the first-order transition is defined by

$$F(R_{\mathrm{inc}}, \delta_{\mathrm{inc}}, \Phi = 0 \text{ or } 2\pi, \beta \text{ arbitrary}) = \\ = F(R_{\mathrm{com}}, \delta_{\mathrm{com}}, \Phi = \pi, \beta = 0). \tag{84}$$

The quantities which are seen in experiments at 38 K are the jumps in the CDW amplitudes R and ρ_{F} on TCNQ and TTF chains, respectively, and the jump in the wavenumber in the a-direction, which is equal to δ_{inc} $(T = 38\,\mathrm{K}) \equiv \delta_0$. After a straightforward calculation we obtain that

$$\frac{R^2_{\mathrm{com}} - R^2_{\mathrm{inc}}}{R^2_{\mathrm{com}}} \approx \frac{b}{2} - \frac{\lambda_a^2\tilde{a}_{\mathrm{Q}}}{\lambda_{a/2}\tilde{a}_{\mathrm{F}}}\Big[\frac{-b\lambda_a}{A_0(1 + \lambda^2_{a/2}/8\lambda_a(-\tilde{a}_{\mathrm{F}}))}\Big]^{1/2} \tag{85}$$

and

$$(a\delta_0)^2 \approx \frac{(-A_0)b}{\lambda_a}\left(1 + \frac{\lambda_{a/2}^2}{8\lambda_a(-\tilde{a}_F)}\right). \tag{86}$$

Here $A_0 \equiv A(\eta_<)$ (Equation (74a)), and

$$\eta_<^2 \simeq \frac{\tilde{a}_F}{\tilde{a}_Q - (2\lambda_{a/2}^2/\lambda_a)} \simeq \frac{2\lambda_a^2}{\lambda_{a/2}^2}. \tag{87}$$

By $\eta_<$ and $\eta_>$ we denote the values of the parameter η for T approaching 38 K from below and above, respectively. The second equality in Equation (87) follows from Equation (59) for $q_a = a^*/4$ and $\lambda_a/\lambda_{a/2} \ll 1$. The jump in the CDW amplitude on TTF chains is connected with the above quantities R_{com}, R_{inc} and $\delta_0 a$ again via Equation (59), which for T close to 38 K reduces to the relation

$$\frac{\eta_< - \eta_>}{\eta_<} \simeq 1 - \frac{\rho_{\text{F inc}}}{\rho_{\text{F com}}} \cdot \frac{R_{\text{com}}}{R_{\text{inc}}} \simeq \frac{\delta_0 a}{2}. \tag{88}$$

The second equality contains only the directly measurable quantities. The data [10, 146] from Figures 23(a) and (b) are $R_{\text{com}}/R_{\text{inc}} \simeq 1.1$, $\rho_{\text{F com}}/\rho_{\text{F inc}} \simeq 1.25$, and $\delta_0 a \simeq 0.05 \times 2\pi \simeq 0.3$, and the relation (32) fits these numbers very well. It should be noted, however, that only the value of $\delta_0 a$ is experimentally well established [10, 34, 146, 147], while there is a large dispersion of data regarding the ratio of the spot intensities below and above 38 K.

For further comparison with experiments we need the temperature dependence of the intrachain coefficients a_Q, a_F involved in $\tilde{a}_Q$ and $\tilde{a}_F$ of Equations (61a, b). In this respect we note that the longitudinal dimension is absent from the free energy (56). This is justified if the fluctuations can be neglected, as is roughly the case when the interchain couplings λ are large enough. This corresponds to the $\xi_{0\perp} > d_\perp$ limit [127] of Equations (43a, b), when the coefficients a_Q, a_F correspond to the coefficient $A'(T/T_L^0 - 1)$ derived from the microscopic approach for the free energy density (41). The coefficients $\tilde{a}_Q$ and $\tilde{a}_F$ are then given by

$$\tilde{a}_Q = (a'_Q + 2\lambda_a + 2\lambda_{c/2})\left(\frac{T}{T_Q} - 1\right) \tag{89a}$$

$$\tilde{a}_F = \left(a'_F + 2\lambda_a + 2\lambda_{c/2} + \frac{\lambda_{a/2}^2}{\lambda_a}\right)\left(\frac{T}{T_F} - 1\right). \tag{89b}$$

Here T_Q and T_F are defined by Equations (61a, b), while a'_Q and a'_F correspond to the coefficient A' of Equation (49) and could be of the order of few λ_a's.

The connection between the coefficients a_Q, a_F and the microscopic coefficients from the expansion (41) is somewhat more involved when the interchain coefficients λ are small, i.e. correspond to the limit $\xi_{0\perp} < d_\perp$ of Equations (43a, b). The longitudinal fluctuations are then important but can be renormalized out by using the mean-field approximation in the interchain coupling [103–105, 108]. This results in the expression for the free energy in which the longitudinal dimension is absent, as in Equation (56), but a_Q and a_F are renormalized, i.e. built from both A' and ξ_0 in the two-chain analogue of Equation (41) [32, 160]. The corresponding coefficients $\tilde{a}_Q$ and $\tilde{a}_F$ are then

$$\tilde{a}_Q = (2\lambda_a + 2\lambda_{c/2})\left(\frac{T^2}{T_Q^2} - 1\right) \tag{90a}$$

$$\tilde{a}_F = \left(2\lambda_a + 2\lambda_{c/2} + \frac{\lambda_{a/2}^2}{\lambda_a}\right)\left(\frac{T^2}{T_F^2} - 1\right), \tag{90b}$$

where T_Q and T_F are again defined by Equations (61a, b), but with the effective coefficients a_Q and a_F, as explained above.

Since the estimates from Section 4 situate TTF-TCNQ somewhere in between ($\xi_{0\perp} \simeq a$), we consider both limits in parallel. Taking into account that $\lambda_{c/2}$ is of the same order as λ_a, and putting $T = 38$ K into Equations (89) and (90), we conclude from the relation (87) that the value of the ratio $\lambda_{a/2}/\lambda_a$ is of the order $2 \sim 4$ in both these limits. This roughly agrees with the microscopic evaluations [33]. The estimates of other quantities are also roughly independent of the limit which is taken. Thus, with the above values we get that $\lambda_{a/2}/8\lambda_a(-\tilde{a}_F) \simeq 0.2 \sim 0.5$ and $(-A_0)/\lambda_a \simeq 1 \sim 2$. From this, it is easy to see that the second term on the right-hand side in Equation (85) is much smaller than the first. It remains to estimate the value of the parameter b. Taking into account that $(\delta_0 a)^2 \simeq 0.1$, we get from Equation (86) that $b \simeq 0.05$ to 0.1. Such value of b leads to the relative jump in TCNQ displacement (i.e. CDW amplitude) $(R_{\text{com}}^2 - R_{\text{inc}}^2)/R_{\text{com}}^2$ of the order 0.025 to 0.05. This is smaller that the value observed in the early neutron scattering measurements [10] (Figure 24), but is perhaps in a better agreement with the more recent data [149]. Finally, let us mention that it is hardly possible to make a direct quantitative estimate of the interchain anharmonic parameter b. It will be seen in Section 5.3 that, beside $2k_F$ CDWs, it also comprises in an indirect way the bilinearly coupled $4k_F$ CDWs.

It is obvious from Equations (85) and (86) that the first-order transition with the lock-in at $q_a = a^*/4$ is possible only if

$$b > 0. \tag{91}$$

This means that the interchain anharmonic contributions have to dominate over the intrachain quartic term on the TTF chains, given by Equation (75b). Thus, if the term (76) is the most important interchain contribution, it should be repulsive ($b_{1\perp} > 0$), and moreover $b_{1\perp} \gtrsim b_F\eta^3$. One can similarly conclude that the attractive biquadratic coupling also contributes to the validity of the condition (91), and so on. The more detailed analysis of various interchain fourth-order terms was carried out in Reference [160]. It has already been mentioned above that those among the interchain terms which involve both TTF and TCNQ chains are expected to have a decisive role in the commensurate lock-in. The amplitude modulation below 38 K, however, fixes the amplitudes of CDWs on TCNQ and TTF chains in different ways. Namely, the solution (83) corresponds to the configuration in which all TCNQ chains have the same CDW amplitude, while TTF chains have amplitudes $\rho_F, 0, \rho_F, 0, \ldots$, i.e. half of TTF chains are not ordered. This is the consequence of the approximation (72) which favors the configuration which suits the TCNQ chains best. In a more realistic approach, the cosine amplitude dependence in the x-direction is shifted, leading to a finite ordering on all chains. Experimentally, it is well established [19, 21, 23, 55, 142, 143] that all TTF chains participate, together with TCNQ chains, in the long-range order below 38 K.

The results of this section are based on the assumption that the bilinear interchain terms introduced in Equation (56) dominate over the quartic interchain terms in the whole temperature range of three-dimensional ordering. Were the two kinds of terms comparable, the expansion (57) would have to be extended with additional q_a-dependent terms, and the types of three-dimensional ordering not encountered in Section 5.1(b) would appear. In particular, a strong biquadratic Q–F coupling was argued to stabilize [163] the solution with $q_a = a^*/2$ if both amplitudes ρ_F and ρ_Q are of the same order [163]. This solution was associated [163] with the three-dimensional ordering in TSeF-TCNQ [164]. We shall come to this point again in Section 5.4(d).

5.2. Hysteresis in the Temperature Range 38 K < T < 49 K

The three-dimensional configurations (82) and (83) used in the above discussion of the transition at 38 K have a simple sinusoidal form. In other words, complex amplitudes of the order parameter defined by the star in Equation (63) and Figure 25 are taken to be constant. In this section we shall extend the analysis to include the configurations in which these quantities may vary in space, more specifically, in the transverse x-direction. This, more detailed analysis of the temperature range 38 K < T < 49 K is raising some questions of principle [162] related to the unusual experimental observations in this range.

The first question has been already mentioned in Section 5.1(b), and concerns the simultaneous appearance of the discommensurations and the modulation domains due to the quartic Umklapp term (65). The next one concerns, in general, the effects of the discrete lattice on the incommensurate ordering with a varying wavenumber. This problem has been widely studied in recent years in some particular discrete models [165]. It has been shown [166, 167] that the wavevector follows the behavior of the complete or incomplete devil's staircases, i.e. that it is essentially a nonanalytic, step-like function of the external variable (e.g. temperature). Formally, the discrete summations like those in Equations (75) and (76) can be transformed into the continuous integrals by keeping the Umklapp, θ_+, θ_- dependent terms of all orders. Carrying out the continuation of all variables in Section 5.1, we truncated this procedure by retaining the presumably leading quartic phase-dependent term (71). It will be argued later that the discreteness of the lattice has an effct on the sliding of q_a, additional to the lock-in of q_a below 38 K.

Experimentally, the curve $q_a(T)$ is not uniquely defined, as is seen from the neutron scattering data [146] (see also Reference [149]) presented in Figure 26. The heating and cooling curves are shifted by about 1.5 K, and close up into a rather wide hysteresis which characterizes the first-order transition at 38 K. A similar hysteresis in the temperature range 38 K < T < 49 K was observed in the conductivity measurements [143]. The two curves can be bridged only by horizontal q_a = const lines, as is shown in Figure 26. There are strong experimental indications that this $q_a(T)$ hysteresis is not related to the imperfections of the samples, and is

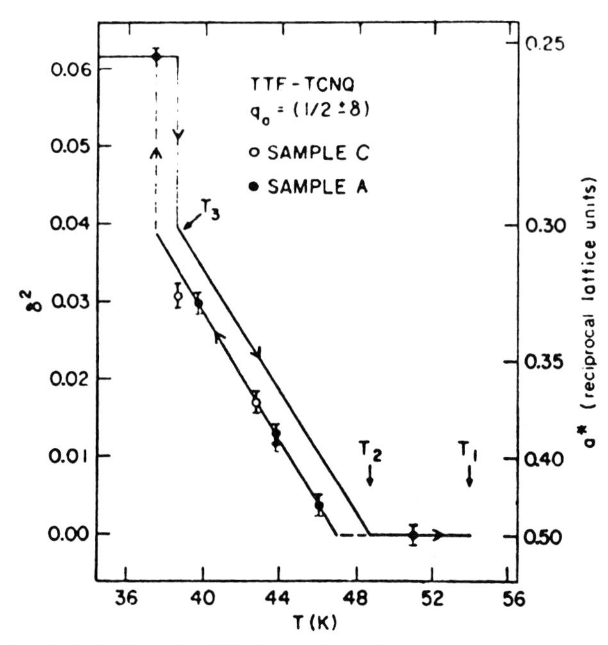

Fig. 26. The hysteresis behavior of $q_a(T)$ observed in the neutron scattering measurements (after Reference [147]). (© The American Physical Society.)

therefore of the microscopic origin [146]. However, it is not likely that this behavior is of the devil's staircase type [166], since then the width of the hysteresis would vary, depending on the order of commensurability through which $q_a(T)$ passes at a given temperature. The aim of the subsequent analysis is to show that the $q_a(T)$ hysteresis of Figure 26 might originate from the coupling between the phase and the modulation present in the free energy functional (80).

We are interested in the configurations which are the (meta)stable minima of the functional (80), and therefore should be found among the solutions of the Euler–Lagrange (EL) equations, defined by $\delta F/\delta\beta(y) = 0$ and $\delta F/\delta\Phi(y) = 0$:

$$\frac{\mathrm{d}^2\Phi}{\mathrm{d}y^2} = \left[\frac{1}{4}\left(\frac{\mathrm{d}\beta}{\mathrm{d}y} - \lambda\right)^2 + \sin^2\frac{\beta}{2} - b\right]\sin\Phi \tag{92}$$

$$\sin\frac{\Phi}{2}\cdot\frac{\mathrm{d}^2\beta}{\mathrm{d}y^2} + \cos\frac{\Phi}{2}\cdot\frac{\mathrm{d}\Phi}{\mathrm{d}y}\left(\frac{\mathrm{d}\beta}{\mathrm{d}y} - \lambda\right) = \sin\frac{\Phi}{2}\cdot\sin\beta. \tag{93}$$

It is useful to note that this system of coupled sine-Gordon equations, together with the expression (80), can be formally viewed as a classical mechanics problem. In this sense Equations (92) and (93) are Newton equations of motion in 'time' y for the material point on the sphere with angular coordinates β and Φ, and with the 'time'-independent Lagrangian given by the integrand in Equation (80). This is a problem with two degrees of freedom, and so the number of integrals of motion may be two or one. The properties of the solutions of EL equations depend in an essential way on whether the number of such integrals is equal or not to the number of degrees of freedom [168]. In the former, integrable case all solutions are periodic functions of time. (Such an example is the sine-Gordon model which is analogous to the problem of a simple pendulum in classical mechanics). More intricate are the latter, nonintegrable situations, in which EL equations also allow the solutions which are erratic in y [168]. However, often it is not possible to prove rigorously that a given system is nonintegrable, i.e. that there are no other integrals of motion beside those which were already found. In our problem only one integral is certainly present, i.e. the value of the mechanical Hamiltonian for a given solution. The useful way in such cases is to analyze directly the solutions of the equations of motion. Numerical calculations clearly show that Equations (92) and (93) have erratic solutions, and therefore present a nonintegrable problem.

In order to discuss the role of erratic solutions in the actual ordering, we have to systematize the configurations in a physical way [162]. Solutions of Equations (92) and (93) may be represented as trajectories in the four-dimensional phase space $(\Phi, \beta, \Phi', \beta')$. Each solution is completely determined by the values of these four functions in some particular point $y = 0$. From this abundant variety of functions we choose those which are expected to represent the domain patterns envisaged in Section 5.1(b). Let $y = 0$ be the center of the wall between the domains with $\Phi \simeq 0$ and $\Phi \simeq 2\pi$. Then $\Phi_0 \equiv \Phi(0) = \pi$, i.e. the center of the wall is amplitude modulated. The preferred value of the phase in this point is $\beta_0 = 0$, as we have seen in Section 5.1(b). The remaining two 'initial' conditions, Φ_0' and β_0', are free. We can vary them and consider the configurations which follow from numerical integrations of Equations (92) and (93). The convenient way to visualize these configurations is to use the planar cross-section Φ, β of the four-dimensional phase space. (Note in this respect that $0 \leq \Phi \leq 2\pi$ due to Equation (70b), while β is defined up to $\mathrm{mod}(2\pi)$).

The y-independent commensurate (Equation (83)) and incommensurate (Equation (82)) configurations discussed in Section 5.1 are obviously the particular solutions of EL equations (92) and (93). Our pair of initial conditions for the former solution is $\Phi_0' = \beta_0' = 0$. Let us now gradually come out from this origin of (Φ_0', β_0') plane, and look for the corresponding solutions. It turns out that only for the pairs of initial values which define a particular curve in the (Φ_0', β_0')plane, the solutions $\Phi(y)$, $\beta(y)$ are periodic, i.e. $\Phi(l) = \pi$, $\beta(l) = 4\pi n$, $\Phi'(l) = \Phi_0'$ and $\beta'(l) = \beta_0$. l denotes the period on the y-scale, and n is an even integer. There is also another family of periodic solutions associated with a curve of initial conditions which does not start from the origin, and for which n is odd. For initial conditions which are outside these two curves, the solutions $\Phi(y)$, $\beta(y)$ are erratic.

Thus we see that the set of erratic solutions is much richer than that of periodic solutions. Still, the more important question is which of these solutions are local functional minima of the free energy, and could therefore be thermodynamically (meta)stable. That requires a step which is not needed in classical mechanics, namely a second-order variation of the free energy. This is not practically realizable for general y-dependent solutions. Instead, we can calculate the free energy per unit length for all solutions, and thus obtain some feeling about the energetic hierarchy of the solutions. The relevant energy scale is given by the difference of the free energies for two homogeneous configurations, (82) and (83), and for a given value of R:

$$\Delta F \equiv F_{\text{com}}(R) - F_{\text{inc}}(R) = B_3 R^4\left(\frac{\lambda^2}{4} - b\right). \tag{94}$$

It turns out that all aforementioned periodic solutions have free energies which are larger than $\min(F_{\text{com}}, F_{\text{inc}})$, and are of the order of ΔF. For example, all solutions with even n have free energies lying in the range $(0, \Delta F)$. On the other hand, the mean energies for all calculated erratic solutions are much larger than $|\Delta F|$.

We thus come to the picture of the 'chaotic' plane (Φ_0', β_0') of configurations which have large free energies and are therefore expected to be thermodynamically unstable. Having much lower free energies, periodic configurations form deep 'canyons' in this plane. Regarding the type of these periodic extrema, we only know that the two limiting homogeneous configurations, (82) and (83), are stable for $\Delta \equiv \lambda^2/4 - b \gtrless 0$, respectively, which according to Equation (94), coincides with $\Delta F \gtrless 0$. Thus, the incommensurate ordering is realized for $\Delta F > 0$. Our further discussion of the incommensurate regime is based on a reasonable assumption that among these periodic solutions there is at least a subset of thermodynamically (meta)stable configurations (presumably those 'close' to the homogeneous configuration (82)).

Regarding the family with even n, one is approaching the incommensurate configuration (82) by going away from the origin of the (Φ_0', β_0') plane, i.e. from the commensurate configuration (83). Then the free energy of the periodic configurations decreases towards F_{inc}, and the period l increases, both in a continuous way. The same type of behavior is also realized in the other family (n odd), which does not have a commensurate end. Thus, both families have the same asymptotic ($l \to \infty$) configuration, i.e. the solution with one domain (82), considered in Section 5.1(b). The numerical results show that – depending on the values of parameters λ, Δ – this limit can be approached in the following two substantially different ways.

The configurations from a given family can be topologically equivalent and pass continuously one into another. Then n does not change, i.e. $n = 0$ and $n = 1$, respectively. Thus, for $n = 0$, one starts from weak oscillations around $\Phi = \pi$, $\beta = 0$, and ends up with the half of the macroscopic ($l \to \infty$) oscillation in Φ, the amplitude of which is equal to π.

Far more interesting is the situation in which n changes within the family of solutions with even or odd n. This is shown in Figure 27 where n changes from 0 to 2, and from 2 to 4. The solutions with successive values of $2n$ are separated by trajectories which are physically not realizable, since Φ extends outside the range $(0, 2\pi)$. The detailed analysis of Equations (92) and (93) shows that this segmentation by separatrices appears provided that [162]

$$\Delta > (1 - \lambda)^2. \tag{95}$$

The rough estimates of Section 5.1(c) suggest that this condition could be fulfilled in TTF-TCNQ.

The presence of the separatrices does not affect the continuity of the free energy and of the period l. However, although energetically degenerate, two configurations from the opposite sides of a given separatrix are topologically incompatible. Indeed, by changing n, we change discontinuously the slope of the soliton lattice in β by 8π

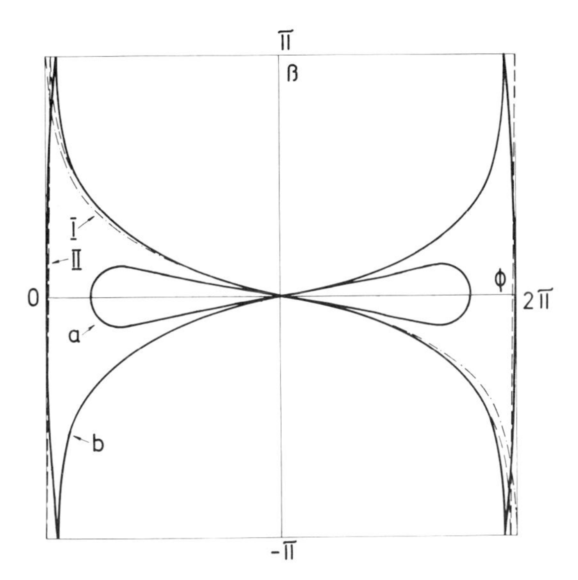

Fig. 27. The periodic configurations $\Phi(y)$, $\beta(y)$ presented in the reduced ($0 \leq \Phi \leq 2\pi$, $-\pi < \beta < \pi$) planar cross-section. Trajectories (a) and (b) belong to $n = 0$ and $n = 2$ subsets respectively. Trajectories I and II separate respectively subsets $n = 0$ and $n = 2$, and subsets $n = 2$ and $n = 4$.

per period l. Looking directly into chain CDWs, it is easy to see that the transition across a separatrix leads to the longitudinal shift of CDWs by a finite length of the order of $\pi/2k_F$ (i.e. half of the Peierls wavelength) along the chains on which $\Phi(x) \approx \pi$. The intermediate states involved in such strong local deformations are highly improbable in comparison with the initial and final configurations, and so represent a high barrier which prevents such topological transitions. In other words, being once in a given periodic configuration, the system is not able to escape from the corresponding segment of the 'canyon' bounded by two separatrices.

The picture of canyons separated in segments allows us to interpret the hysteresis from Figure 26. Let us assume that the condition (95) is realized and that by leaving the commensurate phase at 38 K, the system does not go to the homogeneous incommensurate phase (82), but to some periodic configuration which is 'topologically closer' to the commensurate configuration. Then, in the whole sliding temperature range, the system is captured in the segment to which this configuration belongs. On the other side, the sliding consists of changes in the basic CDW period $2\pi/q_a$, and in the superlattice period x_l (which corresponds to l on the y-scale). Therefore it invokes pinnings on the discrete lattice. These pinning effects may well dominate over the slight differences between the free energies of periodic configurations inside the segment, but still be much weaker than the separatrix barriers. Thus, the system will prefer to keep q_a and x_l fixed by changing the configurations inside the segments. (Note that the coefficient which connects y and x in Equation (79) is temperature dependent, so that l changes with temperature even if x_l is fixed). This freedom is, however, lost when the system comes to one of the separatrices. From here on, the system follows the temperature variation of the separatrix, being forced to change q_a and x_l. This results in the heating and cooling $q_a(T)$ lines of Figure 26. The discontinuities of the devil's staircase type could appear on these lines, but are probably too weak to be observed in TTF-TCNQ. By reversing the direction of the temperature variation, the system can again keep constant periodicities, but only until it arrives to the opposite separatrix. This is the presumable origin of the horizontal bridging lines in Figure 26.

The above interpretation of the $q(T)$ hysteresis is based on the two crucial properties of EL equations (92) and (93). The first one is the nonintegrability, i.e. the presence of chaotic states, due to which the set of physical solutions is reduced to the single-parameter space of periodic configurations. The second property is the appearance of separatrices which leads to the division of this space into segments of topologically equivalent configurations, and to the finite width of the $q_a(T)$ hysteresis on the temperature scale. The system changes the periodicity of the ordered superstructure only after its tendency to accommodate locally the form of the order parameter is completely hindered by the intermediate chaotic and periodic barriers.

Although based on the particular model examined here, this picture may have a more general relevance. We note in this respect that the reduction of Landau expansion to an integrable problem is always an idealization, and that weaker or stronger nonintegrable terms are present in each situation. It would be desirable to analyze the effects of such terms in other interesting solid state examples, especially in those in which hysteresis effects are observed.

5.3. $4k_F$ ORDERING

The $4k_F$ scattering ($q_b = 0.41b^* \equiv 0.59b^*$) was first observed in the X-ray measurements [53, 54] as described in Section 4. In the range of high temperatures the $4k_F$ diffuse scattering dominates in intensity over the $2k_F$ scattering [53] (Figure 11). It is visible even at room temperatures, while $2k_F$ lines disappear above ~150 K. The situation is, however, quite opposite in the regime of three-dimensional ordering ($T < 49$ K), where the intensity of the $4k_F$ spots in the X-ray data is a few times weaker than the intensity of the $2k_F$ spots (Figure 28). In the neutron scattering measurements [147] the faint $4k_F$ spots emerge from the background only below 38 K, with an intensity amounting only to a few percent of the corresponding $2k_F$ intensities.

The $4k_F$ satellites are positioned at $(2q_a, 4k_F, 0)$ for $49\,\mathrm{K} > T > 38$ K, and at $(a^*/2, 4k_F, 0)$ below 38 K, i.e. their wavenumbers are exactly twice the wavenumbers of the $2k_F$ satellites. The corresponding lattice deformation is longitudinally polarized [53, 149], as for the weaker mode of Figure 24(b). In addition, the X-ray measurements [54] also revealed the very weak satellites which appear only below 38 K at $(0, 4k_F, 0)$, and correspond to the longitudinally polarized deformations [169].

We see that the $4k_F$ satellites appear at the outset of the sliding of q_a, i.e. according to the picture developed in the previous sections, with the $2k_F$ activation

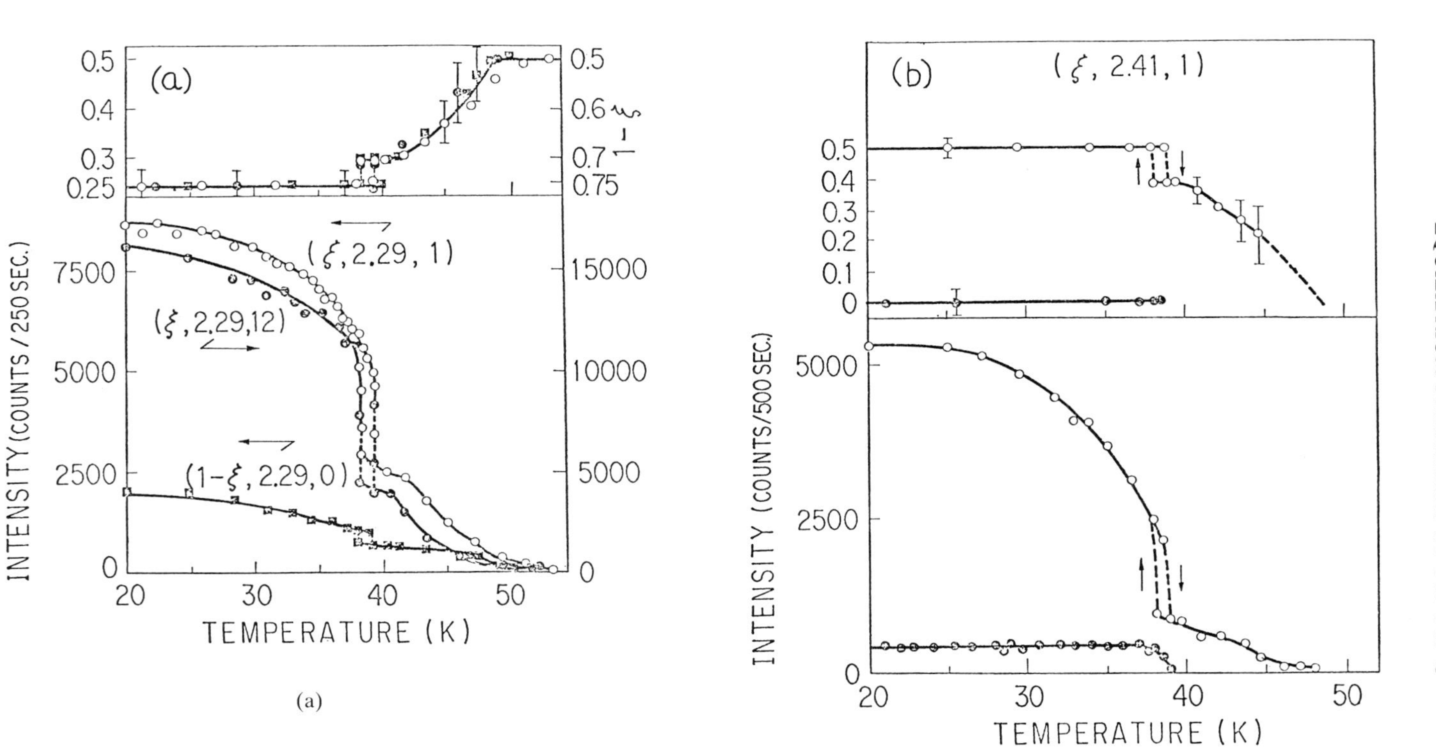

(a)

(b)

Fig. 28. The temperature dependence of the transverse wavenumber and the scattering intensity of $2k_F$ (a), and $4k_F$ (b) ordering, observed in the X-ray measurements. Note that the (○) points belong to the same Brillouin zone, and that the intensity scales in figures (a) and (b) are different (after Reference [54]). (© Physical Society of Japan.)

of the TTF chains. This is supported by the phase diagram of the ternary compounds TTF_{1-x} $TSeF_x$ TCNQ with $0 < x < 1$ [170]. It appears that the three-dimensional $4k_F$ ordering is very sensitive to the alloying of TTF chains and disappears already for $x \gtrsim 0.03$, being replaced by one-dimensional $4k_F$ correlations. The neutron scattering measurements on $TTF_{0.97}$ $TSeF_{0.03}$ TCNQ [171] also give a direct evidence that the $4k_F$ ordering is to be attributed to the TTF chains. The idea which follows naturally from the described experimental situation is that the $4k_F$ static deformations are the secondary effect of the $2k_F$ ordering on the TTF chains. Thus, although the one-dimensional $4k_F$ CDW fluctuations are stronger than the $2k_F$ fluctuations at high temperatures, they do not undergo the independent three-dimensional ordering. Instead, the three-dimensional ordering of the $2k_F$ CDWs on the TTF chains acts presumably as a driving force, which causes the freezing of the $4k_F$ CDWs on those chains (and in turn of the corresponding lattice deformations).

The lowest order possible invariant which couples the $2k_F$ and $4k_F$ CDWs is linear in $4k_F$ components and quadratic in $2k_F$ components [172]. Thus the two lowest order $4k_F$ intrachain contributions to the free energy are

$$\sum_n a_4 |\chi(n)|^2 + a_{24}[\chi(n)\Psi^{*2}(n) + \text{c.c.}]. \tag{96}$$

Here $\chi(n)$ is the complex amplitude of the $4k_F$ deformation

$$u_4(n, j) = \chi(n)\exp(i \cdot 4k_F jb) + \text{c.c.}, \tag{97}$$

and

$$\Psi(n) \equiv \Psi_+ \exp(iqna) + \Psi_-^* \exp(-iqna) \tag{98}$$

has the same meaning for the $2k_F$ deformation (cf. Equation (63)). The coefficient a_4 is related to the $4k_F$ CDW correlation function (36), while a_{24} may also be determined from the microscopic considerations.

After minimizing Equation (96), the component $\chi(n)$ is given by

$$\chi(n) = -\frac{a_{24}}{a_4}\Psi^2(n) = -\frac{a_{24}}{a_4}[\Psi_+^2 \exp(2qnai) + \\ + \Psi_-^{*2}\exp(-2qnai) + 2\Psi_+\Psi_-^*]. \tag{99}$$

The expression (96) then reduces to

$$\sum_n \left(-\frac{a_{24}^2}{a_4}\right)|\Psi(n)|^4, \tag{100}$$

i.e. the $4k_F$ ordering gives an effective attractive intrachain contribution which should be added to the fourth-order terms introduced in Equations (65) and (67). Note that, like all intrachain contributions, the term (100) fulfills the condition (68).

The transverse wavenumbers of the $4k_F$ ordering follow directly from Equation (99). It is seen that the phase modulated $2k_F$ ordering with, for example, a finite component Ψ_+ induces a $4k_F$ component with the wavenumber $+2q_a$, and correspondingly a finite Ψ_- induces the wavenumber $-2q_a$, as is observed in the range $49\,K > T > 38\,K$ [53, 54]. Below 38 K, when $2q_a = a^*/2$, the wavenumbers $2q_a$ and

$-2q_a$ coincide. Then the star ($\pm 2q_a$, $\pm 4k_F$) reduces to two points, and the ($a^*/2$, $4k_F$) component of the $4k_F$ ordering is given by

$$-\frac{a_{24}}{a_4}(\Psi_+^2 + \Psi_-^{*2}). \tag{101}$$

The third term of the right-hand side in Equation (99) indicates that whenever the $2k_F$ ordering is amplitude modulated ($\Psi_+ \neq 0$ *and* $\Psi_- \neq 0$), the $4k_F$ ordering with $q_a = 0$ also appears. The fact that the spots at (0, $\pm 4k_F$, 0) appear below 38 K [54] is thus a natural consequence of the complete amplitude modulation of the $2k_F$ ordering in this temperature range [139] (Equation (83)). The local amplitude modulation above 38 K due to the appearance of the periodic domain patterns (Section 5.2), leads to a similar effect, which, however, should be weaker and therefore hardly observable.

The spots at ($\pm 2q_a$, $\pm 4k_F$, 0) and (0, $\pm 4k_F$, 0) do not belong necessarily to the same phonon mode, as can be shown by a straightforward generalization of the expansion (96) in which more than one $4k_F$ mode is taken into account (see also the next section). Therefore these spots may differ in their polarization, temperature dependence of intensities, etc.

The expansion (96) can also be extended by including the interchain $4k_F$–$4k_F$ and $4k_F$–$2k_F$ terms [172]. Particularly, the bilinear interchain $4k_F$–$4k_F$ interaction

$$\sum_n \chi(n)\chi^*(n + 1) + \text{c.c.} \tag{102}$$

leads, after eliminating $\chi(n)$ components, to an effective biquadratic $2k_F$–$2k_F$ term. Thus the three-dimensional $4k_F$ ordering also affects the coefficients (77), and therefore may help in stabilizing the commensurate phase below 38 K. Some feeling about a relative strength of this indirect mechanism may be obtained from the data on TTF_{1-x} $TSeF_x$ TCNQ alloys [170]. With the disappearance of the three-dimensional $4k_F$ ordering ($\chi \geq 0.03$), the critical temperature of the incommensurate–commensurate transition decreases from 38 K to about 28 K. This gives further support to the assumption that the $2k_F$ components are the driving ones in TTF-TCNQ. The intermediate situation in which the $4k_F$ orderings are treated on equal footing is much more complicated and will not be considered here.

Let us finally note that in the opposite limit of the dominant $4k_F$ ordering the transition at 38 K would have to be interpreted in a way quite different from that of Section 5.1(c). We have already mentioned that at 38 K the fourfold star ($\pm 2q_a$, $\pm 4k_F$) reduces to two points ($a^*/2$, $\pm 2k_F$). Note that this resembles the situation encountered at 49 K for the $2k_F$ CDWs (Section 5.1(b)). The only difference is that the commensurate $a^*/2$ phase for $4k_F$ ordering is present below the transition ($T <$ 38 K), and not above it, as is the case for the $2k_F$ ordering at 49 K. However, the two situations are equivalent from the symmetry point of view, so that one would expect the same type of phase transition in both cases. The fact that the transition at 38 K is of the first order and not of the second order like that at 49 K, indicates again that the limit of the dominant $4k_F$ ordering is not appropriate to the actual behavior of TTF-TCNQ.

5.4. (p, T) PHASE DIAGRAM

5.4(a) *Experimental Situation*

The cascade of phase transitions considered in Sections 5.1–5.3 occurs in a particularly narrow temperature interval of only ~ 15 K. It was pointed out in Section 2.4 that these steep variations give strong evidence for large electron–electron interactions, which lead to the fast temperature dependence of intra- and interchain correlations. Therefore, it has also been interesting to investigate the pressure dependence of the correlations. The experimental analysis of the low-temperature behavior under pressure started with the transport experiments [128, 173, 174], and continued with a series of neutron scattering measurements [175–180]. The results of the latter investigations are summarized in Figure 7. In this rather intricate phase diagram, the following three features can be tentatively distinguished:

(i) The ordering of an additional mode at $T_M(p)$ [180]. $T_M(p)$ starts at 1 bar from the value close above T_F, and increases slowly with p, merging finally into the line $T_Q(p)$ at $p \simeq 4$–5 kbar. No change of q_a is associated with this ordering. The lines $T_Q(p)$ and $T_M(p)$ define the region IV in Figure 7. The presence of two lines, $T_M(p)$ and $T_F(p)$, is consistent with the earlier specific heat data at ambient pressure [156].

(ii) A very steep decrease of the temperatures T_F and T_X as the pressure increases [176, 177], especially in the range 2 kbar $\lesssim p \lesssim 4$ kbar where the sliding $q_a(T)$ phase (II in Figure 7) and the commensurate $a^*/4$ phase (III in Figure 7) disappear, so that for $p \gtrsim 4$ kbar, q_a equals $a^*/2$ (I in Figure 7) in the whole temperature range below $T_Q(p)$.

(iii) The change of character of the critical temperature T_Q in the pressure range 14 kbar $\lesssim p \lesssim 20$ kbar, in which $2k_F = b^* - 4k_F = b^*/3$ and $q_a = 0$ (VI in Figure 7) [175, 176, 179]. Then the line $T_Q(p)$ represents the first-order transition, and varies slightly with pressure [179] (Figure 7). This, however, is at variance with the transport data [128, 173] which show a distinctive maximum of $T_Q(p)$ at about 19 kbar.

In the remaining part of this section we shall mainly discuss point (i) by considering the symmetry properties of molecular displacements which participate in the transitions at T_Q, T_M, and T_F. Some interpretations and controversies regarding points (ii) and (iii) will be briefly reviewed afterwards.

5.4(b) *Symmetry Analysis*

The Landau expansion in Section 5.1 was based on the assumption that there is a one-to-one correspondence between CDWs and the accompanying lattice deformations. However, TTF and TCNQ chains are arrays of tilted, slab-like molecules. It is easy to see that in the scheme of rigid molecular orbitals, a linear change in the overlap between the neighboring orbitals occurs for both the longitudinal (b) and

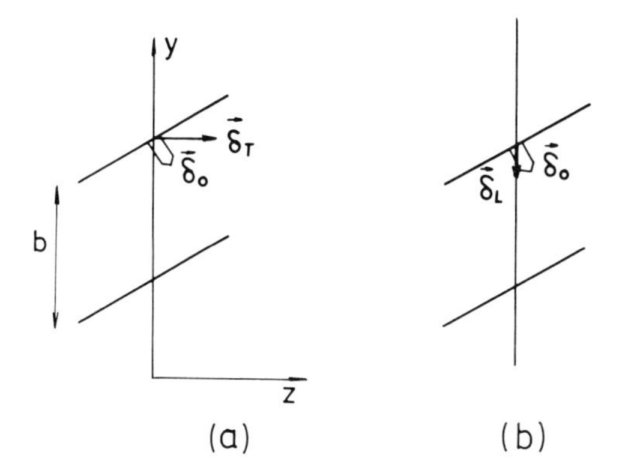

Fig. 29. The change in the longitudinal overlap along $\boldsymbol{\delta}_0$ for the transverse $\boldsymbol{\delta}_T$ (a), and the longitudinal $\boldsymbol{\delta}_L$ (b) relative molecular displacements.

transverse (c) relative molecular displacements (Figure 29). Thus, a CDW on a given chain may induce two translational degrees of freedom through the linear electron–phonon coupling. This corresponds to as many as eight degrees of freedom per unit cell, i.e. to eight different displacive lattice modes. This is still a simplified description in terms of lattice deformations. Namely, we ignored the translations in the *a*-direction since they are not coupled linearly to CDWs, as well as librations and intramolecular deformations. It was already mentioned that none of these modes was seen in the structural measurements.

For a general wavevector $\mathbf{q} = (q_a, q_b = 2k_F, q_c)$ the symmetry group $P2_1/c$ has only one irreducible four-dimensional representation, denoted here by **T**, and defined by the star

$$\pm(q_a, \pm q_b, q_c) \tag{103}$$

(see Figure 25 for $q_c = 0$). Then the eight modes involved in the *b* and *c* translations cannot be distinguished from the symmetry point of view. However, the symmetry splitting into two two-dimensional representations, $\mathbf{T}_+$ and $\mathbf{T}_-$ [154], occurs at the edges of Brillouin zone $\mathbf{q}_0 = (a^*/2, \pm 2k_F, 0)$ which are of particular interest to us. In order to specify this separaion, let us label the translational modes according to the above discussion. It is convenient to start from the modes in which either TCNQ(Q) or TTF(F) chains are displaced, this displacement being either longitudinal (L) or transverse (T), and taking the displacements on the neighboring chains in the *c*-direction either as parallel (i.e. acoustic, A) or as antiparallel (i.e. optic, O). In this way we define altogether eight modes, namely

$$\text{QLO, QLA, QTO, QTA, FLO, FLA, FTO, FTA,} \tag{104}$$

where e.g. QLO denotes the deformation in whch two neighboring TCNQ molecules in the *c*-direction are displaced longitudinally in opposite directions.

The symmetry analysis which takes into account the vector nature of the modes (104) leads to the conclusion [180] that for $\mathbf{q} = \mathbf{q}_0$

$$\begin{array}{ll} \text{QLO, QTA, FLA, FTO} & \text{belong to } \mathbf{T}_+, \\ \text{QLA, QTO, FLO, FTA} & \text{belong to } \mathbf{T}_-. \end{array} \tag{105}$$

This means that for $\mathbf{q} = \mathbf{q}_0$ any eigenmode of the TTF-TCNQ lattice is a linear combination of the modes (104) belonging to only one of the two sets in Equation (105). In other words, the Landau expansion which corresponds to the star $(a^*/2, 2k_F, 0)$ can therefore contain bilinear invariants which couple the modes (104) within the same representation ($\mathbf{T}_+$ or $\mathbf{T}_-$). Similarly, bilinear Lifshitz invariants which are responsible for the sliding of $\mathbf{q}$ in both x and z transverse directions, couple the modes belonging to different sets in the classification (105).

The above results can be easily related to the CDW symmetry analysis used in Reference [154], after specifying the direction of the gradient of the intrachain overlap t for each type of chains. In Figure 30 we assume that this direction is orthogonal to the molecular plane. This figure then shows that, for example, the modes QLO and QTA both correspond to an 'optic' CDW mode. More generally, two different displacive modes belonging to the same family of chains and to the same representation are to be associated with the same CDW mode. Thus, e.g. the $\mathbf{T}_+$ representation comprises the 'optic' CDW mode on TCNQ chains, and the 'acoustic' CDW mode on TTF chains.

In principle, the result (105) allows for the two effects already mentioned in Section 5.1(b), namely, the simultaneous ordering of both chain families at T_Q [154], and the sliding in both x and z transverse directions below T_F [152, 153]. Of course, the symmetry analysis alone does not give any information about the strength of these effects. Physically they are proportional to the small differences of the coupling constants between the pairs of chains at distances $|\mathbf{a}/2 + \mathbf{c}/2|$ and $|\mathbf{a}/2 - \mathbf{c}/2|$, not included in the free energy (56). It is useful in this respect to reproduce the symmetry analysis for the group $Pmcb$, in which the angle (a, c) is equal to 90° (and not 104.5° as in the actual TTF-TCNQ crystal lattice). All invariants which exist for

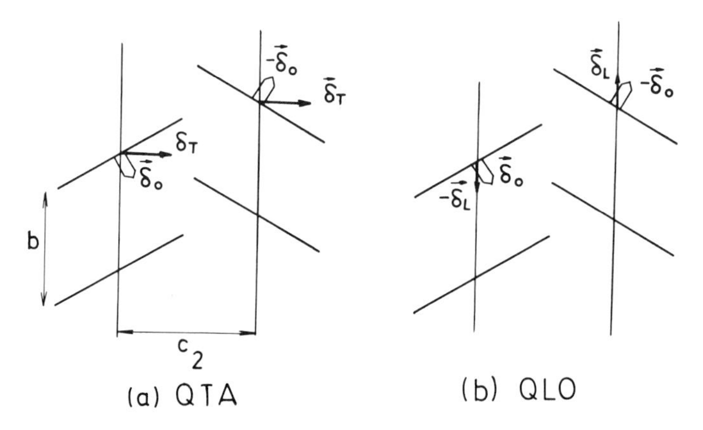

Fig. 30. The overlap variations along δ_0 which allow the CDW to induce the QTA (a) and QLO (b) displacive modes.

the group $P2_1/c$ but vanish for the group *Pmcb*, are to be considered much weaker than the corresponding invariants which exist for both symmetry groups.

For the symmetry group *Pmcb* additional splittings occur for $\mathbf{q} = \mathbf{q}_0$, and the eight modes (104) are separated into the following four irreducible representations [180]:

$$\begin{array}{l} \text{(QLO, QTA)} \\ \text{(FLA, FTO)} \\ \text{(QLA, QTO)} \\ \text{(FLO, FTA)} \end{array} \tag{106}$$

In other words, optic and acoustic CDW modes from different families of chains become decoupled. Furthermore, it can be shown that the Lifshitz coupling in the z-direction between the pairs (QLO, QTA) and (FLO, FTA) (as well as between the other two pairs of modes), also vanishes. The same is valid for the Lifshitz coupling in the x-direciton between the pairs of modes belonging to equivalent chains. Thus, the symmetry analysis of the group *Pmcb* (Equation (113)) suggests that both the simultaneous ordering of TCNQ and TTF chains at 54 K and the sliding in the c-direction are weak effects, in agreement with the already mentioned experimental findings [145–149].

The above comparison between the *Pmcb* and $P2_1/c$ symmetry groups is also helpful for the understanding of lines $T_Q(p)$, $T_M(p)$ and $T_F(p)$ in the (p, T) diagram. First of all, we see that the labels Q and F are well chosen, i.e. that the actual eigenmodes in TTF-TCNQ can be reasonably assigned to the separate chain families. On the other hand, the neutron scattering data suggest that the labels L and T are also well defined for all three lines, at least at their lower pressure ends: the transverse mode condenses on the line $T_Q(p)$ [169, 180], while the longitudinal modes condense on the lines $T_M(p)$ and $T_F(p)$. There is one more assignation based on the inelastic neutron scattering data; the phonon softening at $q_b = 2k_F$ which leads to the phase transition at $T_Q(p)$, is observed on the transverse acoustic branch [149]. Thus, T_Q is the critical temperature for the mode QTA. Finally, the $q_a(T, p)$ dependence is characterized by the absence of any Lifshitz coupling in the z-direction between the lines $T_Q(p)$ and $T_F(p)$, and by the activation of a strong Lifschitz term in the x-direction below $T_F(p)$.

The above facts, combined with the results of the symmetry analysis (Equations (105) and (106)), leave no arbitrariness in the identification of the modes which take part in the transitions $T_M(p)$ and $T_F(p)$. These are QLO and FLO, respectively (possibly combined with FTA). We see that the CDWs which correspond to these modes and to the mode QTA from $T_Q(p)$, are all 'optic', i.e. the CDWs at neighboring chains in the c-direction are always in the opposition of phases. This indicates that on both families of chains the Coulomb CDW–CDW contribution dominates in the coupling constant $\lambda_{c/2}$ [31].

The modes QTA and QLO, which are attributed to the lines T_Q and T_M, belong to the same irreducible representation even in the classification (106). Thus, the symmetry allows the bilinear coupling between these two modes. Consequently, in general, T_M is not a proper phase transition but rather the crossover temperature at which the QLO mode ceases to be driven by the QTA mode, and starts to order by itself [169, 180]. The weaker is the bilinear coupling, the sharper is this crossover. The fact that this line is well defined in experiments indicates that the bilinear coupling between QTA and QLO is rather weak at low pressures. However, it seems that with increasing pressure these two modes gradually hybridize, and that only one of the two new modes is coupled to CDWs. The displacements in the other mode are perhaps directed along the molecular planes, and are therefore coupled very weakly to CDWs. This might explain the merging of the two lines into the single line at $p \gtrsim 5$ kbar. The resulting conclusion, that this line originates completely from the TCNQ chains, should, however, be contrasted to that of Weyl *et al.* [18], who interpreted the thermopower data as the ordering of TTF chains at $T_Q(p)$ for $p \gtrsim 8$ kbar.

5.4(c) *The Commensurate Lock-in of* $2k_F$ *at* $b^*/3$

The pressure range between ~5 and ~12 kbar was not structurally investigated. In this range q_a passes from $a^*/2$ (region I in Figure 7) to a^* (region VI in Figure 7) [178]. Furthermore, at about 14.5 kbar $T_Q(p)$ becomes a first-order transition [128, 173, 178]. Both effects are closely related to the increase of the wavenumber $2k_F$ with pressure, shown in Figure 31 [175, 178]. This increase of $2k_F$ (i.e. of the interchain charge transfer ρ from Equation (1)) is a simple consequence of the

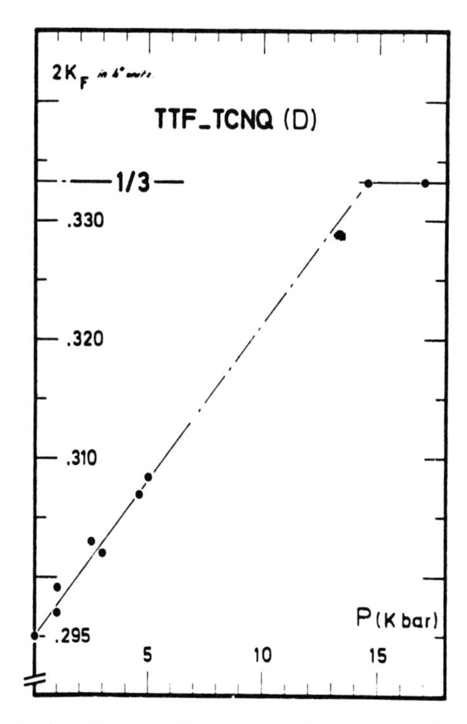

Fig. 31. The neutron scattering data for the pressure dependence of the $2k_F$ wavenumber (after Reference [178]). (© Pergamon Press Ltd.)

increase of bandwidths $4t_Q$ and $4t_F$ due to the decrease of the intermolecular distance b [169]. As $2k_F$ approaches the commensurate value $b^*/3$, the longitudinal Umklapp term of the third order is presumably activated, which, together with the finite bilinear interchain coupling, leads to the first-order transition to the ordered phase. It was shown then that both the increase of the critical temperature due to the Umklapp term and the hysteresis width of the first-order transition are of the same order [122]. The smooth increase of T_Q in Figure 7, and the rather small hysteresis (~ 1 K) observed in the transport measurements [128, 173], suggest however that this Umklapp term is rather weak. On the other hand, it must be strong enough to impose the transverse coherence with $q_a = a^*$, which is, together with $q_a = a^*/3$, the only periodicity compatible with the lock-in at $b^*/3$.

The other interesting feature which could influence the transition at $T_Q\ (p)$ is the Umklapp equivalence of $2k_F$ and $4k_F$ wavenumbers when $2k_F \approx b^*/3$ ($2k_F + 4k_F \approx b^*$). This mixing of the two instabilities which characterize TTF-TCNQ (Sections 3.6 and 5.3) was also invoked [122] to explain the results of transport measurements [128, 173], which, contrary to the structural data of Figure 7, suggest that the increase of T_Q at $2k_F \approx b^*/3$ is much larger (~ 10 K) than the hysteresis width. Unfortunately, it is not known how the high temperature $2k_F$ and $4k_F$ correlations (Figure 11) behave under pressure. As for the low-temperature regime, no $4k_F$ three-dimensional ordering was directly visible in the whole pressure range of Figure 7. This indicates that $4k_F$ correlations are of secondary importance for three-dimensional ordering in region VI of Figure 7 as well.

5.4(d) *The Pressure Dependence of* T_F *and* T_X

In contrast to the smooth increase of the transitions T_Q and T_M with pressure, the two other critical temperatures, $T_F(p)$ and $T_X(p)$ decrease steeply, so that the sliding $q_a(T)$ and the commensurate $q_a = a^*/4$ phases (II and III respectively in Figure 7), are eliminated above 5 kbar [176, 177]. According to the considerations in Section 5.1 and in this section, the two latter phases are characterized by switching on of the bilinear and quartic interactions between TCNQ and TTF chains, and consequently, by the involvement of TTF chains into the long-range order. Thus, the steep decrease of $T_F(p)$ should originate from the strong pressure dependence of either intrachain TTF Coulomb correlations or the above interchain couplings.

The smooth variation of the line $T_Q(p)$ in Figure 7 indicates that intrachain correlations on TCNQ chains are not very sensitive to the decrease of b. It is thus reasonable to expect that the same is valid for TTF chains, and that the lines $T_F(p)$ and $T_X(p)$ are influenced by the pressure dependence of interchain terms. The analysis of the model (56), however, shows that regions II and I from Figure 7 cannot be interpreted by taking into account only the pressure dependence of the bilinear coupling constant $\lambda_{a/2}$ in Equations (59)–(62). Namely, even if this dependence is so strong that $\lambda_{a/2}$ changes the sign [181], this is still incompatible with the existence of the wide range in region I ($p \gtrsim 4$ kbar) in which q_a remains equal to $a^*/2$ in the whole temperature range below $T_Q(p)$. It is therefore necessary to invoke the influence of pressure on the quartic interchain terms.

The detailed analysis of the free energy (56) supplemented by the cubic-linear ($\sim\rho_Q^3\rho_F$) and biquadratic ($\sim\rho_Q^2\rho_F^2$) terms, was undertaken in Reference [169] (see also Reference [182]). It was shown that the temperature T_F decreases fastly under pressure if, for example, the biquadratic term becomes more and more repulsive. The increase of the coefficient $b_{1\perp}$ (Equation (76)), or more precisely of the combination $b_{1\perp}|\lambda_{a/2}|/\lambda_a$, has the same consequence on T_F. With $T_F(p)$ disappearing ($p \gtrsim 4$ kbar), the relatively strong quartic interchain couplings stabilize the ordering of only one (TCNQ) family of chains with $q_a = a^*/2$, i.e. of the solution (58) in Section 5.1(b). This result contradicts the proposition [183] that the biquadratic $\rho_Q^2\rho_F^2$ term leads to the phase with $q_a = a^*/2$ and both families of chains ordered, used also in the interpretation of three-dimensional ordering in TSeF-TCNQ [163], and mentioned in Section 5.1(c).

The line $T_F(p)$ reaches $T_X(p)$ at about 2 kbar, where $T_F(p) \simeq T_X(p) \simeq 35$ K [176, 177]. Furthermore, the lock-in phase transition $T_X(p)$ decreases more slowly than $T_F(p)$, and there exists a pressure range 2 kbar $\lesssim p \lesssim$ 3.2 kbar in which q_a jumps directly from $a^*/2$ to $a^*/4$. This indicates that the condition (91) for the stabilization of the commensurate configuration considered in Section 5.2 remains fulfilled under pressure. This again points towards the importance of the term (76): namely, the increase of the repulsive coupling constant $b_{1\perp}$ may favor both the decrease of the temperature T_F and the fulfilment of the condition (91).

The further increase of pressure leads to the appearance of the new region V (Figure 7) [177, 170], which abruptly replaces region IV at $p \simeq 3.2$ kbar, and likewise abruptly disappears at $p = 4$ kbar. Both nearly vertical lines in Figure 7 represent discontinuous transitions on the pressure scale [177].

The first neutron scattering measurements showed [177] that q_a varies with pressure in the whole of region V. This variation may again be the consequence of the fast pressure dependence of the quartic coupling constants, due to which the condition (91) finally breaks. The later data [179], however, suggest that $q_a(T, p)$-dependence in region V is very weak, so that q_a mainly remains close to $a^*/3$. This possibility opens the question of the eventual lock-in of q_a at $a^*/3$, the origin of which is not clear at present.

5.5. Conclusion

The analysis in Sections 1–4 showed that the $2k_F$ and $4k_F$ CDW instabilities in TTF-TCNQ originate from the electron–electron interactions, and are not of the Peierls type. The coupling of electrons to the lattice, however, determines the structural details of the long-range order. We saw in Section 5.4(b) that only those deformations which are coupled in the lowest order to TCNQ and TTF CDWs, are visible in the structural measurements. Furthermore, the number of displacive degrees of freedom which participate in the static deformations below $T_Q(p)$ is larger than the number of CDW degrees of freedom. The almost degenerate behavior of different displacements belonging to the same chain families is the consequence of the particular symmetry of electron–phonon interactions (Figures 29 and 30). The symmetry analysis of all translational molecular displacements gave a consistent interpretation of most of the transitional lines in the phase diagram from Figure 7. Furthermore, this analysis appeared to be a useful tool in retaining in the Landau functional the terms which are physically the most relevant. This facilitated significantly the analysis of the low-temperature regime in TTF-TCNQ.

The Landau functional introduced in Section 5.1, and extended with $4k_F$ contributions in Section 5.3, proved successful in the explanation of the cascade of phase transitions at ambient pressure. The considerations of Section 5.4 showed that essentially the same model may also account for some complex features of the (p, T) phase diagram. To this end it is, however, necessary to assume that quartic interchain interactions increase strongly under pressure. The important role of these terms in the lock-in of q_a at $a^*/4$ at ambient pressure has already been discussed. Unfortunately, the higher order interchain terms were not analyzed on the microscopic level, so that the reasons for their eventual strong dependence on the interchain distance are not clear at present.

We conclude by pointing out that TTF-TCNQ presents the unique example which shows up most of the features of the incommensurate long-range ordering, which were intensively investigated in many systems in recent years. It comprises the temperature and pressure variation of the incommensurate superstructure (Sections 5.1(b), 5.3, 5.4(d)), the 'simple' incommensurate–commensurate transition which concerns the $2k_F$ component and is usually considered in terms of sine-Gordon model (Section 5.4(c)), and the more specific commensurate lock-in of q_a at $a^*/4$ (Section 5.1(c)). The latter transition goes via the coupling of two nonlinear fields, which represents a particular example of the nonintegrable problem, analogous to those often met in classical mechanics. We argued in Section 5.2 that the nonintegrability causes unusual hysteresis in the sliding temperature range. Finally, the additional complexity of TTF-TCNQ comes from the simultaneous three-dimensional ordering of $2k_F$ and $4k_F$ CDWs (Section 5.3).

The Landau model which served in this section as the basis for the analysis of all these effects, is formulated under criteria which, in the particular case of TTF-TCNQ, have to be taken with some caution (Section 4). It is, however, the only approach known today which successfully links the complex microscopic picture of Sections 1–4 with the impressive body of interesting experimental findings considered in this section.

Acknowledgements

We have benefited from numerous discussions concerning the one-dimensional conductors with J. Friedel, S. Jérome, R. Comès. S. Megtert, V. J. Emery, J. Przystawa and J. R. Cooper. One of us (S.B.) acknowledges especially the collaboration with L. P. Gor'kov on establishing the microscopic picture of TTF-TCNQ. Thanks are due to B. Leontić and M. Šunjić for a careful reading of the manuscript. The work was partially supported by the Yu-U.S. collaboration contract DOE 438.

References

1. S. Barišić, in *Electronic Properties of Inorganic Quasi-One-Dimensional Compounds* (ed. P. Monceau), D. Reidel, Dordrecht, Holland, p. 1 (1985).
2. L. P. Gor'kov, *Lecture Notes in Physics* **96**, 3 (1979); Springer-Verlag, Berlin (eds S. Barišić, A. Bjeliš, J. Cooper, and B. Leontić).
3. D. Jérome and H. J. Schulz, *Adv. in Phys.* **31**, 299 (1982). This paper contains, besides an extensive list of original references, a separate list of conference proceedings, books, and review articles.
4. J. Friedel and D. Jérome, *Contemp. Phys.* **23**, 583 (1982).
5. *Proc. Int. CNRS Colloq. on Physics and Chemistry of Synthetic and Organic Metals*; *J. Physique* **44**, C3 (1983).
6. T. J. Kistenmacher, T. E. Phillips, and D. O. Cowan, *Acta Cryst.* **B30**, 763 (1974).
7. A. J. Berlinsky, J. F. Carolan, and L. Weiler, *Sol. St. Commun.* **15**, 795 (1974).
8. D. Jérome and M. Weger, *Chemistry and Physics of One-Dimensional Metals*, ed. H. J. Keller, Plenum, New York, p. 341 (1977).
9. R. H. Blessing and P. Coppens, *Sol. St. Commun.* **15**, 215 (1974).
10. R. Comès, S. M. Shapiro, G. Shirane, A. F. Garito, and A. J. Heeger, *Rhys. Rev. Lett* **35**, 1518 (1975); *Rhys. Rev.* **B14**, 2376 (1976).
11. D. R. Salahub, R. P. Messmer; and F. Herman, *Phys. Rev.* **B13**, 4252 (1976).
12. P. Grant, R. L. Greene, G. C. Wrighton, and G. Castro, *Phys. Rev. Lett.* 31, 1311 (1973).
13. D. B. Tanner, C. S. Jacobsen, A. F. Garito, and A. J. Heeger, *Phys. Rev.* **B13**, 3381 (1976).
14. I. E. Dzyaloshinskii and E. I. Kats, *Zh. Eksp. Teor. Fiz.* **55**, 338 (1968) [*Sov. Phys. JETP* **28**, 178 (1969)].
15. P. H. Williams and A. N. Bloch, *Phys. Rev.* **B10**, 1097 (1974).
16. S. Barišić, *J. Physique* **44**, 185 (1983); S. Botrić and S. Barišić, *J. Physique* **45**, 185 (1984).
17. P. M. Chaikin, J. F. Kwak, T. E. Jones, A. F. Garito, and A. J. Heeger, *Phys. Rev. Lett.* **31**, 601 (1973).
18. C. Weil, D. Jérome, P. M. Chaikin, and K. Bechgaard, *J. Physique* **43**, 1167 (1982).
19. T. Takahashi, D. Jérome, F. Masin, J. M. Fabre, and L. Giral, *J. Phys. C* ...

20a. G. Soda, D. Jérome, M. Weger, J. Alizon, J. Gallice, H. Robert, J. M. Fabre, and L. Giral, *J. Physique* **38**, 931 (1977).

b. C. Bourbonnais, F. Creuzet, D. Jérome, A. Moradpour and K. Bechgaard, *J. Physique Lett.* **45**, 755 (1984).

21. E. F. Rybaczevskii, L. S. Smith, A. F. Garito, and A. J. Heeger, *Phys. Rev.* **B14**, 2746 (1976).
22. J. C. Scott, A. F. Garito, and A. J. Heeger, *Phys. Rev.* **B10**, 3131 (1974).
23. Y. Tomkiewicz, A. R. Taranko, and J. B. Torrance, *Phys. Rev. Lett.* **36**, 751 (1976); *Phys. Rev.* **B15**, 1017 (1977); Y. Tomkiewicz, B. Welber, P.E. Seiden, and R. Schumaker, *Sol. St. Commun.* **23**, 471 (1977).
24. J. R. Cooper, M. Weger, G. Delplanque, D. Jérome, and K. Bechgaard, *J. Physique Lett.* **37**, 349 (1976).
25. B. Korin, J. R. Cooper, M. Miljak, A. Hamzić, and K. Bechgaard, *Chem. Scr.* **17**, 45 (1981).
26. J. B. Torrance in *Chemistry and Physics of One-Dimensional Metals* (ed. H. J. Keller), Plenum, New York (1977).
27. J. Friedel, *Physics of Metals I: Electrons*, ed. J. M. Ziman, Cambridge Univ. Press (1969); J. Friedel and C. Noguera, *Int. J. Quant. Chem.* **23**, 1209 (1983).
28. A. F. Garito and A. J. Heeger, *Accounts Chem. Res.* **7**, 232 (1974).
29. A. J. Epstein, S. Etemad, A. F. Garito, and A. J. Heeger, *Phys, Rev.* **B5**, 952 (1972).
30. F. Wudl, D. Nalewajek, J. M. Troup, and M. W. Extine, *Science* **222**, 415 (1983).
31. K. Šaub, S. Barišić, and J. Friedel, *Phys. Lett.* **56A**, 302 (1976).
32. A. Bjeliš and S. Barišić, *Lecture Notes in Physics*, **65**, 291 (1977), Springer-Verlag, Berlin, (eds L. Pal, G. Grüner, A. Janossy, and J. Sólyom).
33. C. Hartzstein, V. Zevin, and M. Weger, *J. Physique* **41**, 677 (1980).
34. M. Weger and J. Friedel, *J. Physique* **38**, 241 (1977).

35. G. Shirane, S. M. Shapiro, R. Comès, A. F. Garito, and A. J. Heeger, *Phys. Rev.* **B14**, 2325 (1976); S. M. Shapiro, G. Shirane, A. F. Garito, and A. J. Heeger, *Phys. Rev.* **B15**, 2413 (1977).
36. R. Comès and G. Shirane, *Highly Conducting One-Dimensional Solids* (eds J. T. Devreese, R. P. Evrard, and V. E. van Doren), Plenum, New York, p. 17 (1979).
37. F. E. Bates, J. E. Eldridge, and M. R. Bryce, *Can. J. Phys.* **59**, 339 (1981).
38. D. B. Tanner, K. D. Cummings, and C. S. Jacobsen, *Phys. Rev. Lett.* **47**, 597 (1981).
39. R. Bozio and C. Pecile, *Sol. St. Commun.* **37**, 193 (1981) and Ref. [5], p. 1453.
40. D. Debray, R. Miller, D. Jérome, S. Barišić, L. Giral, and J. M. Fabre, *J. Physique Lett.* **38**, 227 (1977) and S. Megtert, private commun.
41. H. Kuzmany and B. Kundu, *Lecture Notes in Physics* **95**, 259 (1979), Springer-Verlag, Berlin (eds S. Barišić, A. Bjeliš, J. Cooper, and B. Leontić).
42. E. E. Ferguson and F. A. Matsen, *J. Chem. Phys.* **29**, 195 (1958); E. E. Ferguson, *J. Chem. Phys.* **61**, 257 (1964).
43. M. J. Rice, N. O. Lipari, and S. Strässler, *Phys. Rev. Lett.* **39**, 1359 (1977).
44. M. J. Rice, *Phys. Rev. Lett.* **37**, 36 (1976); *Lecture Notes in Physics* **95**, 230 (1979), Springer-Verlag, Berlin, (eds S. Barišić, A. Bjeliš, J. Cooper, and B. Leontić).
45. S. Barišić, J. Labbé, and J. Friedel, *Phys. Rev. Lett.* **25**, 919 (1970); S. Barišić, *Phys. Rev.* **B5**, 932 and 941 (1972).
46. N. O. Lipari, C. B. Duke, and L. Pietronero, *J. Chem. Phys.* **65**, 1165 (1976); N. O. Lipari, M. J. Rice, C. B. Duke, R. Bozio, A. Girlando, and C. Pecile, *Int. J. Quant. Chem. Symp.* **11**, 583 (1977); erratum *ibid.* **12**, 545 (1978).
47. P. A. Lee, T. M. Rice, and P. W. Anderson, *Sol. St. Commun.* **14**, 703 (1974).
48. P. A. Lee and T. M. Rice, *Phys. Rev.* **B19**, 3970 (1979).
49. S. A. Brazovskii and J. E. Dzyaloshinskii, *Zh. Eksp. Teor. Fiz.* **71**, 2338 (1976) [*Sov. Phys. JETP* **44**, 1233 (1976)].
50. A. M. Finkel'stein and S. A. Brazovskii, *J. Phys.* **C14**, 847 (1981); S. A. Brazovskii and A. M. Finkelstein, *Sol. St. Commun.* **38**, 745 (1981).
51. H. Fröhlich, *Proc. R. Soc.* **A223**, 296 (1954).
52. J. Bardeen, *Sol. St. Commun.* **13**, 357 (1973).
53. J. P. Pouget, S. K. Khanna, F. Denoyer, R. Comès, A. F. Garito, and A. J. Heeger, *Phys. Rev. Lett.* **37**, 437 (1976); S. K. Khanna, J. P. Pouget, R. Comès, A. F. Garito, and A. J. Heeger, *Phys. Rev.* **B16**, 1468 (1977).
54. S. Kagoshima, T. Ishiguro, and H. Anzai, *J. Phys. Soc. Japan* **41**, 2061 (1976).
55. S. Etemad, *Phys. Rev.* **B13**, 2254 (1976).
56. R. M. Herman, M. B. Salamon, and G. de Pasquali, *Sol. St. Commun.* **19**, 137 (1976).
57. C. S. Jacobsen, *Lecture Notes in Physics* **95**, 223 (1979), Springer-Verlag, Berlin (eds S. Barišić, A. Bjeliš, J. Cooper, and B. Leontić).
58. S. T. Chui, T. M. Rice, and C. M. Varma, *Sol. St. Commun.* **15**, 155 (1974).
59. J. Sólyom, *Adv. in Physics* **28**, 201 (1979) and references therein.
60. J. E. Hirsch, D. J. Scalapino, R. L. Sugar, and R. Blankenbecker, *Phys. Rev. Lett.* **47**, 1628 (1981); *Rhys. Rev.* **B26**, 5033 (1982); D. J. Scalapino and J. E. Hirsch in Ref [5], p. 1507 and *Phys. Rev.* **B27**, 7196 (1983).
61. S. Mazumdar and A. N. Bloch, *Phys. Rev. Lett.* **50**, 207 (1983); A. N. Bloch and S. Mazumdar in Ref. [5], p. 1273.
62. H. Th. Jonkman, H. J. Zwinderman, and J. J. Kommandeur in Ref. [5], p. 1281.
63. K. Uzelac, *J. Phys.* **A17**, 81 (1984).
64. Yu. A. Bychkov, L. P. Gor'kov, and J. E. Dzyaloshinskii, *Zh. Eksp. Teor. Fiz.* **50**, 738 (1966) [*Sov. Phys. JETP* **23**, 489 (1966)].
65. L. P. Gor'kov and I. E. Dzyaloshinskii, *Zh. Eksp. Teor. Fiz.* **67**, 397 (1974) [*Sov. Phys. JETP* **40**, 198 (1975)].
66. N. Menyhárd and J. Sólyom, *J. Low Temp. Phys.* **12**, 529 (1973); J. Sólyom, *J. Low Temp. Phys.* **12**, 547 (1973).
67. V. N. Prigodin and Yu. A. Firsov, *Zh. Eksp. Teor. Fiz.* **76**, 736 and 1602 (1979) [*Sov. Phys. JETP* **49**, 369 and 813 (1979)].

68. L. Mihály and J. Sólyom. *J. Low Temp. Phys.* **24**, 579 (1976).
69. H. Gutfreund and R. A. Klemm, *Phys. Rev.* **B14**, 1073 (1976).
70. P. A. Lee, T. M. Rice, and R. A. Klemm, *Phys. Rev.* **B15**, 2984 (1977).
71. N. Menyhárd, *Sol. St. Commun.* **21**, 495 (1977).
72. A. J. Schultz, G. D. Stucky, R. H. Blessing, and P. Coppens, *J. Am. Chem. Soc.* **98**, 3194 (1976).
73. D. E. Schaffer, G. A. Thomas, and F. Wudl, *Phys. Rev.* **B12**, 5532 (1975).
74. A. Filhol, G. Bravic, J. Gaultier, D. Chasseau, and D. Vettier, *Acta Cryst.* **B37**, 1225 (1981).
75. V. E. Klimenko, V. Ya. Krivnov, A. A. Ovchinnikov, and J. J. Ukrainskii and A. F. Shvets, *Zh. Eksp. Teor. Fiz.* **69**, 240 (1975), [*Sov. Phys. JEPT* **42**, 123 (1976)].
76. J. B. Torrance and B. D. Silverman, *Phys. Rev.* **B15**, 788 (1977).
77. S. Barišić, *J. Physique C2* **39**, 262 (1978).
78. R. Zallen and E. M. Conwell, *Sol. St. Commun.* **31**, 557 (1979).
79. J. Friedel, *Electron Phonon Interactions and Phase Transitions*, Nato Advanced Study Institute, Series B, Physics, Plenum, New York (1977).
80. T. E. Phillips, T. J. Kistenmacher, A. N. Bloch, and D. O. Cowan, *JCS, Chem. Commun.* 1976, 334 (1976).
81. A. J. Kitaigorodskii, *Molecular Crystals*, Nauka, Moskva (1971).
82. C. Noguera (to be published).
83. S. Barišić and A. Bjeliš, *J. Physique Lett.* **44**, 327 (1983); A. Bjeliš and S. Barišić in Ref. [5], p. 1539.
84. H. Morawitz and G. Siegl, *Phys. Rev.* **B30**, 6180 (1984).
85. P. Županović, S. Barišić and A. Bjeliš, *J. Physique* (1985) to be published.
86. A. J. Epstein, N. O. Lipari, D. J. Sandman, and P. Nielsen, *Phys. Rev.* **B13**, 1569 (1976).
87. R. M. Metzger and C. A. Panetta in Ref. [5], p. 1605 and references therein.
88. J. Hubbard, *Lecture Notes in Physics* **96**, 11 (1979); Springer-Verlag, Berlin (eds S. Barišić, A. Bjeliš, J. Cooper, and B. Leontić); *Phys. Rev.* **B17**, 494 (1978).
89. K. Yamaji, *Sol. St. Commun.* **27**, 425 (1978); J. Kondo and K. Yamaji, *J. Phys. Soc. Japan* **43**, 424 (1977).
90. I. E. Dzyaloshinskii and A. Y. Larkin, *Zh. Eksp. Teor. Fiz.* **61**, 791 (1971) [*Sov. Phys. JETP* **34**, 422 (1972)].
91. V. J. Emery, *Highly Conducting One-Dimensional Solids*, (eds J. T. Devreese, R. P. Evrard, and V. E. van Doren), Plenum, New York, p. 247 (1979).
92. S. Barišić, *Lecture Notes in Physics,* **65**, 85 (1977), Springer-Verlag, Berlin (eds L. Pal, G. Grüner, and A. Janossy); J. Sólyom, *Fizika* **8**, 191 (1976).
93. G. Bergmann and D. Rainer, **Z. Phys. 263**, 59 (1973).
94. B. Horovitz, *Phys. Rev.* **B16**, 3943 (1972).
95. H. Fukuyama, T. M. Rice, C. M. Varma, and B. Halperin, *Phys Rev.* **B5**, 3775 (1974).
96. M. Kimura, *Progr. Theor. Phys. (Osaka)* **49**, 697 (1975); and **53**, 955 (1975).
97. A. Luther and V. J. Emery, *Phys. Rev. Lett.* **33**, 589 (1974).
98. V. J. Emery, A. Luther, and J. Peschel, *Phys. Rev.* **B13**, 1272 (1975).
99. G. E. Gurgenishvili, A. A. Neresyan, G. A. Kharadze, and L. A. Cobanyan, *Physica* **84B**, 243 (1976).
100. A. Luther, *Phys. Rev.* **B14**, 2153 (1976).
101. V. J. Emery, *Phys. Rev.* **B14**, 2989 (1976).
102. P. A. Lee, *Phys. Rev. Lett.* **34**, 1247 (1975).
103. K. B. Efetov and A. J. Larkin, *Zh. Eksp. Teor. Fiz.* **66**, 2290 (1974) [*Sov. Phys. JETP* **39**, 1129 (1974)].
104. P. Maneville, *J. Physique* **36**, 701 (1975).
105. D. J. Scalapino, Y. Imry, and P. Pincus, *Phys. Rev.* **B11**, 2024 (1975).
106. W. Dieterich, *Adv. in Physics* **25**, 615 (1976).
107. G. Toulouse, *N. Cimento* **23B**, 234 (1974).
108. K. Uzelac and S. Barišić, *J. Physique* **36**, 1267 (1975) and *J. Physique Lett.* **38**, 47 (1977).
109. N. Menyhárd, *Sol. St. Commun.* **21**, 495 (1977).
110. A. Blandin, *Cours de 3e Cycle*, Université Paris-Sud, Orsay (1973).

111. A. A. Ovchinnikov, *Zh. Th. Eksp. Phys.* **64**, 342 (1973); A. A. Ovchinnikov and V. A. Onishuk, *Theor. Math. Phys.* **37**, 382 (1978).
112. J. Bernasconi, M. J. Rice, W. R. Schneider, and S. Strässler, *Phys. Rev.* **B12**, 1090 (1975).
113. V. J. Emery, *Phys. Rev. Lett.* **37**, 107 (1976).
114. J. B. Torrance, *Phys. Rev.* **B17**, 3099 (1978).
115. K. B. Efetov, *Zh. Eksp. Teor. Fiz.* **81**, 1099 (1981).
116. A. Bjeliš, K. Šaub, and S. Barišić, *N. Cimento* **23B**, 102 (1974).
117. B. Horovitz, H. Gutfreund, and M. Weger, *Sol. St. Commun.* **11**, 1361 (1972); *Phys. Rev.* **B9**, 1246 (1974).
118. W. L. McMillan, *Phys. Rev.* **B12**, 1197 (1975); R. N. Bhatt and W. L. McMillan, *Phys. Rev.* **B12**, 2042 (1975).
119. N. Menyhárd, *J. Phys.* **C11**, 2207 (1978).
120. H. J. Schulz and C. Bourbonnais, *Phys. Rev.* **B27**, 5856 (1983).
121. C. Bourbonnais and L. G. Caron in Ref. [5], p. 911 and to be published
122. N. L. Bulaevskii, A. J. Buzdin, and D. J. Khomskii, *Sol. St. Commun.* **35**, 101 (1980).
123. P. A. Lee, T. M. Rice, and P. W. Anderson, *Phys. Rev. Lett.* **31**, 462 (1973).
124. M. J. Rice and S. Strässler, *Sol. St. Commun.* **13**, 1389 (1973).
125. A. Bjeliš and S. Barišić, *J. Physique Lett.* **36**, 169 (1975).
126. S. Barišić and S. Marčelja, *Sol. St. Commun.* **7**, 1395 (1969).
127. D. B. Tanner, C. S. Jacobsen, A. F. Garito, and A. J. Heeger, *Phys. Rev.* **B13**, 3381 (1976).
128. A. Andrieux, H. J. Schulz, D. Jérome, and K. Bechgaard, *Phys. Rev. Lett.* **43**, 227 (1979); *J. Physique Lett.* **40**, 385 (1979).
129. B. R. Patton and L. J. Sham, *Phys, Rev. Lett.* **31** (1973) 631 and *Lecture Notes in Physics* **34**, 272 (1975), Springer-Verlag, Berlin (ed. H. G. Schuster).
130. H. Gutfreund, M. Kaveh, and M. Weger, *Lecture Notes in Physics* **95**, 105 (1979), Springer-Verlag, Berlin (eds S. Barišić, A. Bjeliš, J. Cooper, and B. Leontić).
131. A. J. Heeger and M. Weger, *ibid.*, p. 316
132. J. R. Cooper and B. Korin, *ibid.*, p. 181
133. A. A. Gogolin and V. I. Melnikov, *Phys. St. Sol.* (B) **88**, 377 (1978).
134. L. P. Gor'kov and E. J. Rashba, *Sol. St. Commun.* **27**, 1211 (1978).
135. V. Fano, *Phys. Rev.* **124**, 1866 (1961).
136. F. Devreux, C. Jeandey, M. Nechtschein, J. M. Fabre, and L. Giral, *J. Physique* **40**, 671 (1979).
137. Per Bak and V. J. Emery, *Phys. Rev. Lett.* **36**, 978 (1976).
138. T. D. Schultz and S. Etemad, *Phys. Rev.* **B13**, 4928 (1976).
139. A. Bjeliš and S. Barišić, *Phys. Rev. Lett.* **37**, 1517 (1976).
140. L. B. Coleman, M. J. Cohen, M. J. Sandmann, D. J. Yamagishi, A. F. Garito, and A. J. Heeger, *Sol. St. Commun.* **12**, 1125 (1973).
141. J. P. Ferraris, D. O. Cowan, V. V. Walatka, and J. H. Perlstein, *J. Am. Chem. Soc.* **95**, 948 (1973).
142. D. Jérome, W. Müller, and M. Weger, *J. Physique Lett.* **35**, L77 (1974).
143. J. R. Cooper, D. Jérome, M. Weger, and S. Etemad, *J. Physique Lett.* **36**, L219 (1975).
144. F. Denoyer, R. Comès, A. F. Garito, and A. J. Heeger, *Phys. Rev. Lett.* **35**, 445 (1975).
145. S. Kagoshima, H. Anzai, K. Kajimura, and T. Ishiguro, *J. Phys. Soc. Japan* **39**, 1143 (1975).
146. W. D. Ellenson, R. Comès, S. M. Shapiro, G. Shirane, A. F. Garito, and A. J. Heeger, *Sol. St. Commun.* **20**, 53 (1976).
147. W. D. Ellenson, S. M. Shapiro, G. Shirane, and A. F. Garito, *Phys. Rev.* **B16**, 3244 (1977).
148. R. Comès, in *Chemistry and Physics of One-Dimensional Metals* (ed. H. J. Keller), Plenum, New York (1977).
149. J. P. Pouget, S. M. Shapiro, G. Shirane, A. F. Garito, and A. J. Heeger, *Phys. Rev.* **B19**, 1792 (1979).
150. H. Morawitz, *Phys. Rev. Lett.* **34**, 1096 (1975).
151. K. Yamaji, S. Megtert, and R. Comès, *J. Physique* **42**, 1327 (1981).
152. D. Mukamel, *Phys. Rev.* **B16**, 1741 (1977).
153. D. Mukamel and B. Horovitz, *Sol. St. Commun.* **23**, 285 (1977).

154. E. Abrahams, J. Sólyom, and F. Woynarovich, *Phys. Rev.* **B16**, 5238 (1977).
155. R. J. Cooper, J. Lukatela, M. Miljak, J. M. Fabre, L. Giral, and E. Aharon-Shalom, *Sol. St. Commun.* **25**, 949 (1978).
156. D. Djurek, K. Franulović, M. Prester, S. Tomić, L. Giral, and J. M. Fabre, *Phys. Rev. Lett.* **38**, 715 (1977).
157. P. M. Horn, R. Herman, and M. B. Salamon, *Phys. Rev.* **B16**, 5012 (1977).
158. F. C. Frank and J. H. Van der Merwe, *Proc. R. Soc. London* **198**, 205 (1949).
159. M. L. McMillan, *Phys. Rev.* **B14**, 145 and 1496 (1976).
160. S. Barišić and A. Bjeliš, *J. Physique Colloq.* **C7**, 254 (1977).
161. Per Bak, *Phys. Rev. Lett.* **37**, 1071 (1976).
162. A. Bjeliš and S. Barišić, *Phys. Rev. Lett.* **48**, 684 (1982); A Bjeliš and S. Barišić, *Mol. Cryst. Liquid Cryst.* **85**, 1541 (1982).
163. T. D. Schultz, *Sol. St. Commun.* **22**, 289 (1977).
164. C. Weyl, E. M. Engler, K. Bechgaard, G. Jehanno, and S. Etemad, *Sol. St. Commun.* **19**, 925 (1976).
165. For the review see: Per Bak, *Rep. Prog. Phys.* **45**, 587 (1982).
166. S. Aubry, *Ferroelectrics* **24**, 53 (1980); *J. Phys.* **C16**, 2497 (1983).
167. P. Bak and V. L. Pokrovsky, *Phys. Rev. Lett.* **47**, 958 (1981).
168. See, e.g., K. J. Whiteman, *Rep. Prog. Phys.* **40**, 1033 (1977) and R. H. G. Helleman, *Fundamental Problems in Statistical Mechanics* (ed. E. G. D. Cohen), North-Holland, Amsterdam, New York, Vol. 5, p. 165 (1980).
169. S. Megtert, These, Orsay (France) (1984).
170. M. Sakai, S. Kagoshima, and E. M. Engler in Ref. [5], p. 1313.
171. J. P. Pouget and S. Megtert, private communication.
172. K. Sato, S. Iwabuchi, J. Yamauchi, and Y. Nagaoka, *J. Phys. Soc. Japan* **45**, 515 (1978).
173. R. H. Friend, M. Miljak, and D. Jérome, *Phys. Rev. Lett.* **40**, 1048 (1978).
174. C. Weyl, D. Jérome, P. M. Chaikin, and K. Bechgaard, *Phys. Rev. Lett.* **47**, 946 (1981).
175. S. Megtert, R. Comès, C. Vettier, R. Pynn, and A. F. Garito, *Sol. St. Commun.* **31**, 977 (1979).
176. S. Megtert, R. Comès, R. Pynn, and C. Vettier, Proc. of the NATO Advanced Study Institute, *The Physics and Chemistry of Low Dimensional Solids* (ed. L. Alcacer), Tomar, Portugal, p. 113 (1979).
177. C. Fincher, G. Shirane, R. Comès, and A. F. Garito, *Phys, Rev.* **B21**, 5424 (1980).
178. S. Megtert, R. Comès, C. Vettier, R. Pynn, and A. F. Garito, *Sol. St. Commun.* **37**, 875 (1981).
179. S. Megtert, R. Comès, C. Vettier, R. Pynn, and A. F. Garito, *Mol. Cryst. Liq. Cryst.* **85**, 159 (1982).
180. S. Megtert, A. Bjeliš, J. Przystawa, and S. Barišić in Ref. [5], p. 1345 and *Phys. Rev.* **B** (1985), to be published.
181. C. Hartzstein, V. Zevin, and M. Weger, *Sol. St. Commun.* **36**, 545 (1980).
182. Y. Imry, *J. Phys.* **C8**, 567 (1975).
183. C. Hartzstein, V. Zevin, and M. Weger in Ref. [5], p. 1335.

BAND STRUCTURES AND ELECTRONIC PROPERTIES OF METALLIC POLYMER $(SN)_x$ AND ITS INTERCALATES $(SNBr_y)_x$

HIROSHI KAMIMURA

Department of Physics, Faculty of Science, University of Tokyo, Bunkyo-ku, Tokyo 113, *Japan*

and

ATSUSHI OSHIYAMA

Department of Physics, Faculty of Science, University of Tokyo, Bunkyo-ku, Tokyo 113, *Jpan and IBM Research Center, Yorktown Heights, New York* 10598, *U.S.A.*
Present Address: *NEC Fundamental Research Laboratories, Miyazaki, Miyamae-ku, Kawasaki* 213, *Japan*

1. Introduction

In this article we survey the features of the band structures and the Fermi surfaces of metallic polymer $(SN)_x$ and its intercalate $(SNBr_y)_x$, and then we discuss the electronic property and conductivity of these materials.

The metallic polymer $(SN)_x$ is rather unique with properties quite different from both conventional metals and typical one-dimensional conductors such as TTF-TCNQ or KCP. Experiments on the electrical resistivity along the chain axis of $(SN)_x$ [1–5] have shown that this material is metallic down to the liquid helium temperature. The outstanding feature of metallic properties of $(SN)_x$ is that, over a wide temperature region from 10 to 300 K, the resistivity along the chain axis follows a T^2 form in its temperature dependence. Moreover, the anisotropy of the resistivity is quite large and of the order of 10^3 at room temperature [3–5]. This quasi-one-dimensional metallic behavior has been verified by other experiments on the galvanomagnetic effect [6, 7], on plasmon dispersion [8], and on optical reflectivity from the infrared region to 5 eV [9–12]. Nevertheless, $(SN)_x$ becomes superconducting near 0.3 K [13–17] without exhibiting the metal–insulator transition characteristic in one-dimensional conductors. The experiments on the critical magnetic field [18] and on the Meissner effect [19, 20] indicate that $(SN)_x$ is a highly anisotropic type II superconductor. For a review of $(SN)_x$, see, for example, the articles by Kamimura [21], Greene and Street [22], and Yoffe [23].

1.1. Brief survey of brand structure calculations up to self-consistent calculation

Since Parry and Thomas [24] and Kamimura *et al.* [25] reported the first band structure calculation of $(SN)_x$ from opposite standpoints in order to explain the unusual physical properties of $(SN)_x$, a large number of band structure calculations has been attempted [24–40]. Although most of these calculations are in relatively good agreement on the overall structure of valence bands, they differ from one

H. Kamimura (ed.), Theoretical Aspects of Band Structures and Electronic Properties of Pseudo-One-Dimensional Solids, 123–162.

another on the shape of the Fermi surface. The calculated results on the band structure of $(SN)_x$ can be roughly classified into two types. The first type of band structure was proposed by Kamimura *et al.* [25, 32], who performed a tight-binding calculation on a single chain using 2*s*, 2*p* orbitals of nitrogen and 3*s*, 3*p* orbitals of sulfur, assuming that the amount of charge transfer from S to N is very large and that each SN has the ionic configuration close to S^+N^-. The obtained results indicate the occurrence of overlapping conduction bands in which the Fermi level crosses two energy bands (Figure 1(a)). The wave functions of the lower conduction band consist mainly of the π orbitals of sulfur, while the σ orbital constitutes those of the upper conduction band. Further, they considered the effects of interchain interactions on the conduction band structure by the perturbation theory [32]. The calculated Fermi surfaces for the upper conduction band are biconcave sheets

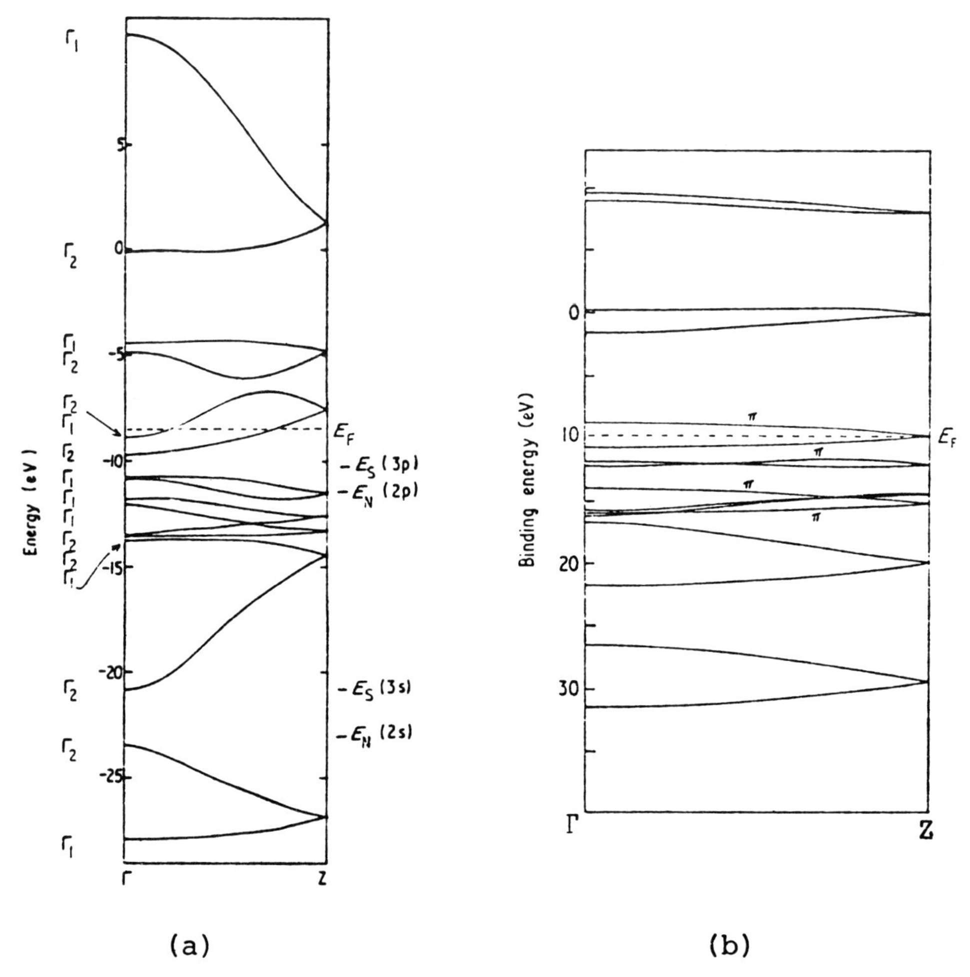

Fig. 1. Band structures for a single chain of $(SN)_x$: (a) that calculated by Kamimura *et al.* [25, 32]; (b) that calculated by Parry and Thomas [24]. The broken curve represents the Fermi level E_F.

located near the center of the Brillouin zone, as shown in Figure 2(a), while those for the lower conduction band are biconvex sheets located near the zone face. The overlapping band structure has been also obtained by Rajan and Falicov [28] using the LCAO tight-binding method on a single chain. The second type of band structure was first reported by Parry and Thomas [24]. They applied the empirical extended Hückel method to a single chain and obtained π conduction bands which cross the Fermi level at the zone edge, as shown in Figure 1(b). Later a three-dimensional orthogonalized plane wave (OPW) calculation was carried out by Rudge and Grant [33]. They used the crystal potential created by superposing sulfur and nitrogen atomic potentials in neutral atomic configurations and a basis set of 550 OPWs at the Γ point. The resulting band structure showed that $(SN)_x$ was expected to be a semimetal owing to interchain interactions. They also calculated the Fermi surface from a pseudopotential interpolation of the OPW results and obtained the smeimetallic Fermi surfaces shown in Figure 2(b). In this Fermi surface the electron pocket is located at the center of the zone face, while the hole pocket separated from the electron one is located near the zone edge. Similar results for the band structure have been also obtained by other authors using the extended Hückel method [27], the empirical pseudopotential method [30], or the extended tight-binding method [37]. The Fermi surfaces slightly different from these results were reported by Ching *et al.* [36] using the LCAO method. They proposed a formation of a closed electron pocket near the center of the zone face and of a closed hole pocket near the zone edge. The disagreement among these calculated results on the

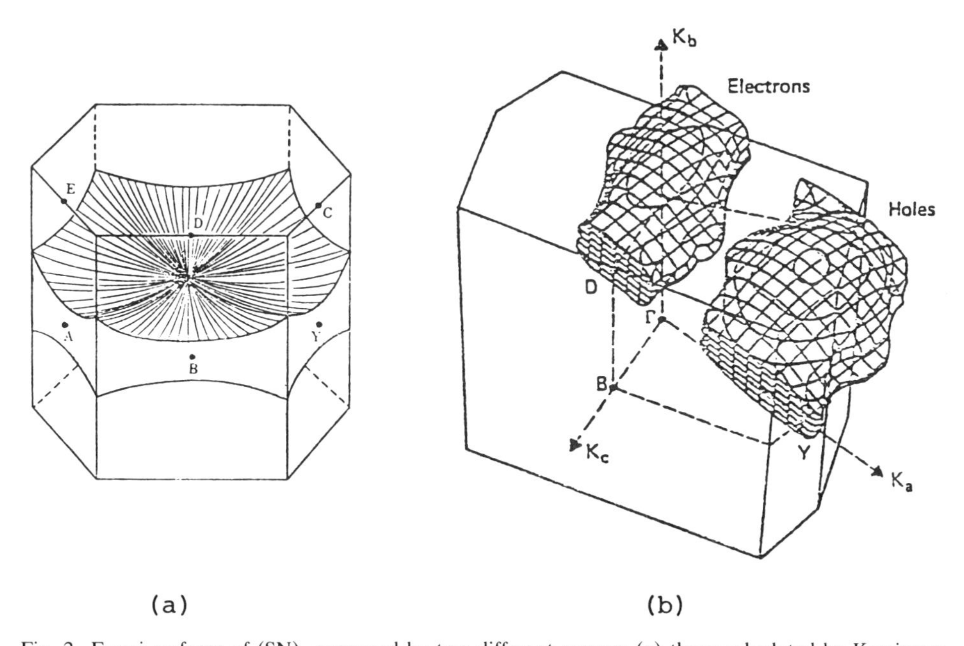

Fig. 2. Fermi surfaces of $(SN)_x$ proposed by two different groups: (a) those calculated by Kamimura *et al.* [32]; (b) those calculated by Rudge and Grant [62]. A second set of Fermi surfaces is not shown in these figures.

shape of the Fermi surface is caused mainly by the uncertainty in the degree of charge transfer between S and N atoms. Namely, in the first type of band structure the charge transfer from S to N is taken to be quite large and, as a result, the antibonding σ type bands cross the Fermi level near the center of the zone, while in the second type the charge transfer is not so large and thus the antibonding σ type bands do not cross the Fermi level.

On the other hand, the experiments on XPS [41–43] and on UPS [43] indicate that $(SN)_x$ is partially ionic. From the measurements of the energy levels of core orbitals in $(SN)_x$, the amount of charge transfer from sulfur atoms to nitrogen atoms is estimated to be 0.4 [43] to 0.5 [42] electrons per SN molecule. In spite of this experimental fact, although Salahub and Messmer [35] performed the self-consistent-field-Xα-scattered-wave calculations for related molecules, all the band calculations of $(SN)_x$ at the earlier stage [24–39] were non-self-consistent and thus they did not include the effect of the charge transfer from sulfur to nitrogen. Therefore, there had remained uncertainty as to the choice of parameters or potentials in these band calculations so that considerable disagreement on the shape of the Fermi surface of $(SN)_x$ was brought about. In order to remove this uncertainty and to obtain a more definite band structure of $(SN)_x$, in 1981 Oshiyama and Kamimura [40] performed a self-consistent band calculation from the first principle. In this article we are mainly concerned with describing the band structures and Fermi surfaces of $(SN)_x$ and $(SNBr_y)_x$ that they obtained.

1.2. Crystal structure of $(SN)_x$

The brass-colored $(SN)_x$ crystals are prepared from the solid-state polymerization of S_2N_2 [44]. Scanning electron microscope pictures show that all these crystals consist

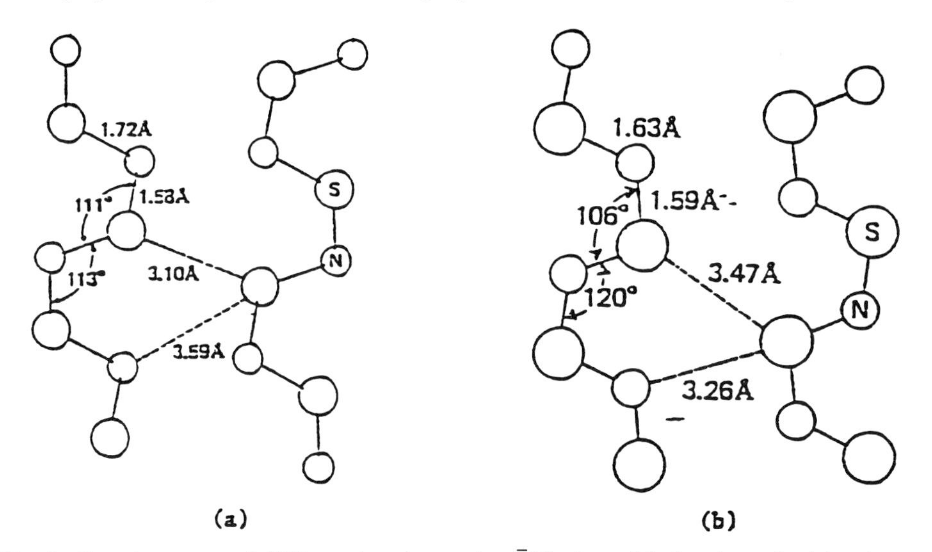

Fig. 3. Crystal structure of $(SN)_x$ projected onto the $(\bar{1}02)$ plane: (a) that determined by electron diffraction [45]; (b) that determined by X-ray [46].

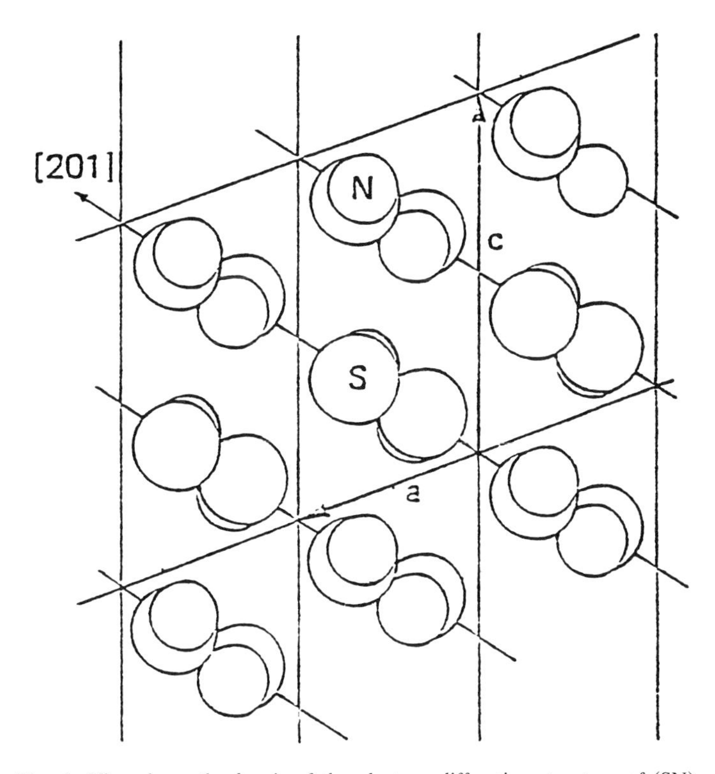

Fig. 4. View down the *b*-axis of the electron diffraction structure of $(SN)_x$.

of *b*-axis oriented fibers with a diameter of about 10^2-10^3 A. The determination of the crystal structure was reported by two different groups: the first reported by Boudelle *et al.* [45] using electron diffraction, and the other more recently reported by Mikulski *et al.* [46] using X-ray diffraction (Figures 3 and 4). The results show that the unit cell of $(SN)_x$ is a monoclinic system (space group $P2_1/c$) containing two SN chains in a cell. Although a slight difference between two results can be seen in the exact location of the four S and four N atoms within the unit cell, the characteristic feature of the crystal structure of $(SN)_x$ is the parallel arrangement of the helical chains in which the interchain distance is much larger than the intrachain bond length. According to this result, $(SN)_x$ can be classified into the category of quasi-one-dimensional conductors.

2. Self-Consistent Numerical Basis Set LCAO Method

In this section we describe briefly the method of solving self-consistently the one-particle equation for $(SN)_x$ within the local density functional (LDF) formalism [47, 48]. The method is based on the LCAO techniques of Ellis, Painter and collaborators [49–51], and on the approximation of expressing crystal charge density as the superposition of the renormalized atomic charge densities in each step of the iterative self-consistent calculation. The basis set of the exact numerical valence orbitals are refined iteratively and employed to solve the one-particle equation which contains the crystal potential obtained in each iteration from the crystal charge density. The effect of the charge transfer from sulfur to nitrogen is clarified by examinining the calculated band structure results in each step of the self-consistent calculation. The calculation was carried out numerically for the Penn structure of $(SN)_x$.

2.1 Local density functional formalism

Hohenberg, Kohn and Sham [47, 48] showed that the ground-state properties of an interacting electron system under an external field were determined by an effective one-particle equation containing an exchange-correlation potential in addition to the usual Hartree potential and the external potential. Such an exchange-correlation potential is given as a functional derivative of the exchange-correlation part of the ground-state energy, E_{xc}, with respect to the total charge density of electrons $\rho(\mathbf{r})$. Identifying the external potential for a polyatomic system as the electron–nuclear and internuclear interactions and varying the total ground-state energy with respect to $\rho(\mathbf{r})$, an effective one-particle equation of the following form is obtained,

$$\left(-\frac{\nabla^2}{2} - \sum_{m,\alpha} \frac{Z_\alpha}{|\mathbf{r} - \mathbf{R}_m - \mathbf{d}_\alpha|} + \int \frac{\rho(\mathbf{r}')}{|\mathbf{r} - \mathbf{r}'|} d\mathbf{r}' + \right.$$

$$\left. + \frac{\delta E_{xc}(\rho(\mathbf{r}))}{\delta\rho(\mathbf{r})}\right) \psi_j(\mathbf{r}) = \varepsilon_j \psi_j(\mathbf{r}). \tag{1}$$

Here Z_α denotes the nuclear charge of the atom situated at site $\mathbf{d}_\alpha + \mathbf{R}_m$, where $\mathbf{R}_m$ is the position vector of the mth unit cell. The eigenfunctions $\psi_j(\mathbf{r})$ are related to the total charge density by

$$\rho(\mathbf{r}) = \sum_{\text{occupied states}} \psi_j^*(\mathbf{r})\,\psi_j(\mathbf{r}) \tag{2}$$

which, in turn, determines self-consistently the LDF potential in Equation (1). Although no satisfactory formulation of $E_{xc}[\rho(\mathbf{r})]$ has been obtained so far for a general $\rho(\mathbf{r})$, the local approxiamtion in the case of slowly varying density has been used for a variety of interacting electron systems [52], including some atoms and molecules, and has yielded reasonably good results. Oshiyama and Kamimura [40] adopted also the usual local approximation for $E_{xc}[\rho(\mathbf{r})]$, and retained only the first term in the gradient expansion of $E_{xc}[\rho(\mathbf{r})]$ as follows;

$$\delta E_{\text{ex}}(\rho(\mathbf{r}))/\delta\rho(\mathbf{r}) \simeq V_{\text{ex}}(\rho(\mathbf{r})) + V_{\text{corr}}(\rho(\mathbf{r})) \tag{3}$$

where the exchange potential V_{ex} has the well-known form of

$$V_{\text{ex}}(\rho(\mathbf{r})) = -[(3/\pi)\,\rho(\mathbf{r})]^{1/3}. \tag{4}$$

For the correlation potential V_{corr}, they used the result of Singwi *et al.* [53] for a uniform electron gas, which can be fitted to a convenient analytical form [54]

$$V_{\text{corr}}(\rho(\mathbf{r})) = -A \ln(1 + B\rho^{1/3}(\mathbf{r})), \tag{5}$$

where $A = 0.0225$ and $B = 33.852$ in atomic unit.

The iterative self-consistent calculation begins with the construction of the crystal potential in each step. As mentioned before, the total charge density $\rho(\mathbf{r})$ is expressed as the sum of renormalized atomic charge densities

$$\rho(\mathbf{r}) = \sum_{m}^{N} \sum_{\alpha}^{s} \rho_{\alpha}(|\mathbf{r} - \mathbf{R}_m - \mathbf{d}_{\alpha}|), \tag{6}$$

where N is the number of unit cells and s is the number of atoms in each cell. Here, the spherical charge density $\rho_{\alpha}(\mathbf{r})$ of atom α is obtained by using the Mulliken gross population analysis [55] for the total charge density calculated in the preceding step of the self-consistent calculation. Thus, the crystal potential, which is refined in each step of the iterative calculation, is obtained by solving Poisson's equation for $\rho(\mathbf{r})$ and by adding the exchange and correlation functionals with this total charge density. The resulting crystal potential contains the electrostatic part (second and third terms in Equation (1)) which causes the long-range Coulomb potential in the case of ionic systems. This 'Madelung-type' potential was treated properly in the case of $(SN)_x$.

The crystal wave functions $\psi_j(\mathbf{k}, \mathbf{r})$ are expanded in terms of $s\eta$ Bloch functions $\varphi_{\lambda\alpha}(\mathbf{k}, \mathbf{r})$ in a standard form:

$$\psi_j(\mathbf{k}, \mathbf{r}) = \sum_{\alpha=1}^{s} \sum_{\lambda=1}^{\eta} c_{j\lambda\alpha}(\mathbf{k})\, \varphi_{\lambda\alpha}(\mathbf{k}, \mathbf{r}) \tag{7}$$

where $\varphi_{\lambda\alpha}(\mathbf{k}, \mathbf{r})$ is defined in terms of the λth basis orbital $\chi_{\lambda}^{\alpha}(\mathbf{r})$ situated at site $\mathbf{d}_{\alpha} + \mathbf{R}_m$:

$$\varphi_{\lambda\alpha}(\mathbf{k}, \mathbf{r}) = N^{-1/2} \sum_{m} \exp(i\mathbf{k} \cdot \mathbf{R}_m)\, \chi_{\lambda}^{\alpha}(\mathbf{r} - \mathbf{d}_{\alpha} - \mathbf{R}_m). \tag{8}$$

As LCAO basis functions numerical atomic-like LDF orbitals were used. These are obtained by solving the following atomic LDF equation:

$$\left(-\frac{\nabla^2}{2} - \frac{Z_{\alpha}}{\mathbf{r}} + \int \frac{\rho_{ss}(\mathbf{r}')}{|\mathbf{r} - \mathbf{r}'|} \mathrm{d}\mathbf{r}' + V_{\text{ex}}(\rho_{ss}^{\alpha}(\mathbf{r})) + \right.$$
$$\left. + V_{\text{corr}}(\rho_{ss}^{\alpha}(\mathbf{r})) \right) \chi_{\lambda}^{\alpha}(\mathbf{r}) = \varepsilon_{\lambda}^{\alpha} \chi_{\lambda}^{\alpha}(\mathbf{r}), \tag{9}$$

where the single-site charge density $\rho_{ss}^{\alpha}(\mathbf{r})$ is obtained in each step of the iteration by averaging spherically the total charge density $\rho(\mathbf{r})$ in Equation (6) around the site α. Thus, the basis orbitals $\chi_{\lambda}^{\alpha}(\mathbf{r})$ have the appropriate nodal behavior as well as the characteristic cusp of an exact atomic solution at the nucleus. On the other hand, the long-range nature of the free-atom eigenfunctions were controlled so as to avoid nonorthogonality between the Bloch functions generated from them. For this purpose, when the calculated orbitals $\chi_{\lambda}^{\alpha}(\mathbf{r})$ have long-range tails, the external potential well [51] was introduced in Equation (9) and the basis orbitals were calculated again. This technique for obtaining an appropriate basis set does not appreciably change the shapes of the basis orbitals around the site α, while it cuts off the long-range tails.

For the case of $(SN)_x$, the effect of employing an extended basis set which includes a d-type virtual orbital was examined by using Gaussian orbitals [37]. According to the results of this work, the contribution of the d orbital in any given energy state was less than 5%. From this result Oshiyama and Kamimura [40] have taken $1s$, $2s$, $2p$, $3s$, $3p$ orbitals of sulfur and $1s$, $2s$, $2p$ orbitals of nitrogen as a numerical basis set.

Once the crystal potential and the basis set are obtained, the usual linear secular equation

$$\sum_{\lambda'}^{n}\sum_{\alpha'}^{s}(H_{\lambda\alpha,\lambda'\alpha'}(\mathbf{k}) - S_{\lambda\alpha,\lambda'\alpha'}(\mathbf{k})\,\varepsilon_j(\mathbf{k}))\,c_{j\lambda'\alpha'}(\mathbf{k}) = 0 \tag{10}$$

is solved, where $H_{\lambda\alpha,\lambda'\alpha'}(\mathbf{k})$, and $S_{\lambda\alpha,\lambda'\alpha'}(\mathbf{k})$ are the Hamiltonian and overlap matrix elements, respectively. In calculating the matrix elements, a large number of multi-center integrals must be carried out, since the Bloch functions $\varphi_{\lambda\alpha}(\mathbf{k},\mathbf{r})$ are expressed in terms of the single-site basis orbitals $\chi_\lambda^\alpha(\mathbf{r})$. For the purpose of performing these integrals, Oshiyama and Kamimura [40] used a three-dimensional Diophantine integration scheme [49, 56] with the numerical form of the potentials and Bloch functions.

The solutions of the secular equation (10) provide eigenvalues $\varepsilon_j(\mathbf{k})$ and eigenfunctions $\psi_j(\mathbf{k},\mathbf{r})$ which lead to the crystal charge density

$$\tilde{\rho}(\mathbf{r}) = \frac{N\Omega}{(2\pi)^3}\sum_j{}'\int_{\mathrm{BZ}}\psi_j^*(\mathbf{k},\,\mathbf{r})\,\psi_j(\mathbf{k},\,\mathbf{r})\,\mathrm{d}\mathbf{k}, \tag{11}$$

where Ω denotes the unit cell volume and the j summation is carried out over the occupied states: $\varepsilon_j(\mathbf{k}) \leq$ the Fermi level E_F. With use of the Mulliken gross population analysis [55] the charge density is expressed in the form of equation (6), where the renormalized atomic charge density $\tilde{\rho}_\alpha(\mathbf{r})$ is given by

$$\tilde{\rho}_\alpha(\mathbf{r}) = \sum_j{}'\sum_{\lambda,\lambda'}^{\eta}\sum_{\alpha'}^{s}\int_{\mathrm{BZ}}\mathrm{Re}(c_{j\lambda'\alpha'}^*(\mathbf{k})\,S_{\lambda'\alpha',\lambda\alpha}(\mathbf{k})\,c_{j\lambda\alpha}(\mathbf{k}))\,|\chi_\lambda^\alpha(\mathbf{r})|^2\mathrm{d}\mathbf{k}. \tag{12}$$

Using the resulting crystal charge density for the input charge density in the subsequent iteration, we can proceed to the next step of the iterative self-consistent calculation.

2.2. EFFECTS OF CHARGE TRANSFER

In order to examine the effect of the charge transfer on the band structure, in this section we investigate the results of the band structure calculation in various steps of the iterative self-consistent calculation. In Figure 5 the band structures at the first, second, and fourth steps of the iteration are shown. The band structure along the

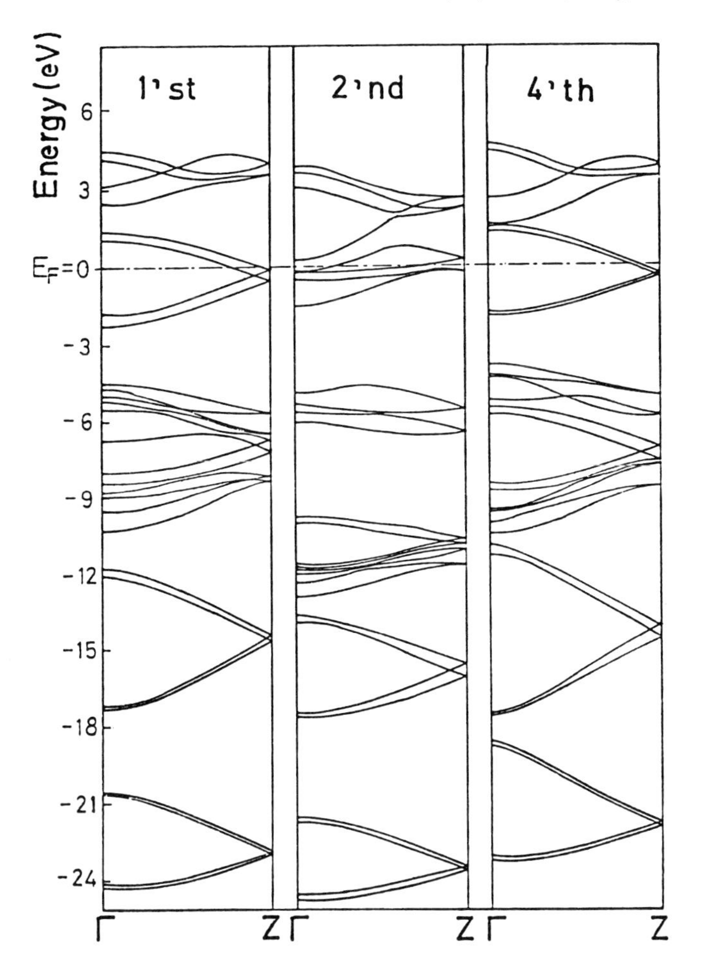

Fig. 5. Band structure along the intrachain direction ΓZ of $(SN)_x$ in the first, second, and fourth steps of the self-consistent band calculation. Here, the chain curve represents the Fermi level E_F [40]. (© The Institute of Physics)

intrachain direction Γ*Z* is shown in this figure. As the initial step, the band calculation started with the charge density with no charge transfer. The obtained band structure is similar to those of the non-self-consistent three-dimensional LCAO calculations by Ching *et al.* [36] and Batra *et al.* [37] and the amount of the charge transfer in the output charge density is about 1.2 electrons per SN molecule. Thus, the results of the first step are non-self-consistent at all. In the second step the charge density in which the charge transfer is about 0.8 electron is used as an input, and the overlapping conduction band structure is obtained. The antibonding σ type bands in the first step are lowered by the effect of the charge transfer so that the overlapping conduction bands appear. The results in the second step are also non-self-consistent since 0.01 electron are transferred from nitrogen to sulfur in the output charge density. In the fourth step the input charge transfer from sulfur to

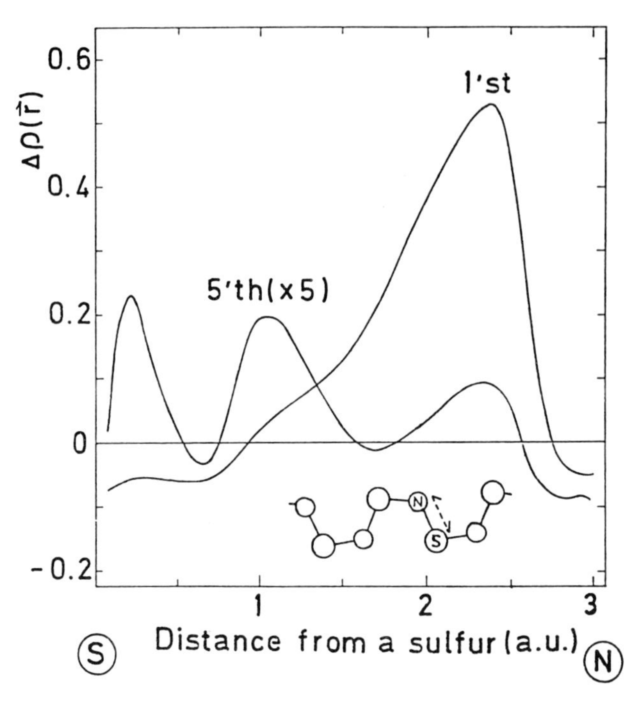

Fig. 6. Plot of $\Delta\rho(\mathbf{r}) = [\bar{\rho}(\mathbf{r}) - \rho(\mathbf{r})]/\rho(\mathbf{r})$ along a intrachain S–N bond, which is represented by the broken curve in the inset, during the first and fifth steps in the iterative self-consistent calculation [40]. (© The Institute of Physics)

nitrogen is 0.3 electron per SN molecule, while the output charge transfer is 0.5 electron. The difference between input and output charge densities along a S–N bond direction is plotted in Figure 6. In the fifth step sufficient convergence for the charge density and the band structure is obtained. The final value of the charge transfer from sulfur to nitrogen is found to be 0.4 electron per SN molecule. This value is in good agreement with the experimental value obtained by XPS and UPS measurements. The final results for the band structure of $(SN)_x$ is presented in the following section.

The iterative self-consistent calculation mentioned in this section reveals that the overlapping conduction band structure proposed by Kamimura *et al.* [25, 32] and by Rajan and Falicov [28] can be realized when the amount of charge transfer from S to N is not less than 0.8 electron per SN molecule, while the semimetallic Fermi surfaces can be realized otherwise.

3. Band Structure and Fermi Surface of $(SN)_x$

In this section we describe the features of the band structure, the Fermi surfaces, and the density of states of $(SN)_x$ calculated by Oshiyama and Kamimura [40]. In Figure 7 we show the calculated band structure along various symmetry directions in

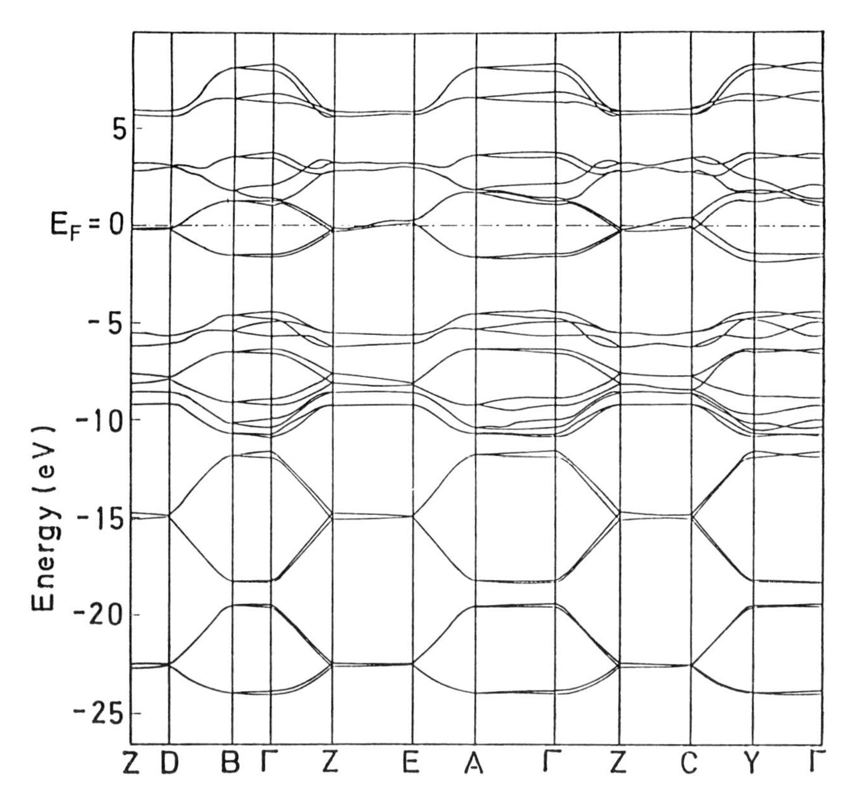

Fig. 7. Energy band structure of $(SN)_x$ along various directions in the Brillouin zone. The chain curve represents the Fermi level E_F [40]. (© The Institute of Physics)

the Brillouin zone, together with the Fermi level E_F. The Brillouin zone of $(SN)_x$ is shown in Figure 8. Reflecting the quasi-one-dimensional feature of the crystal structure, the energy bands are highly dispersive along the chain direction, while those along the interchain direction are quite flat. The widths of the conduction bands along three different directions are about 1.5 eV for the chain direction k_b, 0.36 eV for the direction k_a, and 0.04 eV for the direction k_c. The upper conduction bands are somewhat bent downwards near the Γ point, since the low-lying unoccupied bands are considerably lowered owing to the effect of the charge transfer from sulfur to nitrogen. The band structures obtained by the non-self-consistent LCAO calculations by Ching *et al.* [36] and by Batra *et al.* [37] do not show such features for the conduction band and also for the low-lying unoccupied bands. Nevertheless, in the case of $(SN)_x$ for which the charge transfer is found to be 0.4 electrons in the self-consistent calculation, there appear no overlapping conduction bands.

In spite of the overall quasi-one-dimensional feature of the band structure, the slight dispersion along the direction k_a causes the crossing of the Fermi level E_F with the four conduction bands along this direction. Therefore, the semimetallic Fermi surfaces, consisting of electron and hole pockets, are obtained. This feature of the Fermi surfaces is consistent with that of the non-self-consistent OPW cal-

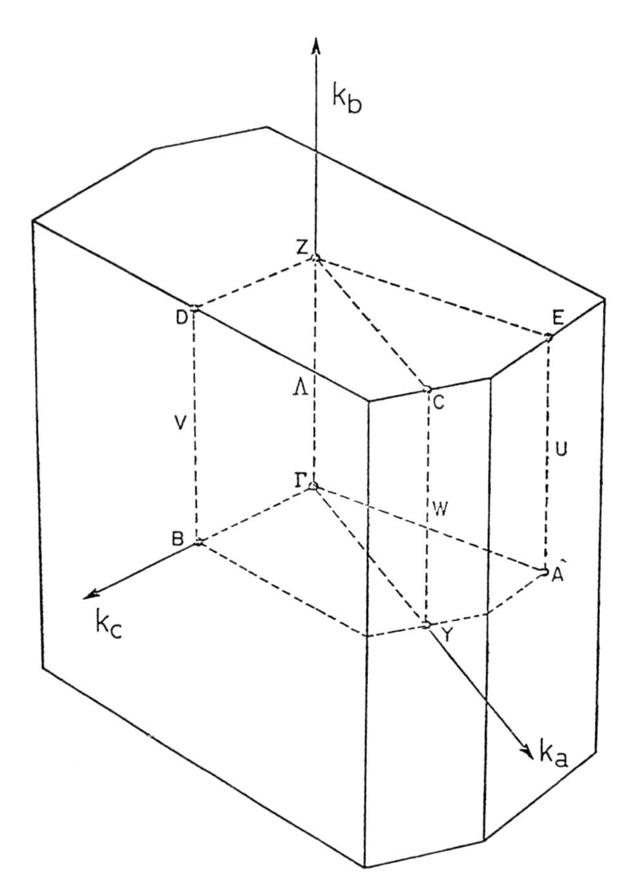

Fig. 8. Brillouin zone of $(SN)_x$.

culation by Rudge and Grant [33]. The Fermi surfaces obtained by the self-consistent calculation are shown in Figure 9. They consist of two electron and two hole pockets which are located around the top surfaces of the Brillouin zone (plane *ZDCE*). Two electron pockets and one of the two hole pockets have the shape of flattened tubes which have an open orbit along the k_c direction, while the other hole pocket has the shape of a flattened ellipsoid. Two electron pockets located at the center of the zone face correspond to the upper two conduction bands. Owing to the presence of the twofold screw axis in the crystal structure, these two pockets are connected with each other on the top zone plane *ZDCE*. Similarly, two hole pockets corresponding to the lower two conduction bands are also connected with each other on the same plane *ZDCE*. Moreover, these four electron and hole pockets have some plane-like regions which are spanned by the wavevector of $q = 0.2\pi/b$. Thus a weak Kohn anomaly in the phonon dispersion is expected to occur along the chain direction. In fact, Pintschovius *et al.* [57] observed a Kohn anomaly

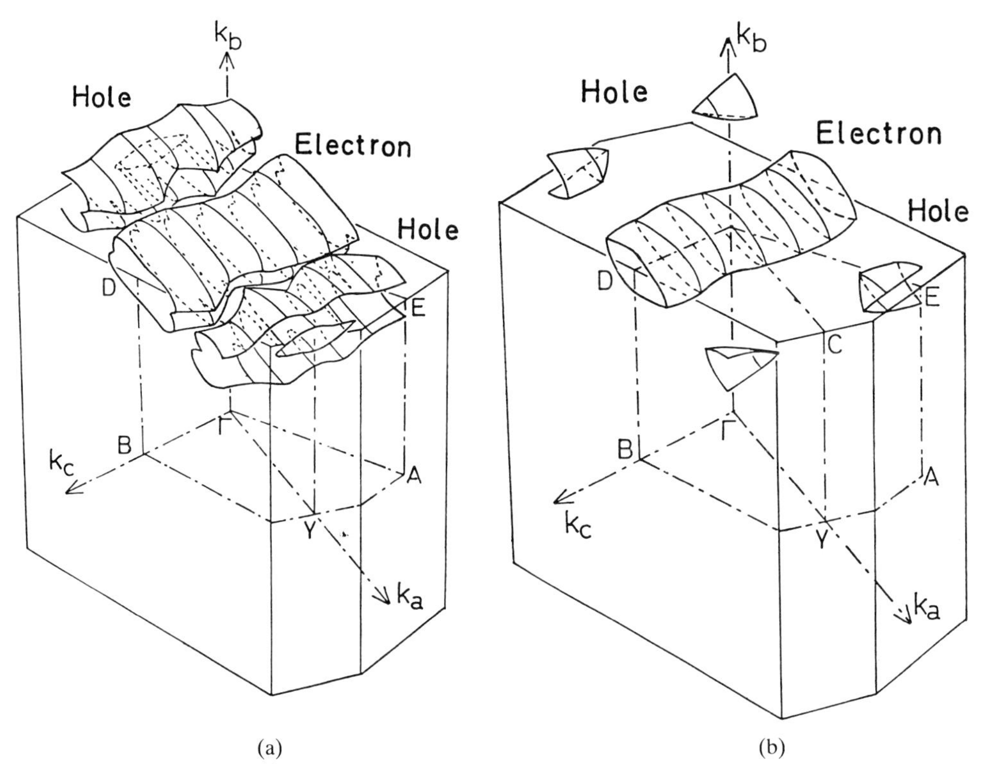

Fig. 9. Fermi surfaces of $(SN)_x$. The outermost surfaces are shown in (a) and the innermost ones shown in (b). The two electron pockets are connected with each other on the plane $ZDCE$ and on the line BD. The two hole pockets are also connected with each other on the same plane and on the line AE [40]. (© The Institute of Physics)

for $(SN)_x$. The observed anisotropy in the transport coefficients can be attributed to this highly anisotropic shape of Fermi surfaces.

The calculated density of states is shown in Figure 10. In this figure we also plot the XPS data obtained by two experimental groups [43, 58]. The agreement in peak positions between the calculated density of states and the XPS data is fairly good. The calculated Fermi level density of states $D(E_F)$ is found to be 0.14 states (eV spin molecule)$^{-1}$. This value is close to the experimental value of 0.18 states (eV spin molecule)$^{-1}$ obtained from the specific heat measurement [59]. From these two values of the Fermi level density of states $D(E_F)$ and $D_{exp}(E_F)$ and using a relation $D_{exp}(E_F) = (1 + \lambda)D(E_F)$, we can estimate the electron–phonon coupling constant λ to be $\lambda = 0.29$. This agrees well with the value $\lambda = 0.31$ estimated from the superconducting transition temperature and the Debye temperature [13].

On the basis of these band structure results, Oshiyama and Kamimura [40] further calculated the imaginary part of the dielectric function of $(SN)_x$ and compared it with the experimental results of the optical reflectivity from the visible region to 25 eV. The results are presented in the following section.

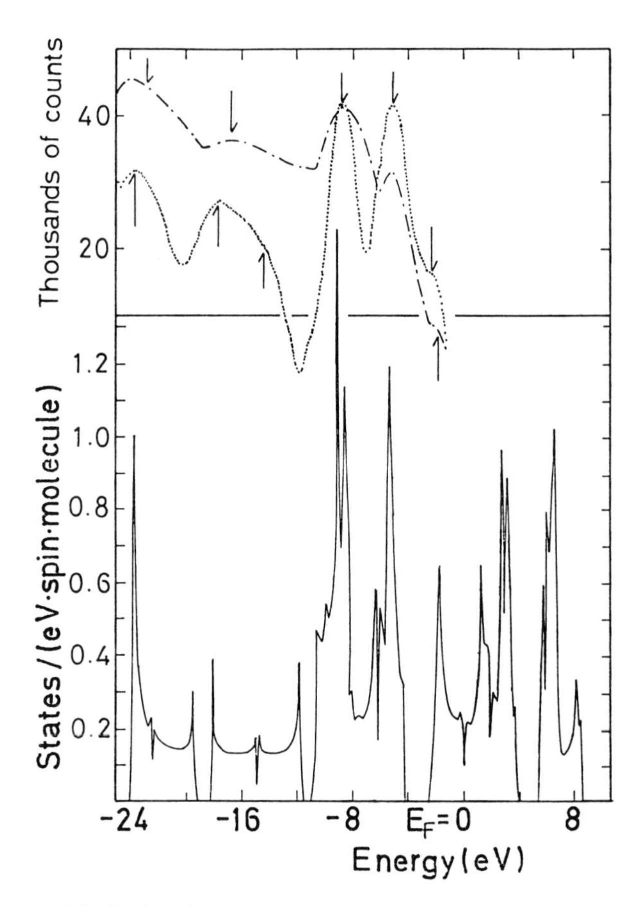

Fig. 10. Comparison of the density of states obtained by the present calculation (full curve) [40] with the XPS results reported by two different experimental groups, the chain curve obtained by Ley [58] and the dotted curve obtained by Mengel *et al.* [43]. (© The Institute of Physics)

4. Optical Absorption Spectra of $(SN)_x$

The optical reflectivity spectrum of $(SN)_x$ films from visible region to 27 eV was first measured by Bordas *et al.* [60] using the synchrotron radiation technique. The obtained spectrum for unpolarized light was compared with the theoretical results based on the one-dimensional semi-empirical tight-binding calculation by Kamimura *et al.* [25, 32]. Later, Mitani *et al.* [61] measured the polarized spectra up to 25 eV for oriented $(SN)_x$ films and $(SN)_x$ single crystals. They compared the results of oriented $(SN)_x$ films with the theoretical results by Grant *et al.* [62] who calculated the dielectric function on the basis of the non-self-consistent OPW calculation by Rudge and Grant [33]. The agreement between the experimental and calculated results was not satisfactory. Although the overall agreement in the spectrum for light polarized parallel to the chain axis was reasonably good, there remained a considerable disagreement in the spectrum for light polarized perpendicular to the chain axis. Then Oshiyama and Kamimura [40] investigated the optical absorption spectra of $(SN)_x$ on the basis of their self-consistent band structure, and found a fairly good agreement with observed results, as will be seen below.

Generally, the interband contribution to the imaginary part of the dielectric function $\epsilon_2(\omega)$ is given by the following formula [63]:

$$\epsilon_2^{\mu}(\omega) = \frac{1}{\pi}\left(\frac{e}{m\omega}\right)^2 \sum_{i,i'} \int_{BZ} |P_{ii'}^{\mu}(\mathbf{k})|^2 \, \delta(\varepsilon_{i'}(\mathbf{k}) - \varepsilon_i(\mathbf{k}) - \hbar\omega)\, d\mathbf{k}, \tag{13}$$

where the index μ corresponds to the direction of a polarization vector of the incident light. Here, $\varepsilon_i(\mathbf{k})$ and $P_{ii'}^{\mu}(\mathbf{k})$ are the energy of the Bloch state $|i\mathbf{k}\rangle$ and the matrix element of the μth component of the momentum operator between the Bloch states $|i\mathbf{k}\rangle$ and $|i'\mathbf{k}\rangle$, respectively. Further, the summation over energy bands and the integration over the Brillouin zone have to be taken over the energy levels which satisfy a relation

$$\varepsilon_i(\mathbf{k}) \leq E_F < \varepsilon_{i'}(\mathbf{k}). \tag{14}$$

In the LCAO formulation the matrix element $P_{ii'}^{\mu}(\mathbf{k})$ can be written, by using Equations (7) and (8), as

$$P_{ii'}^{\mu}(\mathbf{k}) = -i\hbar \sum_{\alpha,\alpha'} \sum_{\lambda,\lambda'} c_{i'\lambda'\alpha'}^{*}(\mathbf{k})\, c_{i\lambda\alpha}(\mathbf{k}) \times$$

$$\times \sum_m \exp(i\mathbf{k}\cdot\mathbf{R}_m) \int \chi_{\lambda'}^{\alpha'}(\mathbf{r} - \mathbf{d}_{\alpha'} - \mathbf{R}_m)\, \nabla^{\mu} \chi_{\lambda}^{\alpha}(\mathbf{r} - \mathbf{d}_{\alpha})\, d\mathbf{r}. \tag{15}$$

The advantage of the numerical basis set LCAO method is that all the values required to calculate the matrix element (15) are given in numerical form. Namely, the basis orbitals $\chi_{\lambda}^{\alpha}(\mathbf{r})$, which consist of the radial wave functions and the appropriate spherical functions, have been obtained from the numerical calculation in Section 2 in addition to their derivatives $\nabla^{\mu}\chi_{\lambda}^{\alpha}(\mathbf{r})$. Further, the eigenvectors, $c_{i\lambda\alpha}(\mathbf{k})$, are given in the form of numerical values for a set of reciprocal lattice vectors of the Brillouin zone. Therefore, the three-dimensional integral in Equation (15) is carried out by analytical integration over angular variables and one-dimensional numerical integration over radial variables. Moreover, since the calculated basis orbitals $\chi_{\lambda}^{\alpha}(\mathbf{r})$ have no long-range tails, the summation over the atomic sites in Equation (15) converges rapidly. In the case of $(SN)_x$ the lattice sum up to 11 a.u. was sufficient.

Oshiyama and Kamimura [40] performed the calculation of $\epsilon_2^{\mu}(\omega)$ for three different polarization vectors of incident light, one parallel to the chain axis b and the other two, c and y, perpendicular to it. The results are shown in Figures 11(a), (b), and (c). The insets in Figures 11(b) and (c) indicate the direction of the polarization of light. The experimental values of $\epsilon_2^{\mu}(\omega)$ by Mitani *et al.* [61] for two different polarization vectors of light are also plotted by dotted lines. These were obtained from the reflectivity data by using the Kramers–Kronig transformation. One of the characteristic features of the calculated spectra for three different polar-

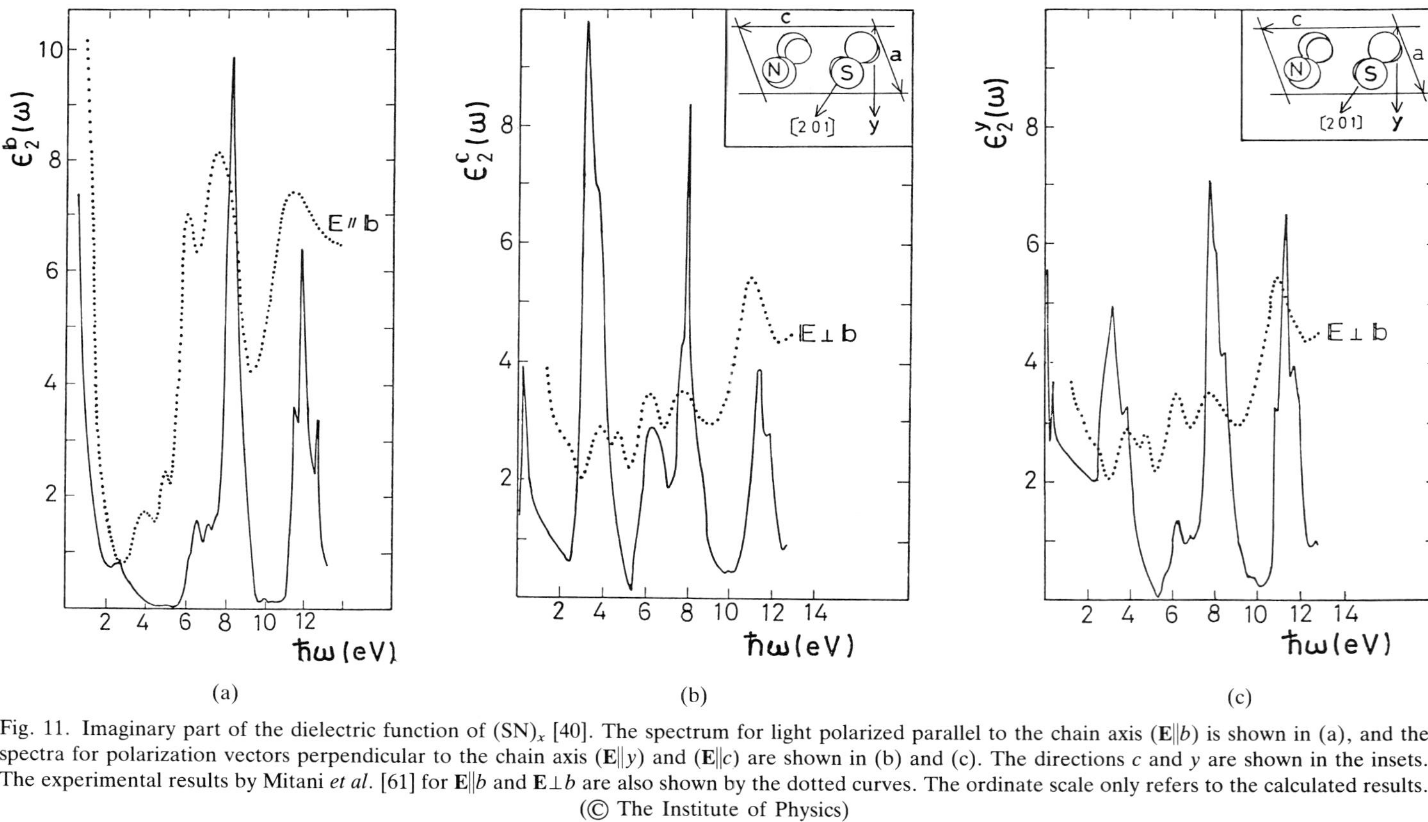

Fig. 11. Imaginary part of the dielectric function of $(SN)_x$ [40]. The spectrum for light polarized parallel to the chain axis ($\mathbf{E}\|b$) is shown in (a), and the spectra for polarization vectors perpendicular to the chain axis ($\mathbf{E}\|y$) and ($\mathbf{E}\|c$) are shown in (b) and (c). The directions c and y are shown in the insets. The experimental results by Mitani *et al.* [61] for $\mathbf{E}\|b$ and $\mathbf{E}\perp b$ are also shown by the dotted curves. The ordinate scale only refers to the calculated results. (© The Institute of Physics)

ization vectors is the strong interband absorption below 1 eV. This can be assigned to the transitions between two conduction bands intersecting the Fermi level along the direction *ZE* in Figure 7. Since the energy of the interband transitions overlaps the Drude region of the reflectivity spectra, the conventional Drude–Lorentz analysis of the reflectivity data below 4 eV reported previously [9–12] should be re-examined. The transitions to the low-lying unoccupied bands whose energies are considerably lowered by the effect of the charge transfer from sulfur to nitrogen contribute significantly to the strong peak around 8 eV in the calculated spectrum for light polarized parallel to the chain axis. The agreement in peak positions between the calculated and experimental results for light polarized parallel to the chain axis is reasonably good. Further, the agreement in intensity of the peaks around 8 and 12 eV is also good. Nevertheless, the intensity of the peak around 6.5 eV in the calculated spectrum is found too weak in comparison with the experimental result. Although the intensity of this peak can be enhanced by introducing the finite interband relaxation time, its calculated value is still smaller than the observed value. The reason for this disagreement is not clear. As for the spectra for light polarized perpendicular to the chain axis, the main peaks in the experimental result are well reproduced by the calculated result.

5. Fermi Surfaces and Optical Absorption Spectra of $(SNBr_y)_x$

So far we have discussed the electronic structure of $(SN)_x$. We would like now to turn our attention to the effect of bromination on the electronic structure of this material. In comparison with the data of pristine $(SN)_x$, brominated sample exhibits the drastic change of both magnitude and temperature dependence of the resistivity, while the plasma edge remains essentially unchanged. In order to clarify the origin of this unusual property, Oshiyama and Kamimura [40] calculated the Fermi surfaces and optical absorption spectra of $(SNBr_y)_x$ on the basis of the rigid band model.

We begin with a brief description of the experimental results on the geometrical positions of the bromines in $(SNBr_y)_x$. When bromines are doped in $(SN)_x$, a crystal expands considerably in volume with no measurable expansion along the chain direction. According to the X-ray and electron diffraction results obtained by Gill *et al.* [39], Bragg diffraction spots apepar in approximately the same positions in $(SN)_x$ and $(SNBr_y)_x$, although the diffraction pattern of $(SNBr_y)_x$ shows considerably streaking features in the direction perpendicular to the chain axis. From these results, Gill *et al.* proposed a structural model in which the bromines mostly reside in the interfiber regions of $(SN)_x$. In order to investigate the molecular structure of the bromines in $(SNBry)_x$, the experiment on Raman scattering was reported by Igbal *et al.* [64, 65]. Although their assignment of the observed peaks was not definite, they proposed the presence of Br_3^- ions and Br_2 molecules in addition to Br_5^- ions. This result was supported by the measurement of the oscillatory part of the extended X-ray absorption (EXAFS) by Morawitz *et al.* [66]. The results on the EXAFS, which contained information on near neighbor shells and coordination numbers, suggested the presence of Br_3^- ions or larger Br_n units. From these experimental results, Oshiyama and Kamimura [40] proposed a charge transfer model in which electrons are transferred from polymeric SN chains to acceptors consisting of Br_3^- and Br_5^- ions, although the structure of the bromines in $(SNBr_y)_x$ was not determined definitely.

If all the bromines enter to $(SNBr_y)_x$ as Br_3^- ions, the amount of charge transfer from the SN chains to the bromines is 0.13 electron per SN molecule for the case of $y = 0.4$. This amount of charge transfer is expected to be unable to cause a significant change in the original band structure of $(SN)_x$. Namely, from the obtained band structure results at each step of the self-consistent band calculation (Figure 5), we can see that the conduction band structure of $(SN)_x$ is not changed significantly unless 0.8 electron per SN molecule are transferred from a sulfur atom to a nitrogen atom. This suggests, in turn, that the band structure of $(SN)_x$ is not affected significantly if the charge transfer from the SN chains to the bromines is about 0.1 electron per SN molecule. Thus, a rigid band seems very reasonable as a first step towards understanding the electronic properties of $(SNBr_y)_x$.

We now discuss the variation of the Fermi surfaces upon bromination in terms of the rigid band model by Oshiyama and Kamimura [40]. The shift of the Fermi level upon bromination gives rise to the variation of the Fermi surfaces. On the basis of the obtained density of states of $(SN)_x$, the Fermi level is calculated as a function of the degree of charge transfer from the SN chains to the bromines per SN molecule, $\Delta\tilde{n}$. The Fermi surfaces thus obtained are shown in Figures 12(a), (b), and (c). Upon bromination, the hole portion of the Fermi surfaces expands while the electron portion shrinks and finally vanishes. For the case of $\Delta\tilde{n} = 0.053$ electron per SN molecule, the Fermi surfaces consist of the two electron and two hole pockets which take the shape of flattened tubes. When the bromine concentration increases so that $\Delta\tilde{n}$ becomes 0.080 electron per SN molecule, one of the electron pockets disappears and the shape of the other electron pocket changes from a flattened tube to a flattened ellipsoid. The remaining electron pocket finally disappears when the charge transfer $\Delta\tilde{n}$ is about 0.1 electron per SN molecule. In this case there remain only two plane-like hole Fermi surfaces. For the case of $(SNBr_y)_x$ with a value of $0 \leq y \leq 0.4$, the electron pockets are still expected to exist in the form of flattened tubes or ellipsoids, since the experimental results on the structure of the bromines in $(SNBr_y)_x$ indicate that the bromines reside in the interfiber regions or in the $(SN)_x$ lattice as Br_3^- ions, Br_5^- ions, or Br_2 molecules. To this extent of bromination ($0 \leq y \leq 0.4$), the total concentration of electrons and holes remains approximately unchanged because of the symmetric structure of the conduction bands near the Fermi level so that we can expect a slight change of the plasma edge upon bromination. On the other hand, the drastic change of both magnitude and temperature dependence of the resistivity upon bromination is unable to be explained if the usual scattering mechanism such as phonon scattering dominantly contributes to the resistivity, for the effective masses of the electron and the hole are approximately equal to one another. Electron–hole scattering, however, is a possible mechanism to solve this problem, as the relaxation time due to such interband carrier–carrier scattering not only depends on the density of states in each band but is also influenced signifi-

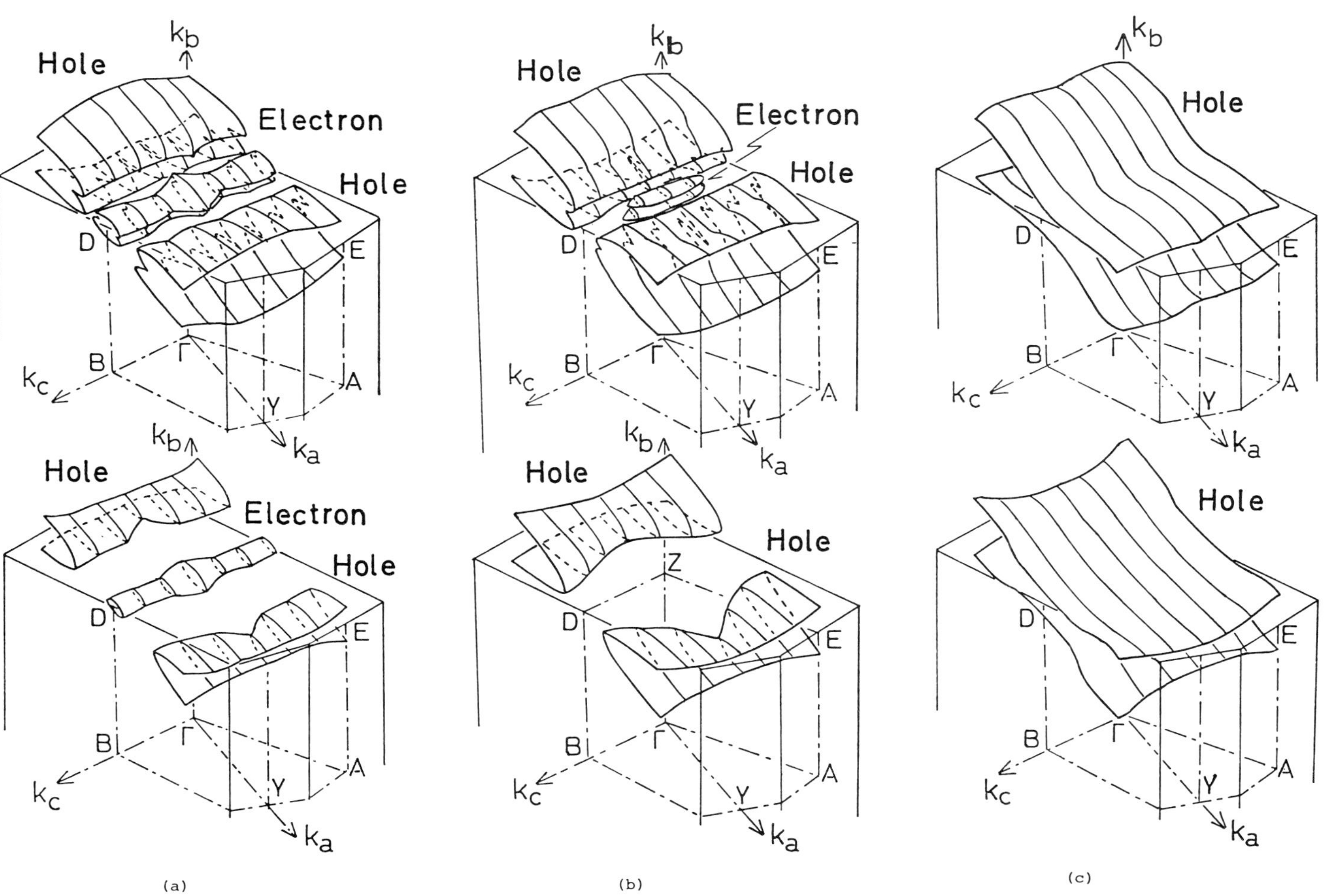

Fig. 12. Fermi surfaces of $(SNBr_y)_x$ for the value of $\Delta\bar{n}$ in electrons per molecule of (a) 0.053 (b) 0.080, and (c) 0.107. The outermost surfaces are shown at the top and the innermost ones shown at the bottom of each figure [40]. (© The Institute of Physics)

cantly by the geometrical configuration of each Fermi surface. This point was discussed in detail by Oshiyama [67] and we shall briefly describe his theory in the following section.

In closing this section we discuss the effect of bromination on the optical absorption spectra. The calculation of the interband contribution to the imaginary part of the dielectric function is carried out using the same method as employed in Section 4 with the rigid shift of the Fermi level. In order to clarify the effect of bromination, the calculation is performed for the value of $\Delta\tilde{n} = 0.25$ electron per SN molecule, which is a somewhat larger value than that of $(SNBr_{0.4})_x$. The calculated spectra below 8 eV for three different directions of polarization of light are shown in Figures 13(a), (b), and (c). The dashed curves in these figures represent the corresponding spectra of pristine $(SN)_x$ which were obtained in Section 4. As seen in these figures, the significant change due to bromination appears in the spectra below 1 eV; that is, the large interband contribution below 1 eV in the case of $(SN)_x$ disappears upon bromination. This can be explained by the fact that the transition between two conduction bands crossing the Fermi level in $(SN)_x$ disappears by the rigid shift of the Fermi level. As regards experimental works, there exists only a preliminary report by Mitani *et al.* [61] on the reflectivity of brominated $(SN)_x$.

6. Electrical Resistivity in an Interacting Two-Carrier System

In this and following sections we investigate the electrical resistivity along the chain axis of $(SN)_x$ and $(SNBr_y)_x$, on the basis of the band structures and the Fermi surfaces of these materials reviewed in preceding sections. The electrical resistivity along the chain axis of $(SN)_x$ is typically of the order of a few hundred $\mu\Omega$ cm at room temperature (RT) with resistivity ratio of RT to 4.2 K, $\rho(RT)/\rho(4.3\,\mathrm{K})$, of 100–1000 [2–5, 68], whereas the resistivity perpendicular to the chain axis is of the order of 0.1 Ω cm at room temperature [3–5, 68]. The resistivity along the chain axis exhibits metallic behavior down to the liquid helium temperature, but is quite different in its temperature dependence from that of ordinary metals. That is, over a wide temperature region from 10 to 300 K, the resistivity of $(SN)_x$ follows a T^2 form [2–5, 68], which does not change at all above and below the Debye temperature 170 K, as shown in Figure 14. This experimental finding suggests that the carrier–carrier scattering plays an important role in governing the resistivity of $(SN)_x$. In fact, this stimulated the first theoretical work on this problem by Oshiyama *et al.* [69]. They calculated the resistivity due to electron–electron Umklapp scattering on the quasi-one-dimensional band model which consists of two pairs of plane-like Fermi surfaces. Although they found out the characteristic temperature dependence of the resistivity such as T^{ν}, where $2 \leqq \nu \leqq 3$, the band model which they adopted was not applicable directly to the case of $(SN)_x$ since the self-consistent band calculation indicated the existence of the highly anisotropic electron and hole pockets.

When bromines are doped in $(SN)_x$, the resistivity along the chain axis decreases by about an order of magnitude at room temperature [39, 68, 70]. In connection

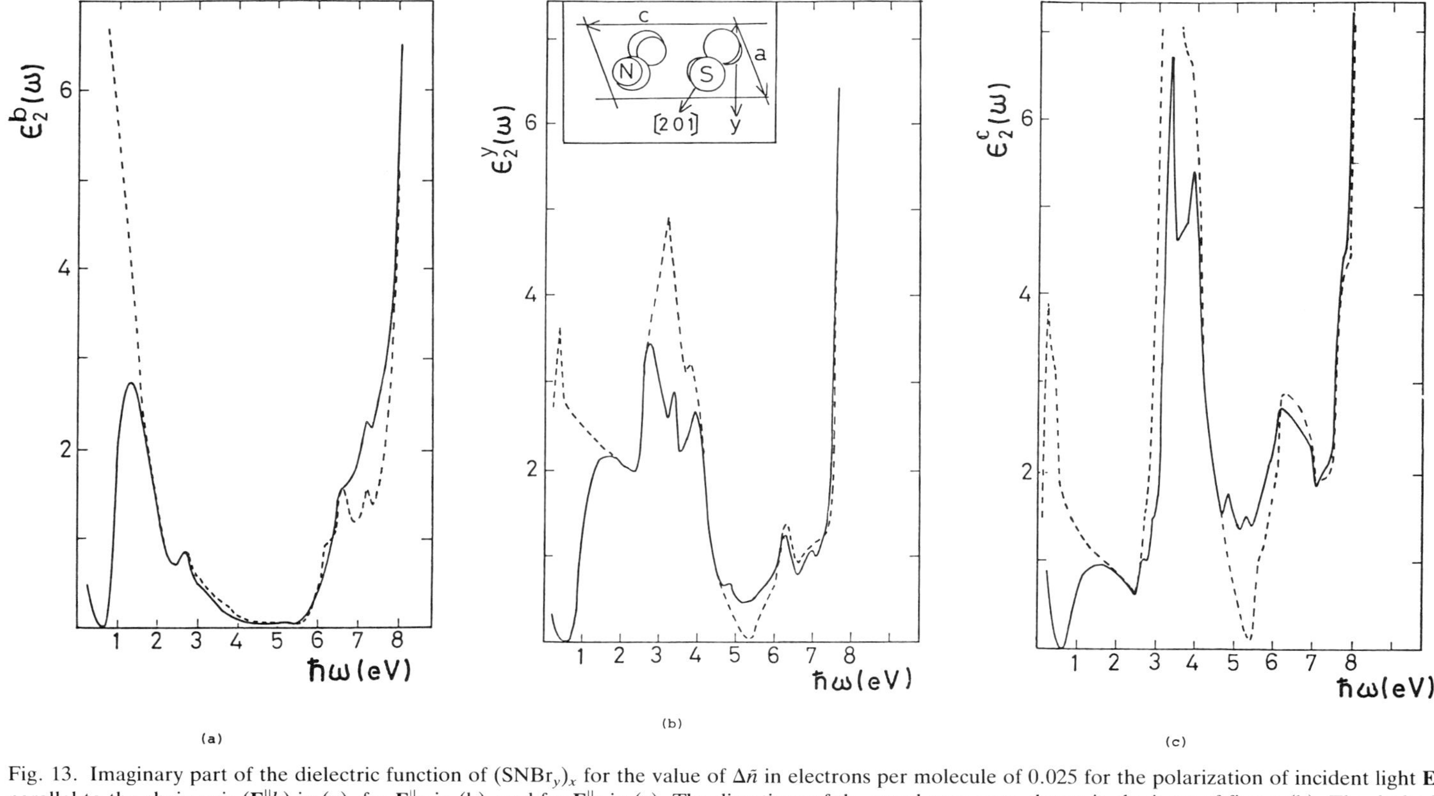

Fig. 13. Imaginary part of the dielectric function of $(SNBr_y)_x$ for the value of $\Delta\bar{n}$ in electrons per molecule of 0.025 for the polarization of incident light **E** parallel to the chain axis (**E**$\|b$) in (a), for **E**$\|y$ in (b), and for **E**$\|c$ in (c). The directions of the c and y axes are shown in the inset of figure (b). The dashed curve represents the corresponding spectra calculated for pristine $(SN)_x$.

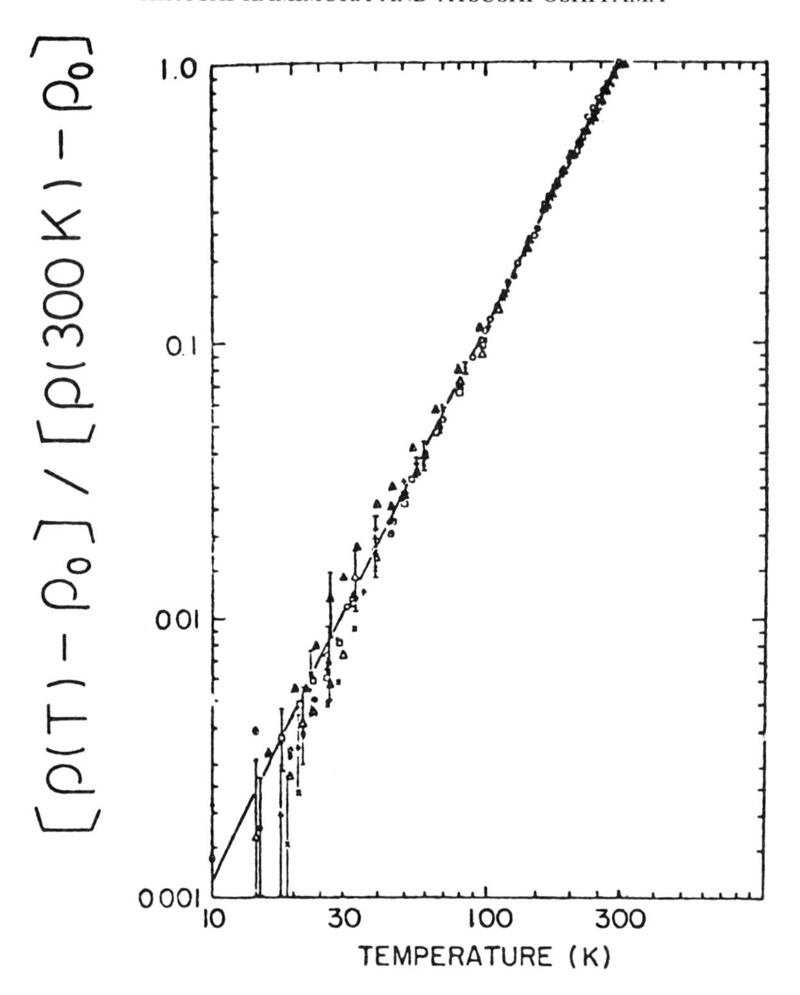

Fig. 14. Log–log plot of the intrinsic resistivity along the chain axis of $(SN)_x$ versus temperature [2]. Here $\rho(T)$ represents the measured resistivity and ρ_0 the residual resistivity.

with this decrease of resistivity the thermoelectric power changes its sign from negative to positive values upon bromination [39]. Detailed measurements on the temperature-dependent resistivity of $(SNBr_y)_x$ were made by Kaneto *et al.* [68] and they showed that the temperature dependence of the resistivity becomes weaker than the T^2 law upon bromination and cannot be decribed by a single power law in the whole temperature region, as shown in Figure 15. According to their result, with an increase in bromine concentration the temperature dependence of the resistivity changes from T^2 to T linear form in the temperature region range above 200 K. A similar variation of the temperature dependence of resistivity was also observed in $(SN)_x$ under hydrostatic pressure [4].

The effect of carrier–carrier scattering on the resistivity has been observed not only in $(SN)_x$ but also in other materials such as certain transition metals [71] or

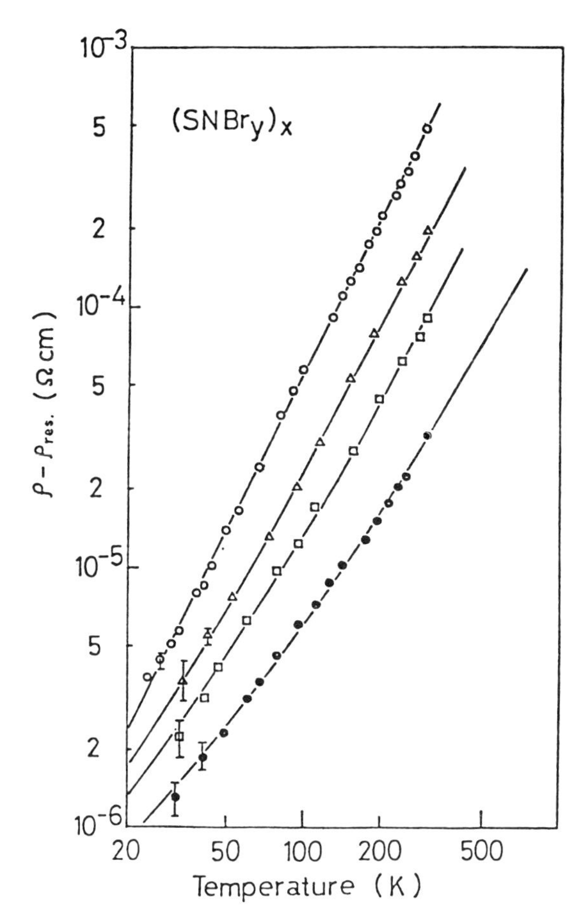

Fig. 15. Log–log plot of the intrinsic resistivity along the chain axis of $(SNBr_y)_x$ versus temperature for four samples: pristine $(SN)_x$ ($y = 0$) ○; $y \cong 0.15$ △; $y \cong 0.28$ □; and $y \cong 0.4$ ● [68]. Here ρ represents the measured resistivity and ρ_{res} the residual resistivity.

some semimetals [72]. In certain magnetic metals paramagnon scattering may give rise to the characteristic T^2 form of resistivity in its temperature dependence [73]. In all these materials, however, the T^2 dependence on resistivity is restricted to lower temperature regions, usually below 20 K. To our knowledge, only $(SN)_x$ and a layered compound TiS_2 [74] exhibit such T^2 dependence on resistivity in a wide temperature region up to 300 K.

Since carrier–carrier scattering is expected to play a dominant role in determing the resistivity of $(SN)_x$, in this section we clarify the characteristic role of the carrier–carrier scattering in the resistivity from a general standpoint before we discuss the transport properties of $(SN)_x$ and $(SNBr_y)_x$. Theoretically the carrier–carrier scattering in single carrier systems has been studied by using the Boltzmann equation [75–77] and by the diagrammatic technique [78]. The exact solution of the Boltzmann equation in a single carrier system was obtained [75–77] and confirmed the view that carrier–carrier normal scattering could not contribute to resistivity because of the total-momentum conservation. This solution was applied to the case of electron–electron Umklapp scattering in simple metals [77]. For two-carrier systems, on the other hand, Baber [79] first predicted the substantial contribution from electron–hole scattering to the resistivity of certain metals. The T^2 form of temperature-dependent resistivity of transition metals was discussed in terms of Baber's model [80]. In this treatment, however, heavy holes were simply regarded as scatterers for light electrons and the current carried by the holes was neglected. Therefore, these works did not clarify the characteristic role of carrier–carrier scattering in the resistivity of multicarrier systems in which different types of carriers could contribute significantly to the total current. Kukkonen and Maldague [81] proposed that the normal scattering between different types of carriers could act so as to relax the relative velocity of two types of carriers towards zero. They used a coupled kinetic equation and applied their solution to TiS_2. Paying attention to the fact that $(SN)_x$ and $(SNBr_y)_x$ were regarded as multicarrier systems containing electrons and holes, Oshiyama [67] investigated both normal and Umklapp scattering mechanisms between different types of carriers in addition to the scattering between the same types of carriers and clarified the characteristic role of each mechanism in the resistivity. Then, in order to reveal the origin of the characteristic behavior in the resistivity of $(SN)_x$ and $(SNBr_y)_x$, he calculated the resistivity due to both carrier–carrier and carrier–phonon scattering mechanisms. Here let us briefly survey his theoretical treatment.

We begin with a discussion on the coupled Boltzmann equation for carrier–carrier scattering and carrier–phonon scattering. As usual, Boltzmann's equation is linealized by expanding the Fermi distribution function $n_i(\mathbf{k})$ about its equilibrium $n_i^0(\mathbf{k}) = (e^t + 1)^{-1}$, where $t = (\varepsilon_{i\mathbf{k}} - E_F)/k_BT$ is the dimensionless energy variable measured from the Fermi level E_F, for a wavevector $\mathbf{k}$ in the ith band; that is,

$$n_i(\mathbf{k}) = n_i^0(\mathbf{k}) - \Phi_{i\mathbf{k}}\left(\frac{\partial n_i^0(\mathbf{k})}{\partial \varepsilon_{i\mathbf{k}}}\right). \tag{16}$$

The resulting linearized Boltzmann equation can be written as

$$\begin{aligned} e v_{i\mathbf{k}} \cdot \mathbf{E}\frac{\partial n_i^0(\mathbf{k})}{\partial \varepsilon_{i\mathbf{k}}} = {} & \frac{2}{k_BT} \sum_{i_2,i_3,i_4} \sum_{\mathbf{k}_2,\mathbf{k}_3,\mathbf{k}_4} W(i\mathbf{k},\, i_2\mathbf{k}_2 \to i_3\mathbf{k}_3,\, i_4\mathbf{k}_4) \times \\ & \times n_i^0(\mathbf{k}) n_{i_2}^0(\mathbf{k}_2)(1 - n_{i_3}^0(\mathbf{k}_3))(1 - n_{i_4}^0(\mathbf{k}_4)) \times \\ & \times (\Phi_{i\mathbf{k}} + \Phi_{i_2\mathbf{k}_2} - \Phi_{i_3\mathbf{k}_3} - \Phi_{i_4\mathbf{k}_4}) \times \\ & \times \delta(\varepsilon_{i\mathbf{k}} + \varepsilon_{i_2\mathbf{k}_2} - \varepsilon_{i_3\mathbf{k}_3} - \varepsilon_{i_4\mathbf{k}_4}) - \\ & - (\Phi_{i\mathbf{k}}/\tau_i)(\delta n_i^0(\mathbf{k})/\delta \varepsilon_{i\mathbf{k}}), \end{aligned} \tag{17}$$

where $W(i\mathbf{k}, i_2\mathbf{k}_2 \to i_3\mathbf{k}_3, i_4\mathbf{k}_4)$ is the transition probability due to carrier–carrier scattering between the states $|i\mathbf{k}, i_2\mathbf{k}_2\rangle$ and $|i_3\mathbf{k}_3, i_4\mathbf{k}_4\rangle$, $\mathbf{E}$ is an electric field and $v_{i\mathbf{k}} = h^{-1}\nabla_\mathbf{k}\varepsilon_{i\mathbf{k}}$ is the group velocity of an electron for the state $\mathbf{k}$ of the ith bands. It can be distinguished from the sign of the velocity whether the carrier of the ith band has an electron-like or a hole-like nature. Further, in the second term of the right-hand side of Equation (17) we consider only an intraband carrier–phonon scattering process for simplicity and use the relaxation time τ_i obtained by the standard variational treatment:

$$\frac{1}{\tau_i} = \frac{4\pi m_i}{n_i \hbar k_B T} \sum_{\mathbf{k},\mathbf{k}'} |g_{i\mathbf{k}-\mathbf{k}'}|^2 N_{\mathbf{k}-\mathbf{k}'}(\nu_{i\mathbf{k}}^{\mu} - \nu_{i\mathbf{k}'}^{\mu})^2 \times$$
$$\times\, n_i^0(\mathbf{k})(1 - n_i^0(\mathbf{k}'))\,\delta(\varepsilon_{i\mathbf{k}} - \varepsilon_{i\mathbf{k}'} + \hbar\omega_{\mathbf{k}-\mathbf{k}'}). \tag{18}$$

Here, μ denotes the direction of the electric field, and $g_{\mathbf{k}-\mathbf{k}'}$, $N_{\mathbf{k}-\mathbf{k}'}$, $\omega_{\mathbf{k}-\mathbf{k}'}$, and n_i are, respectively, the matrix element of carrier–phonon scattering between states of the ith band, the Böse distribution function for a phonon with wavevector $\mathbf{k} - \mathbf{k}'$, the phonon frequency and concentration of the ith carrier. Further, m_i in Equation (18) is the optical mass of the ith carrier defined by

$$\frac{n_i}{m_i} = \frac{1}{4\pi^3} \int (\nu_{i\mathbf{k}}^{\mu})^2 \left(-\frac{\delta n_i^0(\mathbf{k})}{\delta \varepsilon_{i\mathbf{k}}} \right) \mathrm{d}\mathbf{k}. \tag{19}$$

Since it is not feasible to solve exactly the coupled equation (17), we reasonably assume that the angular dependence of the function $\Phi_{i\mathbf{k}}$ is given by

$$\Phi_{i\mathbf{k}} = \nu_{i\mathbf{k}}^{\mu} \varphi_i(t), \tag{20}$$

where φ_i is a general function of energy t. Thus, a solution under the assumption (20) should be regarded as a variational solution in a strict sense. The advantage, however, of this assumption is that we can separate the integration over the **k** space in the first term of the right-hand side of Equation (17) into energy integrals and surface integrals according to

$$\mathrm{d}\mathbf{k} = k_B T \,\mathrm{d}S_{i\mathbf{k}}\, \mathrm{d}t/\hbar |\nu_{i\mathbf{k}}|. \tag{21}$$

The surface integrals in general depend on energy variables, but in the present case, as usual, we can replace them by the surface integrals over the Fermi surface (FS) owing to the Fermi distribution functions in Equation (17). It should be noted that the energy dependence of the surface integral becomes important in determining the temperature dependence of resistivity in the case of carrier–carrier Umklapp scattering in special Fermi surfaces, as pointed out by Oshiyama *et al.* [69]. Here, however, we confine ourselves to the usual case in which the energy dependence of the surface integrals is negligible to the order of $(k_B T)/E_F$. Then, by multiplying both sides of Equation (17) by $(\nu_{i\mathbf{k}} \cdot E)/|\nu_{i\mathbf{k}}|$ and by carrying out the energy and surface integrals, the linearized Boltzmann equation (17) becomes the coupled integral equation:

$$-\frac{n_1 e}{m_1 \cosh(t/2)} = \frac{n_1 \varphi_1(t)}{m_1 \tau_1 \cosh(t/2)} + \frac{(a_{11}^{(0)} + a_{12}^{(1)}) g(t) \varphi_1(t)}{\cosh(t/2)} -$$
$$- 2 \int \frac{F(x-t)}{\cosh(x/2)} [(a_{11}^{(0)} - a_{12}^{(2)} - u_{11}^{(0)}(4) \varphi_1(x) + a_{12}^{(3)} \varphi_2(x)] \,\mathrm{d}x \tag{22a}$$

and

$$-\frac{n_2 e}{m_2 \cosh(t/2)} = \frac{n_2 \varphi_2(t)}{m_2 \tau_2 \cosh(t/2)} + \frac{(a_{22}^{(0)} + a_{21}^{(1)}) g(t) \varphi_2(t)}{\cosh(t/2)} -$$

$$- 2 \int \frac{F(x-t)}{\cosh(x/2)} [(a_{22}^{(0)} - a_{21}^{(2)} - u_{22}^{(0)}(4) \varphi_2(x) + a_{21}^{(3)} \varphi_1(x)] \, \mathrm{d}x \quad (22\mathrm{b})$$

Here, $g(t) = \pi^2 + t^2$ and $F(x) = (x/2)/\sinh(x/2)$. Further, $a_{ii}^{(0)}$, $u_{ii}^{(0)}$, $a_{ii}^{(1)}$, $a_{ij}^{(2)}$, and $a_{ij}^{(3)}$ in Equations (22a) and (22b) are all surface integrals over the Fermi surface. $a_{ii}^{(0)}$ and $u_{ii}^{(0)}$ are concerned with the intraband carrier–carrier scattering; they are given by

$$a_{ii}^{(0)} = \langle W(i\mathbf{k}_1, i\mathbf{k}_2 \to i\mathbf{k}_3, i\mathbf{k}_4)(v_{i\mathbf{k}_1}^{\mu})^2 \rangle \quad (23)$$

and

$$u_{ii}^{(0)} = \langle W(i\mathbf{k}_1, i\mathbf{k}_2 \to i\mathbf{k}_3, i\mathbf{k}_4)(v_{i\mathbf{k}_1}^{\mu} + v_{i\mathbf{k}_2}^{\mu} - v_{i\mathbf{k}_3}^{\mu} - v_{i\mathbf{k}_4}^{\mu})^2 \rangle . \quad (24)$$

Here the angular brackets $\langle \ \rangle$ are defined by the following surface integral:

$$\langle f(i_1\mathbf{k}_1, i_2\mathbf{k}_2, i_3\mathbf{k}_3, i_4\mathbf{k}_4) \rangle = \frac{2(k_B T)^2}{(2\pi)^9} \int \frac{\mathrm{d}S_{i_1\mathbf{k}_1}}{\hbar |v_{i_1\mathbf{k}_1}|} \int \frac{\mathrm{d}S_{i_2\mathbf{k}_2}}{\hbar |v_{i_2\mathbf{k}_2}|} \int \frac{\mathrm{d}S_{i_3\mathbf{k}_3}}{\hbar |v_{i_3\mathbf{k}_3}|} \int \frac{\mathrm{d}S_{i_4\mathbf{k}_4}}{\hbar |v_{i_4\mathbf{k}_4}|} . \quad (25)$$

Further, the surface integral $u_{ii}^{(0)}$ represents the contribution from the intraband carrier–carrier Umklapp scattering along the electric field. Other types of surface integrals are concerned with the interband carrier–carrier scattering and are given by

$$a_{ij}^{(1)} = \langle W(i\mathbf{k}_1, i\mathbf{k}_2 \to j\mathbf{k}_3, j\mathbf{k}_4)(v_{i\mathbf{k}_1}^{\mu})^2 \rangle + + \langle W(i\mathbf{k}_1, j\mathbf{k}_2 \to i\mathbf{k}_3, j\mathbf{k}_4)(v_{i\mathbf{k}_1}^{\mu})^2 \rangle , \quad (26)$$

$$a_{ij}^{(2)} = \langle W(i\mathbf{k}_1, i\mathbf{k}_2 \to j\mathbf{k}_3, j\mathbf{k}_4) v_{i\mathbf{k}_1}^{\mu} v_{i\mathbf{k}_2}^{\mu} \rangle - - \langle W(i\mathbf{k}_1, j\mathbf{k}_2 \to i\mathbf{k}_3, j\mathbf{k}_4) v_{i\mathbf{k}_1}^{\mu} v_{i\mathbf{k}_3}^{\mu} \rangle \quad (27)$$

and

$$a_{ij}^{(3)} = \langle W(i\mathbf{k}_1, i\mathbf{k}_2 \to j\mathbf{k}_3, j\mathbf{k}_4) v_{i\mathbf{k}_1}^{\mu} (v_{j\mathbf{k}_3}^{\mu} + v_{j\mathbf{k}_4}^{\mu}) \rangle - - \langle W(i\mathbf{k}_1, j\mathbf{k}_2 \to i\mathbf{k}_3, j\mathbf{k}_4) v_{i\mathbf{k}_1}^{\mu} (v_{j\mathbf{k}_2}^{\mu} - v_{j\mathbf{k}_4}^{\mu}) \rangle . \quad (28)$$

Here we have considered two types of interband scattering process: $|i\mathbf{k}_1, i\mathbf{k}_2\rangle \leftrightarrow |j\mathbf{k}_3, j\mathbf{k}_4\rangle$ and $|i\mathbf{k}_1, j\mathbf{k}_2\rangle \leftrightarrow |i\mathbf{k}_3, j\mathbf{k}_4\rangle$. It should be noted that the above three surface integrals, $a_{ij}^{(1)}$, $a_{ij}^{(2)}$, and $a_{ij}^{(3)}$ are not independent of each other and satisfy the following relations:

$$a_{ij}^{(3)} = a_{ji}^{(3)} \quad (29\mathrm{a})$$

and

$$a_{ij}^{(3)} = \frac{\theta_i \theta_j m_i (a_{ij}^{(1)} + a_{ij}^{(2)})}{m_j} - u_{ij}^{(1)}, \quad (29\mathrm{b})$$

where θ_i takes the value of $+1$ (or -1) if the carrier of the ith band is an electron (or a hole). Here $u_{ij}^{(1)}$ represents the contribution from interband Umklapp scattering along the electric field and is defined by

$$u_{ij}^{(1)} = \langle W(i\mathbf{k}_1, i\mathbf{k}_2 \to j\mathbf{k}_3, j\mathbf{k}_4) \nu_{i\mathbf{k}_1}^{\mu} [\theta_i \theta_j m_i (\nu_{i\mathbf{k}_1}^{\mu} + \nu_{i\mathbf{k}_2}^{\mu})/m_j - \\ - (\nu_{j\mathbf{k}_3}^{\mu} + \nu_{j\mathbf{k}_4}^{\mu})] \rangle + \langle W(i\mathbf{k}_1, j\mathbf{k}_2 \to i\mathbf{k}_3, j\mathbf{k}_4) \times \\ \times \nu_{i\mathbf{k}_1}^{\mu} [\theta_i \theta_j m_i (\nu_{i\mathbf{k}_1}^{\mu} - \\ - \nu_{i\mathbf{k}_3}^{\mu})/m_j + (\nu_{j\mathbf{k}_2}^{\mu} - \nu_{j\mathbf{k}_4}^{\mu})] \rangle. \qquad (30)$$

Oshiyama [67] showed that the coupled integral equations, (22a) and (22b), can be solved exactly when the carrier–phonon scattering terms are neglected. By using this solution, he obtained an exact formula for resistivity due to carrier–carrier scattering in a two-carrier system in the form of rapidly converging series. According to this exact formula, the interband carrier–carrier normal scattering acts so as to attenuate the relative motion of one carrier in the first band and the other carrier in the second band, while the intraband carrier–carrier normal scattering acts only to renormalize the masses of two types of carriers. Further, he showed that the current carried by the center-of-mass motion of two types of carriers cannot be relaxed towards zero, when carrier–carrier Umklapp scattering and other scattering such as carrier–phonon scattering are absent. However, he pointed out that a semimetal with the equal electron and hole concentrations is a special case in which the center-of-mass motion of electrons and holes carries no net charge so that electron–hole normal scattering gives rise to the finite resistivity by itself.

Further, Oshiyama [67] solved approximately the coupled Boltzmann equations (22a) and (22b) under the simultaneous existence of both carrier–carrier and carrier–phonon scattering by using the variational principle. The resulting electrical resistivity can be written as

$$\rho = \{1 + (M/\tau_{cc})[(\tilde{\tau}_1/n_1 m_1) + (\tilde{\tau}_2/n_2 m_2)]\}\{e^2[n_1 \tilde{\tau}_1/m_1) + \\ + n_2 \tilde{\tau}_2/m_2) + \tilde{\tau}_1 \tilde{\tau}_2 (n_1 \theta_1 + n_2 \theta_2)^2/\tau_{cc}(n_1 m_1 + n_2 m_2)]\}^{-1}, \qquad (31)$$

where

$$1/\tilde{\tau}_i = (1/\tau_i) + (4\pi^2 m_i/3n_i)(\tfrac{1}{4}u_{ii}^{(0)} + \theta_i \theta_j m_j u_{ij}^{(1)}/m_i), \qquad (32)$$

$$1/\tau_{cc} = 4\pi^2 m_1 m_2 |a_{12}^{(3)}|/3M, \qquad (33)$$

and

$$M = (1/n_1 m_1 + 1/n_2 m_2)^{-1}. \qquad (34)$$

Here $\tilde{\tau}_i$ is the combined relaxation time due to carrier–carrier Umklapp scattering along the electric field and due to carrier–phonon scattering, while τ_{cc} is that due to interband carrier–carrier scattering. It should be noted that the carrier–carrier Umklapp scattering along the direction perpendicular to the electric field, if it exists, also contributes to τ_{cc}^{-1}. Equation (31) can be regarded as a generalization of the formula obtained by the simple kinetic equation by Kukkonen and Maldague [81] to the case in which both carrier–carrier Umklapp and normal scattering processes in arbitrary Fermi surfaces exist. This formula (31) still preserves the essential features of the exact solution. That is, in the case of semimetals in which $n_1\theta_1 + n_2\theta_2 = 0$, the electron–hole normal scattering gives rise to the finite resistivity by itself, while in other cases it cannot contribute to the resistivity unless both carrier–phonon and carrier–carrier Umklapp scattering mechanisms are taken into account. Therefore, for the case of semimetals in which electron–hole interaction is sufficiently strong, we can find the drastic change of the scattering mechanism by changing the carrier concentrations from $n_1 = n_2$ to $n_1 \neq n_2$. Further, when the optical masses of two types of carriers are quite different from each other, for example $m_1 \gg m_2$, the resistivity (31) becomes

$$\rho \simeq (m_1/n_1e^2)[(1/\bar{\tau}_1) + (1/\tau_{cc})][1 + m_1\bar{\tau}_2(n_1\theta_1 + \\ + n_2\theta_2)^2/n_1n_2m_2\tau_{cc}]^{-1}. \tag{35}$$

Thus, when m_2 is much heavier than m_1 so that $m_1(n_1\theta_1 + n_2\theta_2)^2\bar{\tau}_2/(n_1n_2m_2\tau_{cc})$ can be neglected in the denominator of Equation (35), the resistivity follows a T^2 form in its temperature dependence for the case of $\tau_{cc}^{-1} \gg \tau_1^{-1}$. This corresponds to the conventional Baber model [78, 79].

7. Electrical Resistivity of $(SN)_x$ and $(SNBr_y)_x$

In this section we discuss the resistivity of $(SN)_x$ and $(SNBr_y)_x$ calculated by the formula (31) and compare it with the observed results. When the formula (31) for the resistivity is applied to the cases of $(SN)_x$ and $(SNBr_y)_x$, we have to consider the variation of the relaxation times, $\bar{\tau}_i$ and τ_{cc} in Equations (32) and (33), upon bromination in addition to the change of the carrier concentrations. In particular, it should be noted that the Fermi level density of states is not only the factor to determine the value of the surface integral $a_{12}^{(3)}$ in Equation (33) which gives the relaxation time τ_{cc} due to the interband carrier–carrier scattering. This integral can also significantly depend on the geometrical configuration of the Fermi surfaces of the two types of carriers because of the crystal momentum conservation.

From this standpoint and also from the standpoint of making the calculation of resistivity easier, we first introduce a simple band model in which the characteristic features of the Fermi surfaces of $(SN)_x$ and $(SNBr_y)_x$ are still preserved. Oshiyama [67] calculated the resistivity along the chain axis on this band model by using formula (31). In his calculation he considered three kinds of scattering mechanisms: electron–hole scattering, electron–phonon scattering, and hole–phonon scattering.

7.1. Model

We begin with a description of a semimetallic band model for $(SN)_x$, which consists of highly anisotropic electron and hole Fermi surfaces. Although the shapes of the real Fermi surfaces of $(SN)_x$ and $(SNBr_y)_x$ are rather complicated, as we have seen in Sections 3 and 5, the essential features of the Fermi surfaces derived by Oshiyama and Kamimura [40] are as follows. First, the Fermi surfaces of these materials are shaped like flattened tubes or flattened ellipoids. Second, with bromination, the hole portion of the Fermi surfaces expands, while the electron portion shrinks and finally vanishes. Taking these features into account, we introduce the following two-band model to calculate the resistivities of $(SN)_x$ and $(SNBr_y)_x$. In the first place, a unit cell is taken to be orthorhombic with lattice constants a, b, and c instead of a monoclinic cell in $(SN)_x$ for simplicity. The lattice constants of the present model are determined so that the unit cell volume of the model coincides with that of $(SN)_x$. They are given in Table I together with the lattice parameters of $(SN)_x$. Then, instead of the four conduction bands of $(SN)_x$ and $(SNBr_y)_x$, the model involves two conduction bands, E_1 and E_2, in which anisotropic electron and hole pockets, respectively, are formed in the shape of flattened tubes or flattened ellipsoids. These conduction bands are described by the following dispersion relations:

$$\begin{aligned}\varepsilon_1(\mathbf{k}) &= \Delta + (b^2\gamma_b/2)[\pi/b - k_b \operatorname{signum}(k_b)]^2 + a^2\gamma_a k_a^2/2 + \\ &\quad + 2c\gamma_c|k_c|/\pi \end{aligned} \tag{36a}$$

for the E_1 band and

$$\begin{aligned}\varepsilon_2(\mathbf{k}) &= -\Delta + 2\gamma_a - (b^2\gamma_b/2)[\pi/b - k_b \operatorname{sign}(k_b)]^2 - \\ &\quad - (a^2\gamma_a/2)[\pi/a - k_a \operatorname{sign}(kk_a)]^2 + 2c\gamma_c|k_c|/\pi \end{aligned} \tag{36b}$$

for the E_2 band, where γ_a, γ_b, and γ_c denote the bandwidths along three different directions, k_a, k_b, and k_c, respectively. These band parameters satisfy the relation, $\gamma_b \gg \gamma_a \gg \gamma_c$, consistent with the band structure of $(SN)_x$. In Equation (36) we adopt the linear dispersion relation along the most nondispersive direction k_c for simplicity. The parameter Δ represents the averaged energy difference between the two bands on the top and bottom surfaces of the Brillouin zone. All the band parameters are determined so as to reproduce the calculated band structure results described in Section 3, and these values are also given in Table I. Further, since we reduce the number of conduction bands to the half of the real case, each band, E_1

TABLE I

Band parameters and lattice constants in the present model. The lattice parameters of $(SN)_x$ are also shown. Here β is the angle between the a and c axes.

Parameters	Model	$(SN)_x$
a (A)	4.03	4.153
b (A)	4.44	4.439
c (A)	7.41	7.637
β (deg)	90	109.7
γ_a (eV)	0.18	—
γ_b (eV)	0.75	—
γ_c (eV)	0.02	—
Δ (eV)	0.09	—

and E_2, is considered to have additional twofold degeneracy. The conduction band structure of the present model is illustrated in Figure 16 with the inset of the Brillouin zone.

For the values of the band parameters given in Table I, an anisotropic electron pocket is formed for band E_1 and an anisotropic hole pocket is formed for band E_2 at the zone face. The detailed shape of these pockets depends on the value of the Fermi level E_F. When the Fermi level crosses two conduction bands at $E_F = \gamma_a + \gamma_c$, the electron concentration n_e is equal to the hole concentration n_h. For the values given in Table I, n_e and n_h are calculated to be $n_e = n_h = 1.17 \times 10^{21}\,\text{cm}^{-3}$ for $E_F = \gamma_a + \gamma_c = 0.2\,\text{eV}$. The Fermi surfaces in this case, which are shown in Figure 17(a), consist of anisotropic electron and hole pockets which have the shape of flattened tubes. The electron pocket is located at the center of the zone face and this is separated along the direction k_a from the hole pocket at the zone edge. This corresponds to the case of $(SN)_x$. Then, as E_F is lowered gradually to $\Delta + 2\gamma_c$, n_e decreases monotonously while n_h increases. They are given, as a function of E_F, by

$$n_e = \frac{2(E_F - \Delta - \gamma_c)}{\pi abc\sqrt{\gamma_a \gamma_b}} \qquad \text{for } E_F \geqq \Delta + 2_c \tag{37a}$$

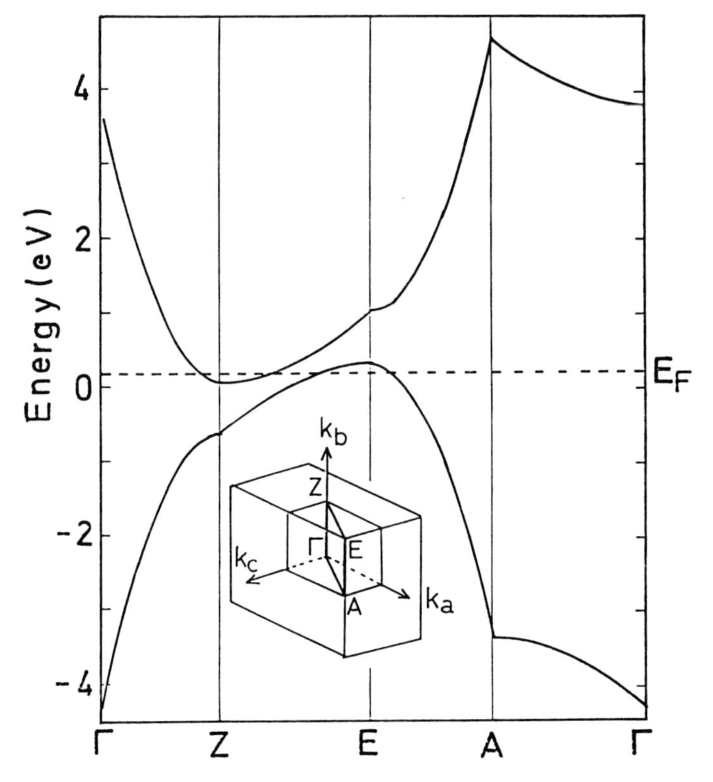

Fig. 16. Conduction band structure of the present band model, together with the Fermi level $E_F = \gamma_a + \gamma_c$.

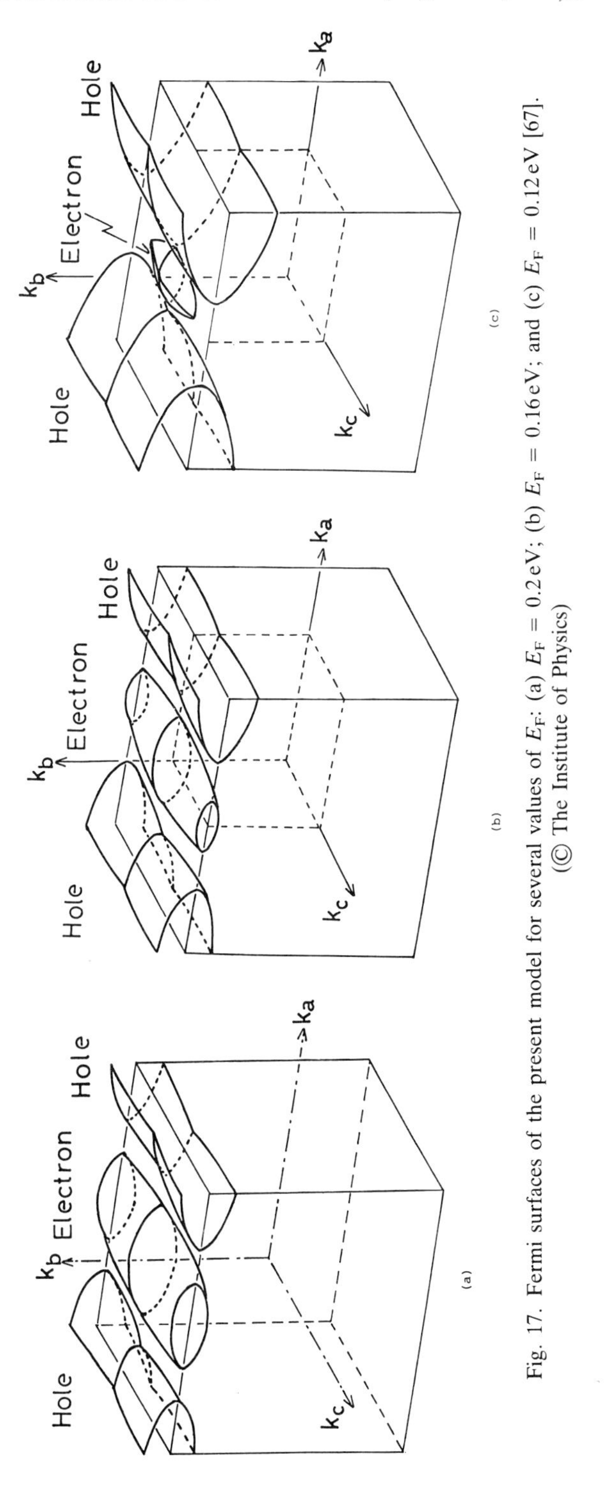

Fig. 17. Fermi surfaces of the present model for several values of E_F: (a) $E_F = 0.2$ eV; (b) $E_F = 0.16$ eV; and (c) $E_F = 0.12$ eV [67]. (© The Institute of Physics)

and

$$n_h = \frac{2(\gamma_c + 2\gamma_a - \Delta - E_F)}{\pi abc\sqrt{\gamma_a\gamma_b}}. \tag{37b}$$

The Fermi surfaces in this case are shown in Figure 17(b). The electron portion of the Fermi surfaces shrinks while the hole portion expands. Both the electron and hole pockets still have the shape of flattened tubes. If the Fermi level E_F decreases further from $\Delta + 2\gamma_c$, the electron pocket takes a closed shape for all directions so that its shape becomes flattened ellipsoids. This feature of the Fermi surfaces is shown in Figure 17(c). The concentration of electrons in this case is given by

$$n_e = \frac{(E_F - \Delta)^2}{2\pi abc\gamma_c\sqrt{\gamma_a\gamma_b}} \quad \text{for } \Delta + 2\gamma_c \geqq E_F \geqq \Delta \tag{37c}$$

while the hole concentration is still given by Equation (37b). This electron pocket finally vanishes at $E_F = \Delta$, while the hole portion of the Fermi surfaces is still expanding. These features of the Fermi surfaces in the present band model exactly corresponds to the case of $(SN)_x$ and $(SNBr_y)_x$.

The charge transfer from SN chains to bromines per SN molecule can be written as

$$\Delta\tilde{n} = abc(n_h - n_e)/4 \tag{38}$$

in the present model, since there are four SN molecules in a unit cell of $(SN)_x$. This quantity $\Delta\tilde{n}$ is the function of E_F through n_e and n_h. In the following subsection we use this quantity $\Delta\tilde{n}$ as a parameter of the charge transfer in order to compare the calculated results with the experimental results of $(SNBr_y)_x$.

7.2. Temperature Dependence and Magnitude of Resistivity

In this subsection the resistivity along the chain axis of $(SN)_x$ and $(SNBr_y)_x$ calculated on the basis of the band model described in the preceding subsection is presented. By using the formula (31) together with Equations (32) and (31), the calculation was carried out for several values of the charge transfer $\Delta\tilde{n}$ [67].

7.2.1. *Electron–Hole Scattering*

First, let us consider the relaxation time τ_{cc}, Equation (31), which is given in terms of the surface integral $a_{12}^{(3)}$, Equation (28). In the first place, two types of electron–hole scattering processes are considered in the present band model. In scattering process (i), one electron and one hole are scattered to another electron and another hole; i.e. $|\text{band } E_1\mathbf{k}_1, \text{band } E_2\mathbf{k}_2\rangle \leftrightarrow |\text{band } E_1\mathbf{k}_3, \text{band } E_2\mathbf{k}_4\rangle$. In scattering process (ii), two electrons (or holes) are scattered to two holes (or electrons); i.e. $|\text{band } E_1\mathbf{k}_1, \text{band } E_1\mathbf{k}_2\rangle \leftrightarrow |\text{band } E_2\mathbf{k}_3, \text{band } E_2\mathbf{k}_4\rangle$. Since the hole pockets are located near the zone edge, $k_a = \pm\pi/a$, these two scattering processes include the Umklapp scattering along the direction k_a perpendicular to the chain axis. Such an Umklapp scattering process also contributes to the relaxation time τ_{cc}.

For the calculation of the transition probability W in the expression (28) of the surface integral, the Thomas–Fermi screened Coulomb interaction was adopted and its matrix elements between plane-wave states were calculated. Thus $W(i_1\mathbf{k}_1, i_2\mathbf{k}_2 \rightarrow i_3\mathbf{k}_3, i_4\mathbf{k}_4)$ can be written as

$$W(1\mathbf{k}_1, 2\mathbf{k}_2 \rightarrow 1\mathbf{k}_3, 2\mathbf{k}_4) = (2\pi/\hbar)[4\pi e^2/(\epsilon_c|\mathbf{k}_1 - \mathbf{k}_3|^2 + q_e^2 + q_h^2)]^2 \times \\ \times \delta[\mathbf{k}_1 + \mathbf{k}_2 - \mathbf{k}_3 - \mathbf{k}_4 - (2\pi l_a/a)\hat{k}_a] \tag{39}$$

for the scattering process (i), and

$$W(1\mathbf{k}_1, 1\mathbf{k}_1 \rightarrow 2\mathbf{k}_3, 2\mathbf{k}_4) = (2\pi/\hbar)[4\pi e^2/(\epsilon_c|\mathbf{k}_1 - \mathbf{k}_3|^2 + q_{eh}^2)]^2 \times \\ \times \delta[\mathbf{k}_1 + \mathbf{k}_2 - \mathbf{k}_3 - \mathbf{k}_4 - (2\pi l'_a/a)\hat{k}_a] \tag{40}$$

for the scattering process (ii), where ϵ_c is an averaged dielectric constant and $\hat{k}_a$ is the unit vector in the direction k_a. Further, the integers l_a and l'_a take the value 0 or ± 1, which are determined uniquely for the four wavevectors of carriers, $\mathbf{k}_1$, $\mathbf{k}_2$, $\mathbf{k}_3$ and $\mathbf{k}_4$, in the first Brillouin zone. Three kinds of screening lengths, q_e^{-1}, q_h^{-1}, and q_{eh}^{-1}, are obtained by taking the long-wavelength limit of the usual polarization function. They are given, in the present model, by

$$q_e^2 = \begin{cases} 2\pi e^2 n_e/(E_F - \Delta - \gamma_c) & \text{for } E_F \geqslant \Delta + 2\gamma_c \\ 4\pi e^2 n_e/(E_F - \Delta) & \text{for } \Delta + 2\gamma_c \geqslant E_F \geqslant \Delta, \end{cases} \tag{41a}$$

$$q_h^2 = 2\pi e^2 n_h/(\gamma_c + 2\gamma_a - \Delta - E_F), \tag{41b}$$

and

$$q_{eh}^2 \simeq 8\pi e^2(n_h - n_e)/\gamma_b. \tag{41c}$$

By inserting Equations (39) and (40) into Equation (28), the surface integral $a_{12}^{(3)}$ is given in the form of eightfold integration which contains a three-dimensional δ function representing the momentum conservation. This eightfold integral is not feasible for arbitrary Fermi surfaces, but in the present band model this integral can be reduced to the threefold integral by elementary calculation. The resulting integral was carried out numerically for several values of $\Delta\tilde{n}$, Equation (38), with the use of the empirical value of $\epsilon_c = 4.5$ [9–12, 25].

By inserting the value of $|\tilde{a}_{12}^{(3)}|$ thus calculated into Equation (33), the temperature-dependent relaxation time due to electron–hole scattering was calculated. When $\Delta\tilde{n}$ is equal to zero, this scattering mechanism causes the finite resistivity ρ_{eh} which follows the T^2 form in its temperature dependence. From Equations (31) and (33) we obtain

$$\rho_{eh} = \{4\pi^2\hbar^4[|a_{12}^{(3)}|/(k_B T)^2]/3n^2e^2b^4\gamma_b^2\}(k_B T)^2, \tag{42}$$

where $n = n_e = n_h$. The coefficient of the T^2 term in the present band model becomes $3.16 \times 10^{-9}\,\Omega\,\text{cm}\,\text{K}^{-2}$ [67]. This value is in good agreement with the experimental value of $5.6 \times 10^{-9}\,\Omega\,\text{cm}\,\text{K}^{-2}$ for $(SN)_x$ obtained by Chiang *et al.* [2]. Therefore, both magnitude and temperature dependence of the resistivity of $(SN)_x$

can be well explained by electron–hole scattering in the highly anisotropic Fermi surfaces.

7.2.2. *Carrier–Phonon Scattering*

We must next consider the relaxation time τ_i, Equation (32), in the band E_i. Since the carrier–carrier Umklapp scattering along the chain axis is inhibited owing to the geometrical configuration of the Fermi surfaces, only the carrier–phonon scattering contributes to the relaxation time $\tilde{\tau}_i$. Thus Oshiyama [67] calculated the relaxation time τ_i due to carrier–phonon scattering using Equation (18).

As for the phonon structure of $(SN)_x$, he simply assumed an isotropic Debye model; that is, he adopted a linear dispersion relation, $\omega_q = s|q|$, where s is an averaged sound velocity. In this model, when temperature is higher than the Debye temperature T_D, $1/\tau_i$ can be obtained as follows:

$$1/\tau_i = C_i^2 D_i(E_F) abc(k_B T)/8\hbar s^2(m_S + m_N), \tag{43}$$

where C_i, m_S, and m_N are the deformation potential constant, the mass of a sulphur atom, and the mass of a nitrogen atom, respectively. Here the Fermi level density of states of the band E_i, $D_i(E_F)$, is simply given, in the present model, by

$$D_1(E_F) = \begin{cases} 1/[\pi abc(\gamma_a\gamma_b)^{1/2}] & \text{for } E_F \geqslant \Delta + 2\gamma_c \\ (E_F - \Delta)/[2\pi abc\gamma_c(\gamma_a\gamma_b)^{1/2}] & \text{for } \Delta + 2\gamma_c \geqslant E_F \geqslant \Delta \end{cases} \tag{44a}$$

and

$$D_2(E_F) = 1/[\pi abc(\gamma_a\gamma_b)^{1/2}]. \tag{44b}$$

As regards the deformation energy constant of $(SN)_x$, its magnitude can be estimated from the dimensionless electron–phonon coupling constant λ [82]. That is, τ_i^{-1} can be related to λ through the electron–phonon matrix element and the Fermi level density of states [83]. Assuming $\tau_1 = \tau_2$, we have

$$1/\tau_i \sim \pi\lambda(k_B T)/2\hbar \tag{45}$$

in the case of four conduction bands. Further, the electron–phonon coupling constant λ is evaluated through the relation $D_{BS}(E_F) = D_{EX}(E_F)/(1 + \lambda)$, where D_{BS} and D_{EX} are the calculated and observed density of states, respectively. From self-consistent band structure results described in Section 3, we have found $D_{BS}(E_F) =$ 0.14 states (eV spin molecule)$^{-1}$, while $D_{EX}(E_F)$ is found to be 0.18 states (eV spin molecule)$^{-1}$ from the specific heat measurement [59]. From these values λ becomes 0.29 in the present case. By using Equations (43), (44), and (45) and the value of $s =$ 1.46×10^5 cm s^{-1} which is evaluated from the Debye temperature 170 K [59] of $(SN)_x$, the deformation potential constants, C_1 and C_2, are estimated to be 2.07 eV. Oshiyama [67] used this value of 2.07 eV for C_1 and C_2 together with $T_D = 170$ K in examining the effect of the carrier–phonon scattering.

By using these values of parameters, the inverse of the relaxation time τ_i^{-1}, Equation (43), at $T = 300$ K was found to be $\tau_1^{-1} = \tau_2^{-1} = 1.72 \times 10^{13}$ s^{-1} for the case of

$E_F \geqq \Delta + 2\gamma_c$ (or $\Delta\tilde{n} \leqq 0.065$). Since the calculated value of τ_{cc}^{-1} at $T = 300\,K$ is $3.64 \times 10^{14}\,s^{-1}$ for the case of $\Delta\tilde{n} = 0$, the contribution from carrier–phonon scattering to resistivity is much smaller than that from electron–hole scattering in the case of $(SN)_x$. When $\Delta\tilde{n} \neq 0$, however, the carrier–phonon scattering is expected to be important to resistivity, since the inverse of the relaxation time τ_{cc}^{-1} due to electron–hole scattering decreases with increasing $\Delta\tilde{n}$ and further the current carried by the center-of-mass motion is unable to be relaxed towards zero by electron–hole normal scattering.

7.2.3. *Resistivity of* $(SN)_x$ *and* $(SNBr_y)_x$

The resistivity of $(SN)_x$ and $(SNBr_y)_x$ was calculated with use of formula (31), considering both electron–hole and electron(hole)–phonon scattering mechanisms [67]. By using the calculated relaxation times, τ_{cc} and τ_i, the carrier concentrations and the optical masses, $m_e = m_h = h^2/(b^2\gamma_b)$, the calculation of the resistivity was carried out for several values of $\Delta\tilde{n}$ corresponding to the amount of charge transfer to the bromines per SN molecule. The calculated resistivity in the temperature

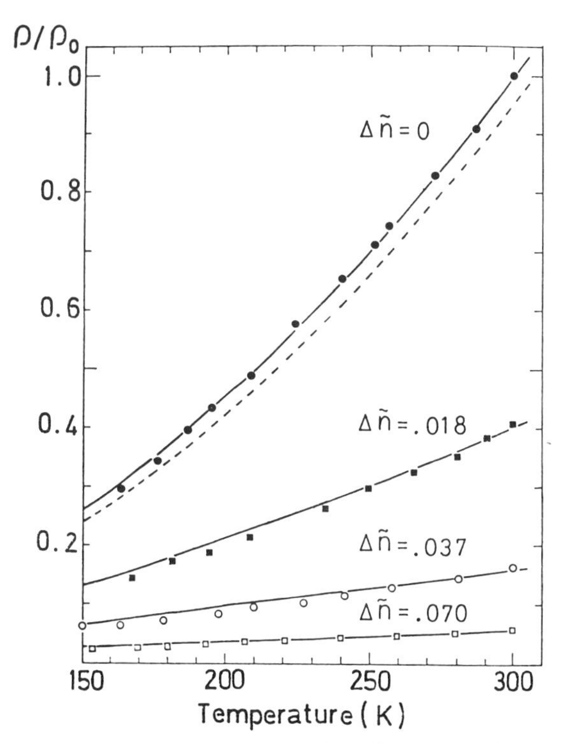

Fig. 18. The temperature dependence of the resistivity along the chain axis for several values of $\Delta\tilde{n}$. The scale of the ordinate is normalized to the magnitude at 300 K of the resistivity for $\Delta\tilde{n} = 0$, ρ_0. The broken curve represents the contribution from electron–hole scattering ρ_{eh}/ρ_0 for $\Delta\tilde{n} = 0$. The experimental data by Kaneto *et al.* [68] on $(SN)_x$ and $(SNBr_y)_x$ are also plotted: pristine $(SN)_x$ ●; $y \cong 0.15$ ■; $y \cong 0.28$ ○; and $y \cong 0.4$ □ [67]. (© The Institute of Physics)

region above $T_D = 170\,K$ is shown in Figure 18 for several values of $\Delta\tilde{n}$ [67]. The broken curve in this figure represents the resistivity ρ_{eh}, Equation (42), due to electron–hole scattering for the case of $\Delta\tilde{n} = 0$. It is seen from this figure that the electron–hole scattering is a dominant mechanism contributing to the resistivity for the case of $\Delta\tilde{n} = 0$: i.e. for the case of $(SN)_x$. Since the contribution to resistivity from electron(hole)–phonon scattering is expected to vary as T^5 in the temperature region much lower than T_D, the observed T^2 form of the resistivity of $(SN)_x$ in the temperature region below 10–300 K can be explained by electron–hole scattering term ρ_{eh}.

On the other hand, for the case of $\Delta\tilde{n} \neq 0$, electron(hole)–phonon scattering plays an important role, and the temperature dependence of the resistivity changes from T^2 to T linear form with the increasing $\Delta\tilde{n}$. This behavior is well explained by the following two facts. First, in the case of $\Delta\tilde{n} = 0$ no net charge is carried by the center-of-mass motion, while in the case of $\Delta\tilde{n} \neq 0$, this motion carries the net charge of $-e(n_e - n_h)$ and can be attenuated not by electron–hole scattering but by electron(hole)–phonon scattering. Second, the inverse of the relaxation time τ_{cc}^{-1} due to electron–hole scattering decreases with increasing $\Delta\tilde{n}$ owing to both the expansion of the hole portion and the shrinkage of the electron portion of the Fermi surfaces. In order to compare the calculated results with the experimental ones, we also plot the experimental data by Kaneto *et al.* [68] on the resistivity of $(SN)_x$, $(SNBr_{0.15})_x$, $(SNBr_{0.28})_x$, and $(SNBr_{0.4})_x$ in Figure 18. The agreement between the calculated and the experimental results is fairly good in spite of the simplification of the present two-band model. From the value of $\Delta\tilde{n}$ in the calculated curve, and that of y in the experimental data on $(SNBr_y)_x$, the net charge of a bromine, $\Delta\tilde{n}/y$, in $(SN)_x$ can be estimated to be 0.12, 0.13, and 0.17 for the three calculated curves: $Br^{-1.2\sim-0.17}$. This value is consistent with the experimental results of the extended X-ray absorption fine structure (EXAFS) [66] and of Raman scattering [64, 65] which indicate the existence of Br_3^- and Br_5^- ions in addition to Br_2 molecules.

In ordinary cases, this drastic change in both the magnitude and the temperature dependence of the resistivity is expected to cause a significant change in the plasma edge. This is not the case in $(SN)_x$ and $(SNBr_y)_x$, as shown below [84]. The motion of electrons and holes in a electron–hole system under the polarized light, $\mathbf{E}(t) = \mathbf{E}_0 e^{-i\omega t}$, is described by the following simple kinetic equations for the total velocities of electrons and holes, $\mathbf{V}_1(t) = \mathbf{V}_1^0 e^{-i\omega t}$ and $\mathbf{V}_2(t) = \mathbf{V}_2^0 e^{-i\omega t}$:

$$-n_1 e\mathbf{E}(t) = m\frac{d\mathbf{V}_1(t)}{dt} + \frac{m\mathbf{V}_1(t)}{\tau_1} + \frac{\tau_{cc}}{M}\left(\frac{\mathbf{V}_1(t)}{n_1} - \frac{\mathbf{V}_2(t)}{n_2}\right) \tag{46a}$$

and

$$n_2 e\mathbf{E}(t) = m\frac{d\mathbf{V}_2(t)}{dt} + \frac{m\mathbf{V}_2(t)}{\tau_2} - \frac{M}{\tau_{cc}}\left(\frac{\mathbf{V}_1(t)}{n_1} - \frac{\mathbf{V}_2(t)}{n_2}\right), \tag{46b}$$

where masses of an electron and a hole are assumed to be equal to each other and are denoted by m. In fact, the effective masses of the electron and the hole in $(SN)_x$ and $(SNBr_y)_x$ are approximately equal to one another because of the symmetric structure of the conduction bands near the Fermi level. By solving Equations (46a) and (46b), the conductivity $\sigma(\omega)$ which is defined by $\sigma(\omega) = (-e\,|\mathbf{V}_1^0| + e\,|\mathbf{V}_2^0|)/|\mathbf{E}_0|$ can be written as

$$\sigma(\omega) = \frac{(\omega_{p1}^2 + \omega_{p2}^2)\,\tau_T}{4\pi(1 - i\omega\tau_T)}\left[1 + \left(\frac{n_1 - n_2}{n_1 + n_2}\right)^2 \frac{1}{\tau_{cc}/\tau - i\omega\tau_{cc}}\right], \tag{47}$$

where

$$\frac{1}{\tau_T} = \frac{1}{\tau_{cc}} + \frac{1}{\tau} \tag{48}$$

and

$$\omega_{pi}^2 = 4\pi n_i e^2/m. \tag{49}$$

Here τ_1 is assumed to be equal to τ_2 and they are denoted by τ. Since τ_{cc} is of the order of 10^{-14} s, the second term in the square bracket in Equation (47) can be neglected for the frequency region around the plasma edge which is determined to be about 5 eV from the reflectivity data [9–12, 39, 70]. Therefore, it is seen from Equation (47) that the plasma frequency in this electron–hole system – that is, $\omega_{p1}^2 + \omega_{p2}^2$ – is expected to be essentially unchanged upon bromination, since the total concentration of carriers, $n_1 + n_2$, remains approximately constant, as was pointed out in Section 5.

8. Summary

Let us summarize the main results with regard to the band structures, the Fermi surfaces, optical properties, and resistivity of $(SN)_x$ and $(SNBr_y)_x$ which we have reviewed in this article.

The self-consistent numerical basis set LCAO method within the local density functional formalism has determined the band structure, the charge transfer, and the Fermi surfaces of $(SN)_x$ from the first principles, and has settled the long-standing dispute on the shape of the Fermi surfaces of $(SN)_x$. The charge transfer from the sulfurs to the nitrogens is 0.4 electron per SN molecule: i.e. $(S^{+0.4}N^{-0.4})_x$. This is in agreement with the experimental value. The band structure has a quasi-one-dimensional feature but interchain interactions give rise to the highly anisotropic Fermi surface. The Fermi surfaces consist of two electron and two hole pockets near the top surfaces of the Brillouin zone. The two electron pockets and one of the hole pockets have the shape of flattened tubes, and the other hole pocket has the shape of a flattened ellipsoid. Although these four pockets have some plane-like regions over their surface area, they have a closed shape owing to interchain interactions. As a result, $(SN)_x$ becomes superconducting without exhibiting the Peierls metal-insulator transition. On the other hand, the large anisotropy in the normal state properties can be attributed to the plane-like feature of the Fermi surfaces.

The calculated density of states of the valence bands agrees well with the XPS data. The Fermi level density of states is also in reasonable agreement with the experimental value obtained from the result of the specific heat measurement. Further, the calculated spectra are in good agreement with experimental optical spectra. Moreover, the Fermi surfaces and the optical absorption spectra of $(SNBr_y)_x$ have been calculated in terms of the rigid band model based on the band structure of $(SN)_x$. The result shows that, upon bromination, the hole portion of the Fermi surfaces expands while the electron portion shrinks. This variation of the Fermi surface, together with the change of the optical spectra, has clarified the effect of bromination on the electronic properties of $(SN)_x$.

The calculation of resistivity along the chain axis of $(SN)_x$ and $(SNBr_y)_x$ has been performed for the Fermi surfaces of these materials. Both the observed T^2 temperature dependence and magnitude of the resistivity of $(SN)_x$ have been well explained by electron–hole scattering. Moreover, it has been shown that electron(hole)–phonon scattering becomes important upon bromination and thus the drastic change of temperature and magnitude of the resistivity upon bromination is brought about. The characteristic feature of the variation of temperature dependence with increasing bromine concentration has been reproduced well by the calculated results. This agreement between the calculated and experimental results, in turn, has confirmed the validity of the obtained self-consistent band structure results of $(SN)_x$.

Acknowledgements

One of the authors (H.K.) would like to acknowledge the collaboration of Dr A. D. Yoffe, A. J. Grant, and F. Lévy in the early stage of the work reviewed in this article when H.K. stayed in the Cavendish Laboratory from 1974 to 1975, and stimulating discussions and encouragement by them on the whole subject of the present article. Both of the authors also would like to thank Dr Y. Natsume, Dr K. Kaneto, and Professors Y. Inuishi, K. Yoshino, T. Mitani, T. Koda, and K. Nakao for their valuable discussion on the subject of this article.

References

1. V. V. Walatka, M. M. Labes, and J. H. Perlstein, *Phys. Rev. Lett.* **31**, 1139 (1973).
2. C. K. Chiang, M. J. Cohen, A. F. Garito, A. J. Heeger, C. M. Mikulski, and A. G. MacDiarmid, *Sol. St. Commun.* **18**, 1451 (1976).
3. H. Kahlert and K. Seeger, *Proc. 13th Int. Conf. Physics Semiconductors, Rome*, Tipographia Marves, Rome, p. 353 (1976).
4. R. H. Friend, D. Jerome, S. Rehmatullah, and A. D. Yoffe, *J. Phys.* **C10**, 1001 (1976).
5. K. Kaneto, Ph.D. Thesis, University of Osaka (1976); K. Kaneto, M. Yamamoto, K. Yoshino, and Y. Inuishi, *J. Phys. Soc. Japan* **47**, 167 (1979).
6. W. Beyer, W. D. Gill, and G. B. Street, *Sol. St. Commun.* **27**, 185 (1978).
7. M. Yamamoto, K. Kaneto, K. Yoshino, and Y. Inuishi, *Sol. St. Commun.* **29**, 541 (1979).

8. C. H. Chen, J. Silcox, A. F. Garito, A. J. Heeger, and A. G. MacDiarmid, *Phys. Rev. Lett.* **36**, 525 (1976).
9. A. A. Bright, M. J. Cohen, A. F. Garito, A. J. Heeger, C. M. Mikulski, P. J. Russo, and A. G. MacDiarmid, *Phys. Rev. Lett.* **34**, 206 (1975).
10. L. Pintschovius, H. P. Geserich, and W. Möller, *Sol. St. Commun.* **17**, 477 (1975).
11. P. M. Grant, R. L. Greene, and G. B. Street, *Phys. Rev. Lett.* **35**, 1743 (1975).
12. W. Möller, H. P. Geserich, and L. Pintschovius, *Sol. St. Commun.* **18**, 791 (1976).
13. R. L. Greene, G. B. Street, and L. J. Suter, *Phys. Rev. Lett.* **34**, 577 (1975).
14. W. D. Gill, R. L. Greene, G. B. Street, and W. A. Little, *Phys. Rev. Lett.* **35**, 1732 (1975).
15. R. Civiak, W. Junker, C. Elbaum, H. I. Kao, and M. M. Labes, *Sol. St. Commun.* **17**, 1573 (1975).
16. W. H. G. Müller, F. Baumann, G. Dammer, and L. Pintschovious, *Sol. St. Commun.* **25**, 119 (1978).
17. L. R. Bickford, R. L. Greene, and W. D. Gill, *Phys. Rev.* **B17**, 3525 (1978).
18. L. J. Azvedo, W. G. Clark, G. Deutscher, R. L. Greene, G. B. Street, and L. J. Suter, *Sol. St. Commun.* **19**, 197 (1976).
19. R. H. Dee, D. H. Dollard, B. G. Turrell, and J. F. Carolan, *Sol. St. Commun.* **24**, 469 (1977).
20. Y. Oda, H. Takenaka, H. Nagano, and I. Nakada, *Sol. St. Commun.* **32**, 659 (1979).
21. H. Kamimura, *Proc. 13th Int. Conf. Physics Semiconductors, Rome*, Tipografia Marves, Rome, p. 51 (1976).
22. R. L. Greene and G. B. Street, *Chemistry and Physics of One-Dimensional Metals* (ed. H. J. Keller), Plenum, New York, p. 167 (1977).
23. A. D. Yoffe, *Chem. Soc. Rev.* **5**, 51 (1976).
24. D. E. Parry and J. M. Thomas, *J. Phys.* **C8**, L45 (1975).
25. H. Kamimura, A. J. Grant, F. Levy, A. D. Yoffe, and G. D. Pitt, *Sol. St. Commun.* **17**, 49 (1975).
26. A. Zunger, *J. Chem. Phys.* **63**, 4854 (1975).
27. W. I. Friesen, A. J. Berlinsky, B. Bergerson, L. Weiler, and T. M. Rice, *J. Phys.* **C8**, 3549 (1975).
28. V. T. Rajan and L. M. Falicov, *Phys. Rev.* **B12**, 1240 (1975).
29. M. Kertesz, J. Koller, A. Azman, and S. Suhai, *Phys. Lett.* **55A**, 107 (1975).
30. M. Schlüter, J. R. Chelikowsky, and M. L. Cohen, *Phys. Rev. Lett.* **35**, 869 (1975) and **36**, 452 (1976).
31. A. A. Bright and P. Soven, *Sol. St. Commun.* **18**, 317 (1976).
32. H. Kamimura, A. M. Glazer, A. J. Grant, Y. Natsume, M. Schreiber, and A. D. Yoffe, *J. Phys.* **C9**, 291 (1976).
33. W. E. Rudge and P. M. Grant, *Phys. Rev. Lett.* **35**, 1799 (1975).
34. C. Merkel and J. Ladik, *Phys. Lett.* **56A**, 395 (1976).
35. D. R. Salahub and R. P. Messmer, *Phys. Rev.* **B14**, 2592 (1976).
36. W. Y. Ching, J. G. Harrison, and C. C. Lin, *Phys. Rev.* **B15**, 5975 (1977).
37. I. P. Batra, S. Ciraci, and W. E. Rudge, *Phys. Rev.* **B15**, 5858 (1977).
38. L. Mihich, *Sol. St. Commun.* **28**, 521 (1978).
39. W. D. Gill, W. Bludau, R. H. Geiss, P. M. Grant, R. L. Greene, J. J. Mayerle, and G. B. Street, *Phys. Rev. Lett.* **38**, 1305 (1977).
40. A. Oshiyama and H. Kamimura, *J. Phys. C: Solid State Phys.* **14**, 5091 (1981).
41. P. Mengel, P. M. Grant, W. E. Rudge, and B. H. Schechtman, *Phys. Rev. Lett.* **35**, 1803 (1975).
42. W. R. Salaneck, J. W.-P. Lin, and A. J. Epstein, *Phys. Rev.* **B13**, 5574 (1976).
43. P. Mengel, I. B. Ortenburger, W. E. Rudge, and P. M. Grant, *Proc. Int. Conf. Organic Conductors and Semiconductors, Siofok*, Springer-Verlag, p. 591 (1976).
44. For a review see G. B. Street and R. L. Greene, *IBM J. Res. Dev.* **21**, 98 (1977).
45. M. Boudeulle, Ph.D. Thesis, University of Lyon (1974) and *Cryst. Struct. Commun.* **4**, 9 (1975); M. Boudeulle and P. Michel, *Acta Cryst.* **A28**, 199 (1972).
46. M. J. Cohen, A. G. Garito, A. J. Heeger, A. G. MacDiarmid, C. M. Mikulski, M. S. Saran, and J. Kleppinger, *J. Am. Chem. Soc.* **98**, 3844 (1976).
47. P. Hohenberg and W. Kohn, *Phys. Rev.* **B136**, 864 (1964).
48. W. Kohn and L. J. Sham, *Phys. Rev.* **A140**, 1133 (1965).
49. D. E. Ellis and G. S. Painter, *Phys. Rev.* **B2**, 2887 (1970).

50. G. S. Painter and D. E. Ellis, *Phys. Rev.* **B1**, 4747 (1970).
51. F. W. Averill and D. E. Ellis, *J. Chem. Phys.* **59**, 6412 (1973).
52. B. Y. Tong and L. J. Sham, *Phys. Rev.* **144**, 1 (1966); N. D. Lang, *Sol. St. Phys.* **28**, 225 (1973); T. Ando, *Phys. Rev.* **B13**, 3468 (1976); O. Gunnarsson and B. I. Lundqvist, *Phys. Rev.* **B13**, 4274 (1976).
53. K. S. Singwi, A. Sjolander, P. M. Tosi, and R. H. Land, *Phys. Rev.* **B1**, 1044 (1970).
54. L. Hedin and B. I. Lundqvist, *J. Phys.* **C4**, 2064 (1971).
55. R. S. Mulliken, *J. Chem. Phys.* **23**, 1841 (1955).
56. H. Conroy, *J. Chem. Phys.* **47**, 5307 (1967).
57. L. Pintschovius, H. Wendel, and H. Kahlert, *Proc. Int. Conf. Organic Conductors and Semiconductors, Siofok*, Springer-Verlag, p. 589 (1976); L. Pintschovius and R. Pynn, *Proc. Int. Conf. Quasi-One-Dimensional Conductors, Dubrovnik*, Springer-Verlag, p. 421 (1978).
58. L. Ley, *Phys. Rev. Lett.* **35**, 1796 (1975).
59. J. M. E. Harper, R. L. Greene, P. M. Grant, and G. B. Street, *Phys. Rev.* **B15**, 539 (1977).
60. J. Bordas, A. J. Grant, H. P. Hughes, A. Jakobson, H. Kamimura, F. A. Levy, K. Nakao, Y. Natsume, and A. D. Yoffe, *J. Phys.* **C9**, L277 (1976).
61. T. Mitani, H. Mori, S. Suga, T. Koda, S. Shin, K. Inoue, I. Nakada, and H. Kanzaki, *J. Phys. Soc. Japan* **47**, 679 (1979); T. Mitani, private communication.
62. P. M. Grant, W. E. Rudge, and I. B. Ortenburger, *Proc. Int. Conf. Organic Conductors and Semiconductors, Siofok*, Springer-Verlag, p. 575 (1976).
63. H. Ehrenreich and M. H. Cohen, *Phys. Rev.* **115**, 786 (1959).
64. Z. Iqbal, R. H. Baughman, J. Kleppinger, and A. G. MacDiarmid, *Sol. St. Commun.* **25**, 409 (1978).
65. Z. Iqbal, J. Sharma, R. H. Baughman, M. Akhtar, and A. G. MacDiarmid, *Proc. Int. Conf. Quasi-One-Dimensional Conductors, Dubrovnik*, Springer-Verlag, p. 432 (1978).
66. H. Morawitz, W. D. Gill, P. M. Grant, G. B. Street, and D. Sayers, *Proc. Int. Conf. Quasi-One-Dimensional Conductors, Dubrovnik*, Springer-Verlag, p. 390 (1978).
67. A. Oshiyama, *J. Phys. C: Solid State Phys.* **14**, 5109 (1981).
68. K. Kaneto, K. Yoshino, and Y. Inuishi, *Proc. Symp. Design of Inorganic and Organic Materials of Technological Importance, Kyoto* (1979).
69. A. Oshiyama, K. Nakao, and H. Kamimura, *J. Phys. Soc. Japan* **45**, 1136 (1978).
70. C. K. Chiang, M. J. Cohen, D. L. Peebles, A. J. Heeger, M. Akhtar, J. Kleppinger, A. G. MacDiarmid, J. Milliken, and M. J. Moran, *Sol. St. Commun.* **23**, 607 (1977).
71. J. T. Schriempf, *J. Phys. Chem. Solids* **28**, 2851 (1967); *Phys. Rev. Lett.* **19**, 1131 (1967).
72. R. Hartman, *Phys. Rev.* **181**, 1070 (1969).
73. R. Jullien, M. T. Béal-Monod, and B. Coqblin, *Phys. Rev.* **B9**, 1441 (1974).
74. A. H. Thompson, *Phys. Rev. Lett.* **35**, 1786 (1975).
75. G. A. Brooker and J. Sykes, *Phys. Rev. Lett.* **21**, 279 (1968).
76. H. H. Jensen, H. Smith, and J. W. Wilkins, *Phys. Rev.* **185**, 323 (1969); H. Smith and J. W. Wilkins, *Phys. Rev.* **183**, 624 (1969).
77. W. E. Lawrence and J. W. Wilkins, *Phys. Rev.* **B7**, 2317 (1973).
78. L. P. Gor'kov and I. E. Dzyaloshinskii, *JETP Lett.* **18**, 403 (1973).
79. W. G. Baber, *Proc. R. Soc. London* **A158**, 383 (1937).
80. M. J. Rice, *Phys. Rev. Lett.* **20**, 1439 (1968).
81. C. A. Kukkonen and P. F. Maldague, *Phys. Rev. Lett.* **37**, 782 (1976); P. F. Maldague and C. A. Kukkonen, preprint.
82. W. L. McMillan, *Phys. Rev.* **167**, 331 (1968).
83. J. J. Hopfield, *Comm. Sol. St. Phys.* **3**, 48 (1970); *Superconductivity in d- and f-Band Metals*, (ed. D. H. Douglass), Am. Inst. Phys., New York, p. 358.
84. A. Oshiyama, Ph.D. Thesis, University of Tokyo (1980).

THE QUASI-ONE-DIMENSIONAL CHALCOGENIDES AND HALIDES OF TRANSITION ELEMENTS

D. W. BULLETT

School of Physics, University of Bath,
Claverton Down, Bath BA2 7AY, England.

1. Introduction

In this article we survey the pseudo-one-dimensional compounds of the transition elements, looking in particular at the consequences for the band structures and electronic properties of the connectivity of atoms in linear chains in these compounds. The unique transport properties of one such material, niobium triselenide, have attracted so much recent experimental and theoretical attention [1–18] that they demand a separate, detailed discussion in the following article by Shima and Kamimura, while the characteristic features of the cyanoplatinate family of compounds are presented elsewhere [19–23].

1.1. One-dimensional compounds and metal-metal bonding

A wide variety of quasi-one-dimensional compounds are now known, especially among the oxides, chalcogenides and halides of the 4d and 5d transition and rare earth elements [24–28]. The simplest description of their geometrical structures is usually in terms of the condensation of small clusters of atoms into infinite chains [24]. Thus in the trichalcogenides, MX_3, trigonal prismatic MX_6 units (Figure 1) condense into linear chains by sharing triangular faces (whereas in the layered dichalcogenides MX_2 reviewed in the related books of Series A [29], octahedral or trigonal prismatic MX_6 units condense into infinite two-dimensional sheets). In many cases these one-dimensional compounds would be regarded as metal-rich according to the traditional rules of valence electron transfer from metal atoms to the surrounding anions; the 4d and 5d elements are able to use the remaining valence electrons to form metal–metal (M–M) bonds, either within or between the clusters. Very often the chemical and electronic properties of these materials are dominated by the influence of these M–M bonding interactions. As Simon [24] has pointed out, the basic structural principle is simple: when the number of nonmetal atoms in a compound is not sufficient to completely surround the metal cluster, the latter clusters link up (or 'condense') via direct M–M bonds.

1.2. Electron band structure methods

This article reviews the results of electronic structure calculations without presenting the various methods of calculation in any detail. The inputs to such calculations

H. Kamimura (ed.), Theoretical Aspects of Band Structures and Electronic Properties of Pseudo-One-Dimensional Solids, 163–230.

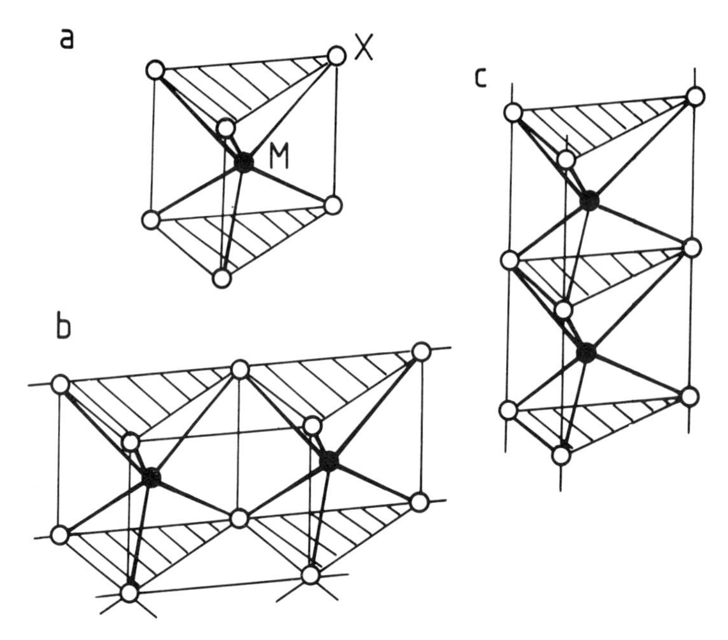

Fig. 1. (a) Trigonal prismatic MX_6 clusters, condensed (b) laterally into the two-dimensional layers of the $NbSe_2$ structure, and (c) longitudinally into the one-dimensional columns of the $NbSe_3$ structure.

are the chemical composition, crystal structure, and some approximation to the potential which an electron feels at any point within the unit cell. The outputs are the one-electron energy levels and crystal wave functions. In principle these determine all the physical properties of the system, so that from the results one should be able to understand and predict whether this particular atomic arrangement should be especially stable, whether the compound will behave as a semiconductor or a metal, whether a metallic compound is likely to be a particularly good superconductor, whether the structure is likely to go through soft-mode phase transitions under variations of external conditions (such as temperature and pressure) or internal changes of composition, etc.

Band structure methods differ principally in the extent to which they contain adjustable or undetermined parameters [30, 31]. Completely ab initio calculations at the Hartree–Fock level or beyond are economically practicable only for very simple structures. The extended Hückel model [23, 32, 33] aims to approximate these calculations by assuming that matrix elements of the total energy operator H between any pair of atomic orbitals φ_i and φ_j are directly proportional to the overlap of the two orbitals

$$\int \varphi_i^* H \varphi_j \, \mathrm{d}V = \frac{I_i + I_j}{2} K \int \varphi_i^* \varphi_j \, \mathrm{d}V,$$

where the constant of proportionality involves a semi-empirically chosen parameter K (usually in the range $1 < K < 2$) and the ionization potentials I_i and I_j to remove an electron from each orbital in the isolated atom. Given this assumption one can easily deduce the electronic structure of very complicated atomic arrangements containing perhaps fifty or more atoms in the unit cell.

Less drastic assumptions are made in those methods [29, 34–46] that approximate the exchange correlation part of the potential influencing an electron in the solid by the equivalent exchange correlation potential in a free electron gas of the same local density. The Coulomb part of the potential is found by solving Poisson's equation for a given charge distribution $\rho(\mathbf{r})$, often constructed by spherically averaging the overlapping charge clouds of individual atoms. Added to this electrostatic potential, the local density approximation for the exchange correlation potential provides a well-defined effective field within which each electron moves. Self-consistent calculations adopt an iterative approach forcing the charge distribution $\rho(\mathbf{r})$ from which the effective potential is constructed to be consistent with the output charge distribution formed by occupying the calculated valence band of wave functions.

Various procedures are available for calculating the wave functions and energies in a given one-electron potential. In the simplest approximation [6, 22, 30] we evaluate the interactions between any pair of atomic orbitals in the system by direct numerical integration and use these matrix elements to set up the secular determinant for Bloch waves in the crystal, neglecting all three-centred effects. This two-centred approximation can often provide an invaluable visual framework in which to picture the competing interactions that make up the bonding in the solid [30, 31]. Where the two-centred approximation is thought too severe, we can adopt a three-dimensional Diophantine integration scheme in which a set of pseudo-random integration sampling points within a unit cell, and an associated set of weight functions, are chosen to evaluate the matrix elements between numerical potentials and Bloch functions without neglecting any multicentred terms [34–36]. In the self-consistent-field Xα scattered-wave method [37] the pictorial advantages of atomic orbital overlaps are discarded in favour of a rigorous wave matching condition, to make expansions in energy-dependent partial waves continuous at a spherical boundary around each atom or group of atoms.

The so-called linear methods of band theory [38, 41] combine the advantages of the methods employing fixed basis functions with those of the partial wave methods: they use energy-independent basis functions derived from the partial waves to provide solutions of high accuracy. For fairly complex structures the most popular of these methods is the so-called 'linear muffin-tin-orbital' or LMTO method [42–46] in the atomic sphere approximation that treats the crystal potential as spherically symmetric within each Wigner–Seitz sphere.

2. Metal–Metal Pairing in NbI_4

2.1. THE NbI_4 CRYSTAL STRUCTURE

One of the simplest examples of M–M pairing of transition element atoms in a one-dimensional chain is provided by NbI_4 [47–51] (and by its isomorphous cousin TaI_4

[48]). Fine needle-like crystals of NbI_4 may be prepared [47, 48] by heating niobium penta-iodide to 270–300 °C at one end of a sealed tube and collecting the evolved iodine in the other end of the tube at about 35 °C. NbI_4 is a quasi-one-dimensional semiconductor at atmospheric pressure; the electrical resistivity $R_{\|}$ along the needle axis is about $3 \times 10^4\,\Omega\,m$ at room temperature but the resistivity $R_{\perp}$ perpendicular to this direction is about five times large [50].

The structure of the low-temperature ($T < 348\,°C$) form, α-NbI_4, has been determined by Dahl and Wampler [49]. The crystals are orthorhombic, space group $Cmc2_1$ (C_{2v}^{12}), with eight formula units per cell. The most significant aspect of the structure for present purposes is the occurrence of infinite chains of NbI_6 octahedra sharing opposite edges (Figure 2). Iodine atoms are arranged in the unit cell in an approximately hexagonal close-packed array, with one in four of the available octa-

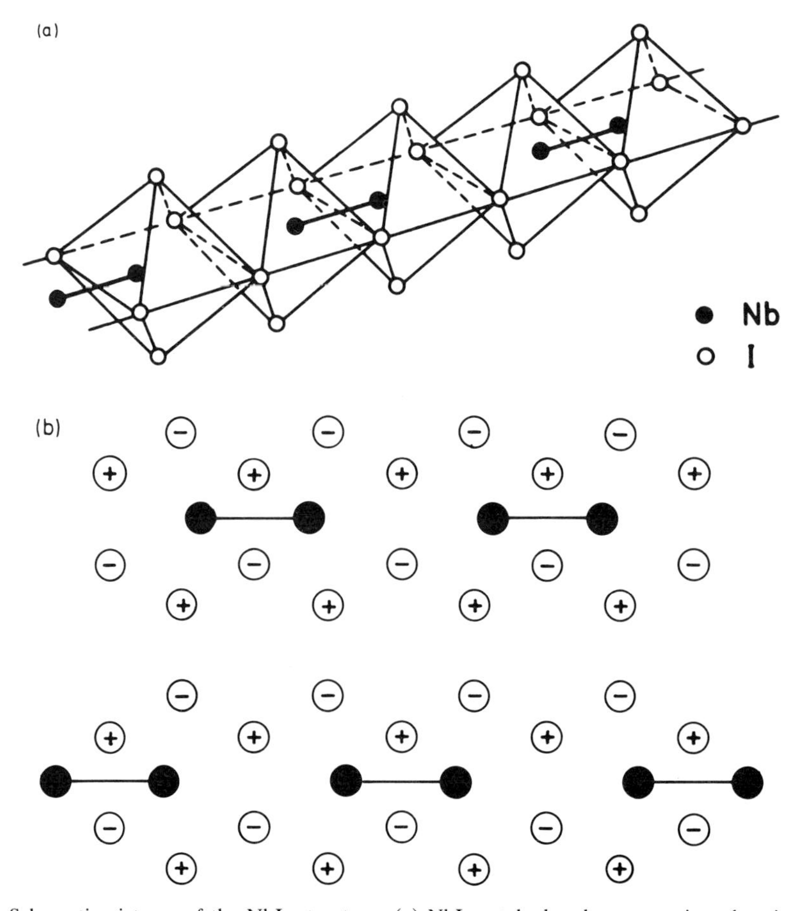

Fig. 2. Schematic pictures of the NbI_4 structure. (a) NbI_6 octahedra share opposite edges in a linear chain; Nb atoms are displaced from the octahedron centres to form metal–metal bonds along the chain axis. (b) Nb atoms occupy alternate rows of interstitial sites in alternate layers of a close-packed iodine lattice. (Bullett [51].) (© American Chemical Society.)

hedral interstices occupied by niobium atoms. The transition element forms linear chains by filling alternate rows of interstices in alternate layers. Niobium atoms are displaced 0.26 Å from the centre of the iodine octahedra, to give alternate short (3.3 Å) and long (4.3 Å) Nb–Nb distances (cf. 2.86 Å in niobium metal) along the chain.

2.2. Metal–metal Bonding and the Electronic Structure of NbI_4

The driving force responsible for this metal–metal pair bonding reveals itself very clearly when electronic structure calculations are carried out in an atomic orbital basis [51]. We start from *s*, *p*, and *d* valence level orbitals on the transition element atom in the appropriate electron configuration, and *s* and *p* orbitals on the halogen. Making the two-centre approximation we can then calculate the rate of decay of the matrix element between each pair of orbitals in the system, as a function of the separation between atomic sites; in practice these interactions are negligibly small beyond about 5 Å. As an example, Figure 3 shows the matrix elements $dd\sigma$, $dd\pi$, and $dd\delta$ for the three possible orientations of niobium *d* orbitals quantized along

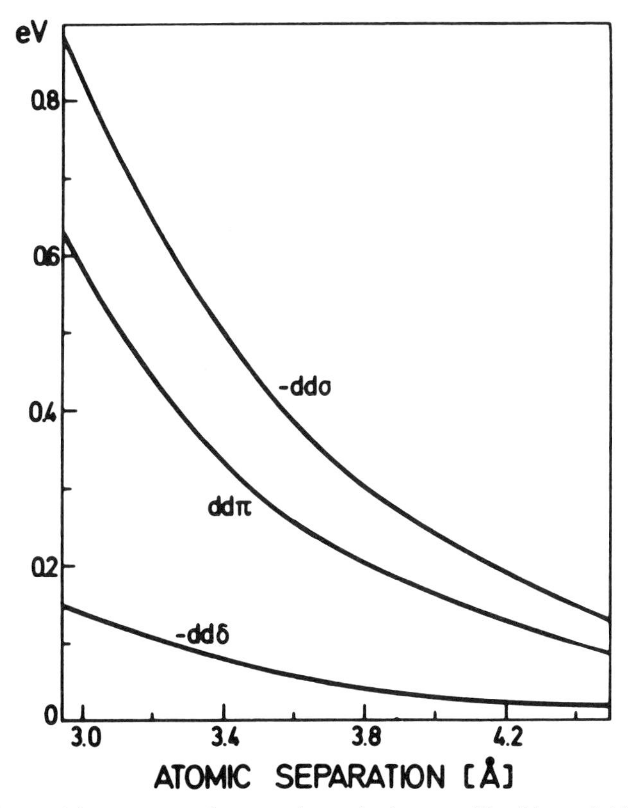

Fig. 3. Rate of decay of the two-centre interatomic matrix elements $dd\sigma$, $dd\pi$, and $dd\delta$ as a function of the separation between two Nb atoms. (Bullett [51].) (© American Chemical Society.)

their mutual axis, as a function of Nb–Nb distance. For any given arrangement of atoms in the unit cell, these two-centre interactions may then be used to set up the secular matrix in Bloch form, and the band structure energies $E_n(\mathbf{k})$ and crystal wave functions $\psi_n(\mathbf{k})$ may be deduced by matrix diagonalization.

The calculated density of electron states [51] in the highest valence and lowest conduction bands is shown in Figure 4, both for the full three-dimensional unit cell and for an isolated chain taken from the NbI_4 structure. Either case produces essentially the same result. Centred at −16.6 eV and of width ~1.2 eV is the narrow band of states formed by iodine 5*s* orbitals. The main valence band centred at −8 eV is composed predominantly of iodine 5*p* orbitals. The *p* bandwidth of 3.6 eV for the single chain increases slightly to 4.1 eV when interactions between chains are included, and at the same time the sharp one-dimensional band-edge singularities are

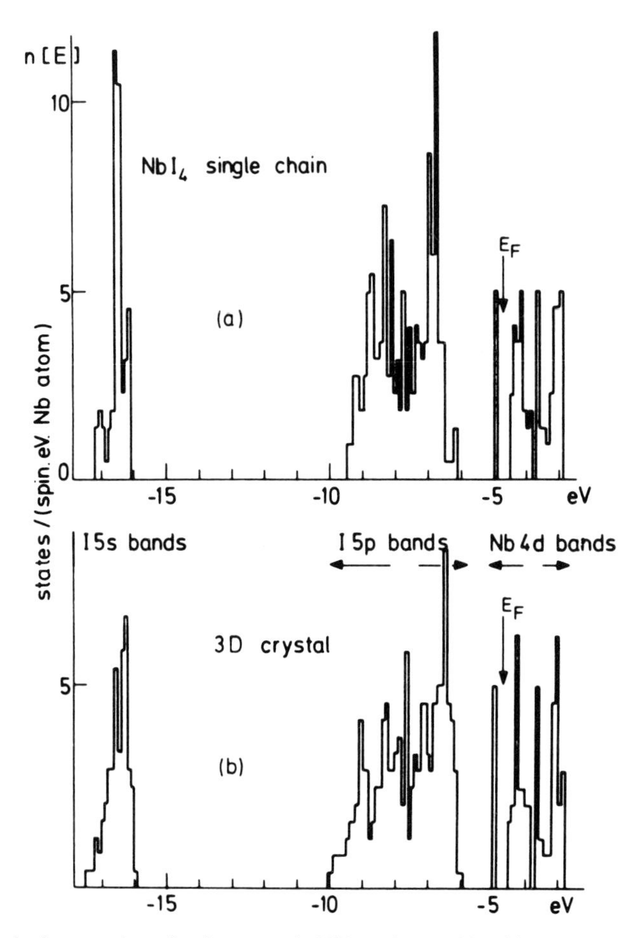

Fig. 4. Density of valence and conduction states in NbI_4, calculated for (a) an isolated chain and (b) for the full three-dimensional crystal (Bullett [51]). Isolated atom energy levels used in the calculation were $I_p = -7.7$ eV, $Nb_d = -5.0$ eV. The Fermi level E_F falls within the gap which separates M–M bonding states at −4.9 eV from higher Nb *d* states. (© American Chemical Society.)

partially smoothed out in the three-dimensional result. Next in order of energy come the group of narrow bands which are essentially the transition element *d* bands after hybridization with the ligand *p* orbitals. Interchain Nb–Nb distances are so large (~7 Å) that there is very little difference between the *d* bands calculated for an isolated chain or for coupled chains. The important point is that one *d* band is split off at a distinctly higher binding energy than the others, and forms the highest occupied band of M–M bonding states. Calculated energy gaps between highest occupied and lowest unoccupied states are 0.41 eV for the single chain and 0.38 eV for the full three-dimensional structure.

It is easy to see how the doubling of the unit cell along the chain direction, via pairing of Nb atoms, establishes a filled band of M–M bonds below the other empty *d* bands. Figure 5 compares the details of the one-dimensional energy bands near the Fermi level for the two chain geometries: 5(a) is for the observed single-chain geometry of NbI_4 while 5(b) shows the results for an idealized chain in which Nb atoms sit exactly at the centres of the coordinating iodine octahedra. The *d* states split in the octahedral field into the familiar triplet and doublet, separated in this case by ~1.2 eV. Within the triplet group of bands for the idealized chain, the d_{z^2}

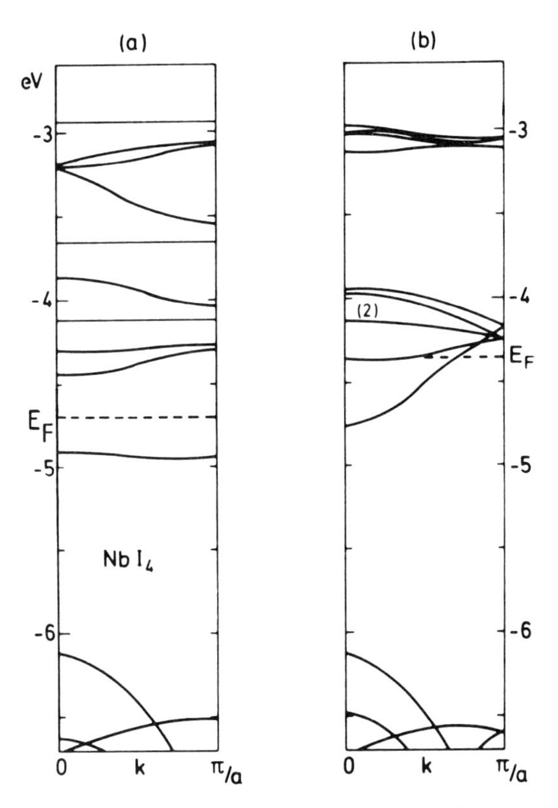

Fig. 5. One-dimensional energy bands near the Fermi level in NbI_4: (a) observed single chain of NbI_4; (b) NbI_4 chain containing no Nb–Nb pairing. (Bullett [51].) (© American Chemical Society.)

orbitals pointing along the chain axis form the main component of the broadest band, extending from -4.8 to -3.9 eV. This bandwidth may be readily understood from Figure 3, which shows that at a separation of $a/2 = 3.83$ Å, $dd\sigma$ is equal to -0.28 eV, and a purely d_{z^2} band in the regular one-dimensional chain would have a width of $4|dd\sigma| = 1.1$ eV. Hybridization with Nb s and p orbitals and with iodine p orbitals has the effect of reducing this slightly. In the distorted chain of the real NbI_4 structure, Nb–Nb separations are alternately 3.3 and 4.3 Å, and the d_{z^2} band splits into bonding and antibonding sub-bands. The alternating $dd\sigma$ interactions are then -0.56 and -0.16 eV, so that purely d_{z^2} sub-bands would extend in the ranges $E_d - 0.56 \pm 0.16$ and $E_d + 0.56 \pm 0.16$ eV. Although complicated rehybridization processes occur in the distorted octahedral environment, we can identify the lower sub-band with that at -5.0 eV in Figure 5(a). Complete occupation of this sub-band stabilizes the structure by lowering the average energy of the occupied d states. An analogous instability lies at the heart of the physical properties of many of the materials mentioned in this article.

Formally each Nb atom in NbI_4 donates one of its five valence electrons to each of four iodine atoms and uses the remaining electron to form a single pair bond to a neighbouring metal atom. In order to assess quantitatively the extent to which valence and conduction bands are really 'p bands' and 'd bands' one needs a procedure to divide the charge in each wave function between the different atomic sites. For an LCAO wave function an elementary assignment is to divide all overlap charges equally between the corresponding atoms [30]. The local atomic densities of states thus provide an assessment of ionicity; in the case of NbI_4 [51] approximately 0.5 electron is transferred from the Nb atom to each iodine atom, $Nb^{2+}(I^{0.5-})_4$.

The electrical properties of NbI_4, and their pressure dependence, have been studied by several groups. At atmospheric pressure Kepert and Marshall [50] recorded an activation energy for conduction of 0.12 eV. If the Fermi level lies near the middle of the semiconducting gap, this would suggest an energy gap ~0.24 eV, slightly smaller than the calculated band structure estimate. High-pressure studies by Kawamura *et al.* [52, 53] show the resistivity decreasing monotonically with pressure. By 130 kbar the activation energy has shrunk to 0.02 eV, and beyond about 150 kbar NbI_4 shows metallic bahaviour, with a typical positive linear temperature dependence of resistivity. An increase in effective dimensionality under pressure is typical of the behaviour of quasi-low-dimensional materials. Presumably under increasing pressure the separate chains are forced closer together, in this case broadening the iodine valence bands and thus tending to push up the Nb–Nb bonding band. At the same time the compression of each octahedron within a chain forces the transition element back towards the centre of the octahedron, until eventually indirect band overlap occurs between the lowest Nb–Nb bonding band and the next higher d band.

Similar crystal structures are adopted by niobium tetrabromide [54, 55] and tetrachloride [55, 56], and by the tantalum tetrahalides. The smaller the size of the anion, the stronger are the M–M bonds as the paired metal atoms may come closer together. Thus in the series NbI_4, $NbBr_4$, $NbCl_4$ the activation energies for conduction increase in the sequence 0.12, 0.37, 0.44 eV [50].

3. Transition Metal Trichalcogenides

The trichalcogenides of the group IV and group V transition elements MX_3 all cling to a common structural element, the trigonal prismatic column, which bequeaths on them, to a greater or lesser extent, quasi-one-dimensional properties. Whereas the group IV trichalcogenides [57–61] are isostructural, from TiS_3 right through to $ThTe_3$, small structural differences among the group V family are accompanied by major qualitative changes in their electronic structure and properties [62–64].

None of these materials is structurally one-dimensional to the same degree as the NbI_4 arrangement that we have just discussed; all contain substantial structural M–X cross-linking between chains and one could argue that they are as close to being layer structures as true one-dimensional materials. Most of these compounds crystallize monoclinically in fibrous strands or filamentary ribbon-shaped platelets, with the chain direction along the crystallographic b-axis (see Figure 6). Each column is displaced from its two immediate neighbours by half the trigonal prism height along the b-axis. Thus the columns are bound together in layers and the crystals cleave easily between the layers, which are held together only by weak van der Waals-type bonding between adjacent layers of chalcogen atoms.

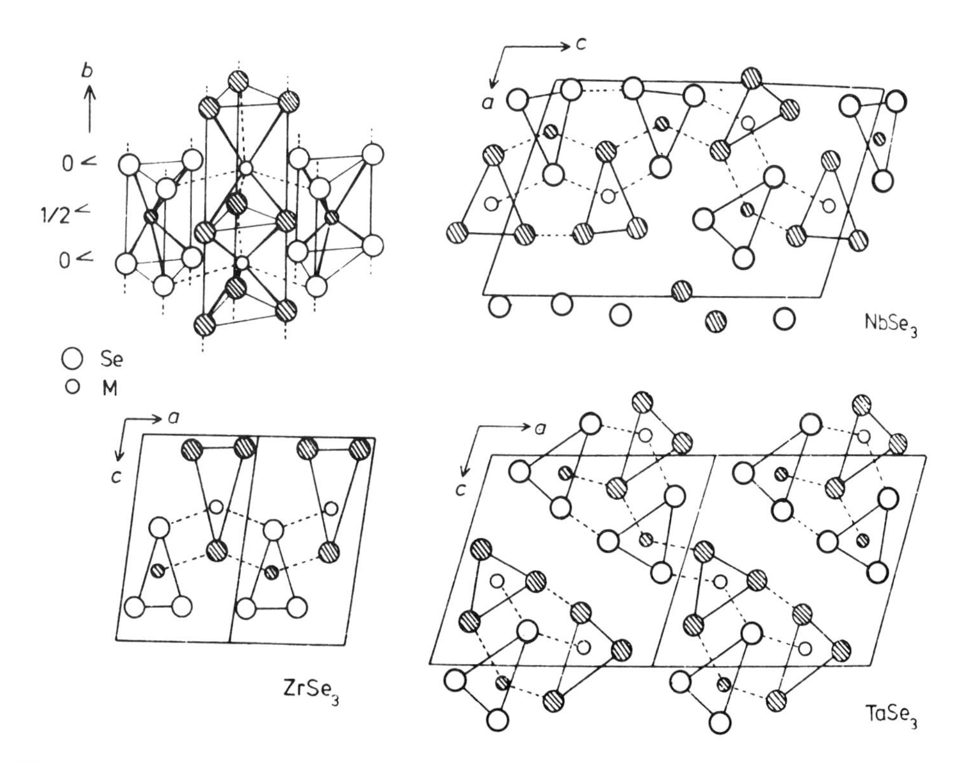

Fig. 6. Atomic arrangements in MSe_3 crystals: stacking of prisms along the b-axis and projections in the ac plane of the $ZrSe_3$, $NbSe_3$, and $TaSe_3$ structures. Open atoms at 0, shaded atoms at $\frac{1}{2}b$. [62]. (© The Institute of Physics.)

3.1. Trichalcogenides of Group IV

3.1.1. MX_3 *Structure for Group IV Transition Metals*

Brattås and Kjekshus [58] reported the easy synthesis of TiS_3, ZrS_3, $ZrSe_3$, $ZrTe_3$, HfS_3, $HfSe_3$ and $HfTe_3$ by direct reactions between the elements, and subsequently reported the crystal structures of these phases [59]. Despite numerous attempts they were unable to prepare phases corresponding to the formulae $TiSe_3$ and $TiTe_3$; presumably the mismatch of atomic radii in these two cases is just too great.

The monoclinic unit cell (space group $P2_1/m = C_{2h}^2$) contains two formula units, in

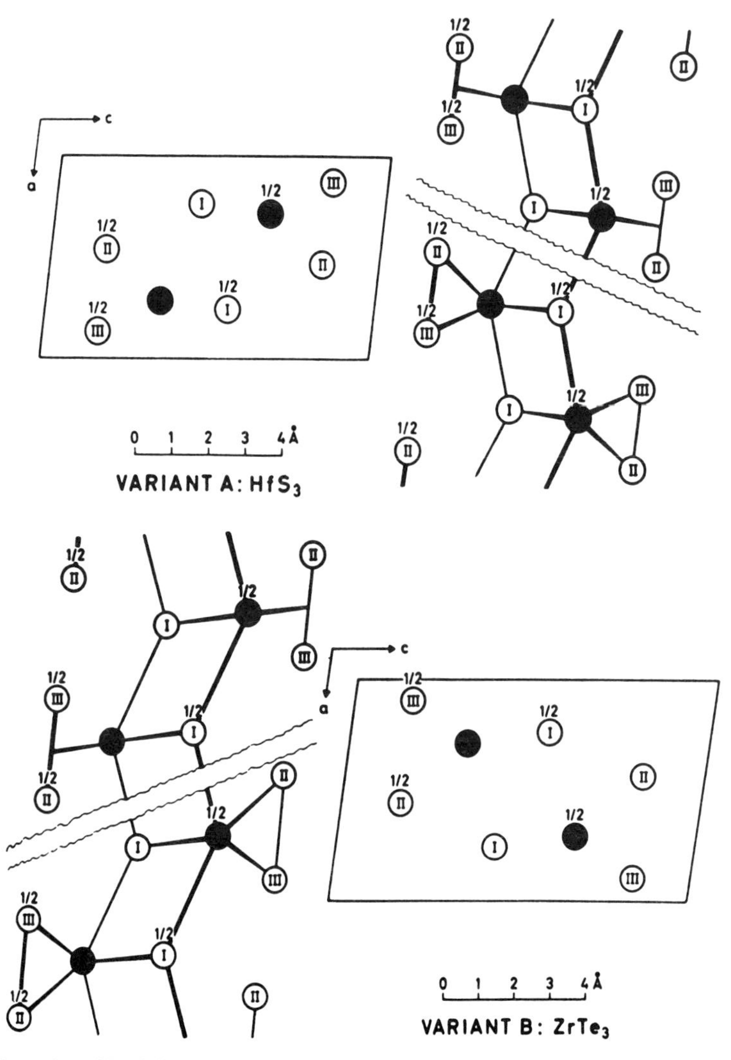

Fig. 7. Variants A and B of the $ZrSe_3$ type structure in {010} projection, illustrated by data obtained for HfS_3 and $ZrTe_3$, respectively. Filled and open circles represent metal (M) and nonmetal (X) atoms, respectively. [59]. (© Acta Chemica Scandinavica.)

sites related through a centre of inversion. There are three inequivalent X atoms, according to the structural formula $MX_IX_{II}X_{III}$. The local atomic coordination can be seen in Figure 7. Each M atom is surrounded by six near X (two X_I, two X_{II} and two X_{III}) atoms at the corners of a triangular prism and by two X_I atoms outside two of the rectangular faces of the prism. X_I atoms are coordinated by four M atoms at the corners of a deformed tetrahedron. $X_{II}(X_{III})$ atoms are coordinated by one near $X_{III}(X_{II})$ atom and two M atoms. The short interatomic distances between X_{II}–X_{III} pairs are in reasonable accord with the expectation values for the corresponding X–X single bonds. X–X pairs also occur within the trigonal prisms of the MX_3 compounds of group V transition metals, but there they can be present to a different extent on different columns, thus $NbSe_3$ contains three types of columns containing strong, intermediate and weak Se–Se pairing [62–64].

Twinning is frequently encountered among the MX_3 crystals, but in addition to this phenomenon came the closely connected observation that there are two variants of the $ZrSe_3$-type structure, denoted A and B by Furuseth *et al.* [59]. These are almost mirror images of each other, as illustrated by Figure 7. The B-type variant was measured for TiS_3, $ZrTe_3$, and $HfSe_3$, and some of the family have been prepared as both A- and B-type samples. To a very good approximation the positional parameters of the A and B types obey the relations

$$x_A = 1 - x_B,\ y_A = y_B, \text{ and } z_A = z_B,$$

simulating the behaviour of a mirror plane of symmetry parallel to (100) at $x = \frac{1}{2}$. In the B structure, the two interchain M–X distances, which are almost identical in the A structure, become highly differentiated and the shorter of the two interchain distances is less than the average M–X distance within the chain. Tables I and II record the unit cell dimensions and interatomic distances for compounds with one or the other variant of the $ZrSe_3$ structure.

3.1.2. *Optical Properties*

Several studies of the optical properties of these materials have been reported [65–72]. While most of these were in the visible wavelength region, reflectivity

TABLE I

Lattice parameters for compounds with the $ZrSe_3$ type structure (from Brattås and Kjekshus [58]). (© Acta Chemica Scandinavica.)

Compound	a (Å)	b (Å)	c (Å)	β(°)
TiS_3	4.958	3.401	8.778	97.32
ZrS_3	5.124	3.624	8.980	97.28
$ZrSe_3$	5.411	3.749	9.444	97.48
$ZrTe_3$	5.894	3.926	10.100	97.82
HfS_3	5.092	3.595	8.967	97.38
$HfSe_3$	5.388	3.722	9.428	97.78
$HfTe_3$	5.879	3.902	10.056	97.98

TABLE II

Interatomic distances (Å) in the structures of TiS_3, ZrS_3, $ZrSe_3$, $ZrTe_3$, HfS_3, and $HfSe_3$ (from Furuseth *et al.* [59], and Krönert and Plieth [57] for $ZrSe_3$). (© Acta Chemica Scandinavica.)

Compound		TiS_3 (type B)	ZrS_3 (type A)	$ZrSe_3$ (type A)	$ZrTe_3$ (type B)	HfS_3 (type A)	$HfSe_3$ (type B)
2 $M-X_I$	(intrachain)	2.496	2.602	2.717	3.030	2.588	2.752
2 $M-X_{II}$	"	2.667	2.602	2.733	3.163	2.612	2.935
2 $M-X_{III}$	"	2.358	2.605	2.743	2.771	2.590	2.586
1 $M-X_I$	(interchain)	2.416	2.724	2.870	2.829	2.697	2.624
1 $M-X_I$	"	2.855	2.707	2.868	3.467	2.698	3.100
1 $X_{II}-X_{III}$	(intrachain)	2.038	2.090	2.344	2.761	2.102	2.333

measurements have recently been extended to the vacuum ultraviolet region (with photon energies up to 14 eV, and in some cases to 30 eV by using synchrotron radiation as a soft X-ray source [67, 68]). The wavelength derivative of the reflectivity of ZrS_3 has been measured using wavelength modulated spectroscopy [72]. The energy distributions of electrons in core and valence bands have been probed by photoemission spectroscopy, using either AlKα [73–75] or synchrotron [76, 77] radiation.

Some of the results of optical measurements are summarized in Table III. All of these compounds are diamagnetic semiconductors, but resistivity measurements by McTaggart [78] suggest that the group IV tritellurides ($ZrTe_3$, $HfTe_3$) may be metallic; Bayliss and Liang [71] have observed rather different reflectivity behaviour in $ZrTe_3$ as compared to $ZrSe_3$. The strong polarization dependence observed in the optical properties indicate that the electronic structures of $ZrSe_3$-type materials are highly anisotropic. As an example of this dichroism, Figure 8 compares the basal plane normal incidence reflectivity spectra for the $\mathbf{E}\perp b$ and $\mathbf{E}\|b$ orientations,

TABLE III

Optical band gaps (eV) of MX_3 compounds of the group IV metals

TiS_3	ZrS_3	HfS_3	$ZrSe_3$	$HfSe_3$	
0.9	2.2	2.8	1.2		Grimmeiss *et al.* [65] (optical absorption)
0.83	1.91	1.95	1.11	1.02	Brattås and Kjekshus [58] (diffuse reflectance spectroscopy)
	~2.8	3.1			Schairer and Schafter [66] (optical absorption)
	2.1[a], 2.5, 3.18				Perluzzo *et al.* [69] (thermomodulation spectroscopy)
	2.0[a], 2.5, 3.18				Nee *et al.* [72] (wavelength modulated reflectance)

[a] Indirect gap in ZrS_3.

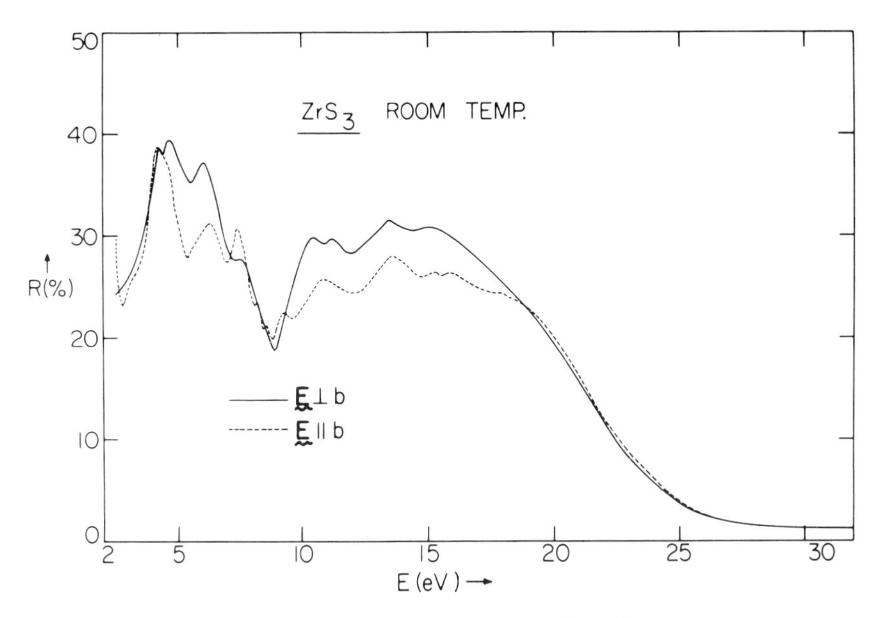

Fig. 8. Basal plane near-normal-incidence reflectivity spectrum (2.5–30 eV) of ZrS_3 at room temperature for the $\mathbf{E}\perp b$ (solid line) and $\mathbf{E}\| b$ (broken line) orientations (Khumalo *et al.* [68].) (© North-Holland Publishing Co.)

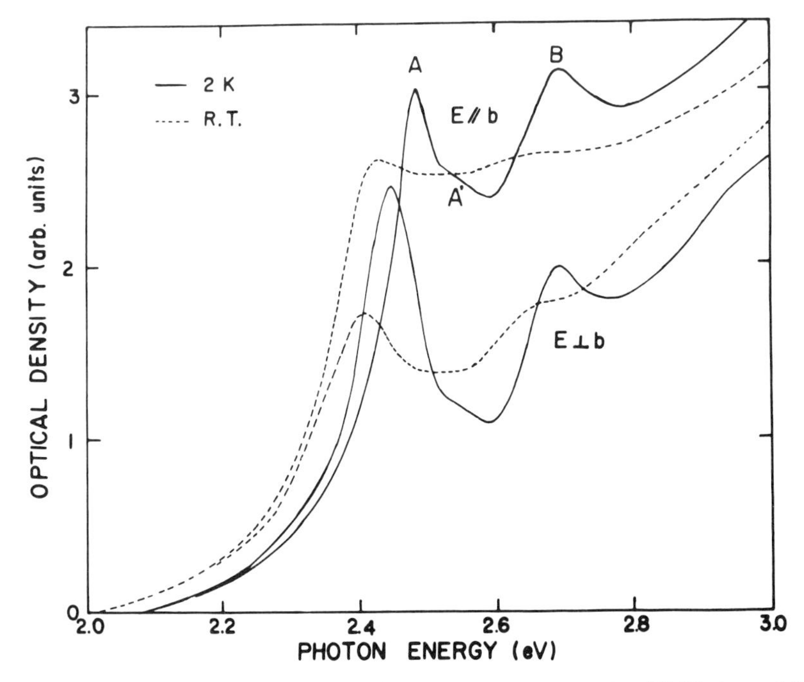

Fig. 9. Absorption spectrum of ZrS_3 single crystal at room temperature and at 2 K (Kurita *et al.* [70]), for light polarized parallel and perpendicular to the monoclinic *b*-axis. (© North-Holland Publishing Co.)

as measured [68] for ZrS_3. Khumalo *et al.* used a Kramers–Kronig analysis of these data to deduce the polarization dependence of the real and imaginary parts of the dielectric function.

Figure 9 shows the absorption spectrum measured [70] for ZrS_3, where the effect of dichroism is again seen; the features are broadly similar to those observed earlier by Schairer and Schafer [66]. Kurita *et al.* [70] conclude that in ZrS_3 the direct energy gap is 2.56 eV with an exciton ionization energy $E_b = 0.08$ eV for $\mathbf{E}\|b$, and 2.57 eV with $E_b = 0.12$ eV for $\mathbf{E}\perp b$. In $ZrSe_3$ the direct transition is allowed only for $\mathbf{E}\|b$ (with the first exciton peak at 1.81 eV) and forbidden for $\mathbf{E}\perp b$. Both ZrS_3 and $ZrSe_3$ have an indirect band gap. For $ZrSe_3$ the exciton indirect gap is at 1.535 eV and shows a negligible polarization dependence, whereas the exciton indirect gap of ZrS_3 is 2.085 eV for $\mathbf{E}\perp b$ and 2.055 eV for $\mathbf{E}\|b$.

3.1.3. *Electronic Structure Calculations*

Band structure calculations have so far been reported only for $ZrSe_3$ [62, 79], but we may regard this compound as typical of the whole series. In the atomic-orbital-based calculation [62] $ZrSe_3$ was found to have a semiconducting direct gap of 1.6 eV and indirect gap of 1.5 eV (slightly larger than the indirect gap of 1.0 eV calculated by the same method [80] for $ZrSe_2$). The calculated density of states is reproduced in Figure 10. The main valence band of Se 4*p* states extends ~6 eV below the valence band maximum, while the Se 4*s* states form the multipeaked bands between about 11 and 16 eV below E_F. These *s* bands are broadened considerably relative to the corresponding layered dichalcogenide as a result of the closer approach of some Se atoms. The 2.34 Å Se_{II}–Se_{III} distance [57] (compared to 3.77 Å for the Se–Se intralayer spacings in $ZrSe_2$) gives rise to bonding and antibonding *s* states below and above the Se_I contribution. The overall width of the *p* band is rather similar to that of $ZrSe_2$, except that because of the Se–Se pairing one of the 'Se *p* states' is forced up into the conduction bands as an unoccupied *p* antibonding state. Thus in the trichalcogenides of the group IV transition metals the energy gap occurs after eight, and not nine, *p* bands per MX_3 formula unit, and $ZrSe_3$ is a semiconductor. When the valence band has to accommodate further electrons, as in the group V trichalcogenides, separate MX_3 chains approach, align and distort [62] so that some of the states which would otherwise have been X–X antibonds within a chain become X–X or M–X bonds between chains.

Myron *et al.* [79] have presented the calculated electron bands of $ZrSe_3$ along certain high symmetry lines and their results are reproduced, together with the Brillouin zone for the simple monoclinic lattice, in Figure 11. They used a non-self-consistent and nonrelativistic version of the KKR method, including angular momentum states up to $l = 2$ on Zr sites and up to $l = 1$ on Se sites. The crystal potential was constructed from overlapping neutral Hartree–Fock–Slater atomic charge densities using the local density approximation of Kohn and Sham for exchange; no attempt was made to correct for nonspherical corrections to the potential outside the inscribed atomic spheres, so that individual bands may shift appreciably when the model is improved. The main features are very similar to

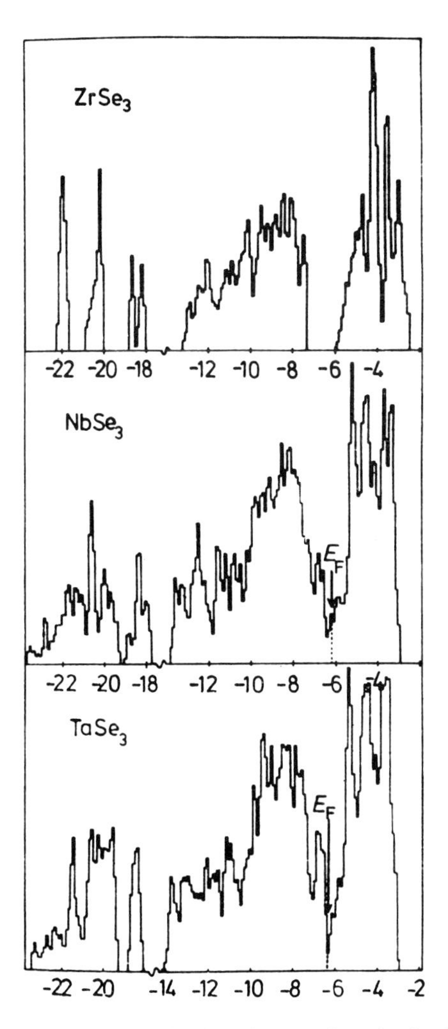

Fig. 10. Electron density of states calculated for the valence and conduction bands of $ZrSe_3$, $NbSe_3$, and $TaSe_3$. (Bullett [62].) (© The Institute of Physics.)

those of the atomic-orbital-based calculation [62] (which uses a similar potential). The direct band gap calculated between occupied and unoccupied states is 1.8 eV at A and 2.7 eV at Γ, but an indirect gap of 1.3 eV is present from the valence band maximum at Γ to the conduction band minimum at A. Although, even if these were fully self-consistent calculations, the one-electron energy gap calculated in the local density approximation should not be interpreted too literally as the experimental semiconducting gap in the quasiparticle excitation spectrum, it is nevertheless encouraging to see this degree of agreement between calculated and experimental [58, 65, 66, 73] values. The main difference between the results of the two calculations is the stronger dispersion in the lowest KKR conduction bands, and consequently the

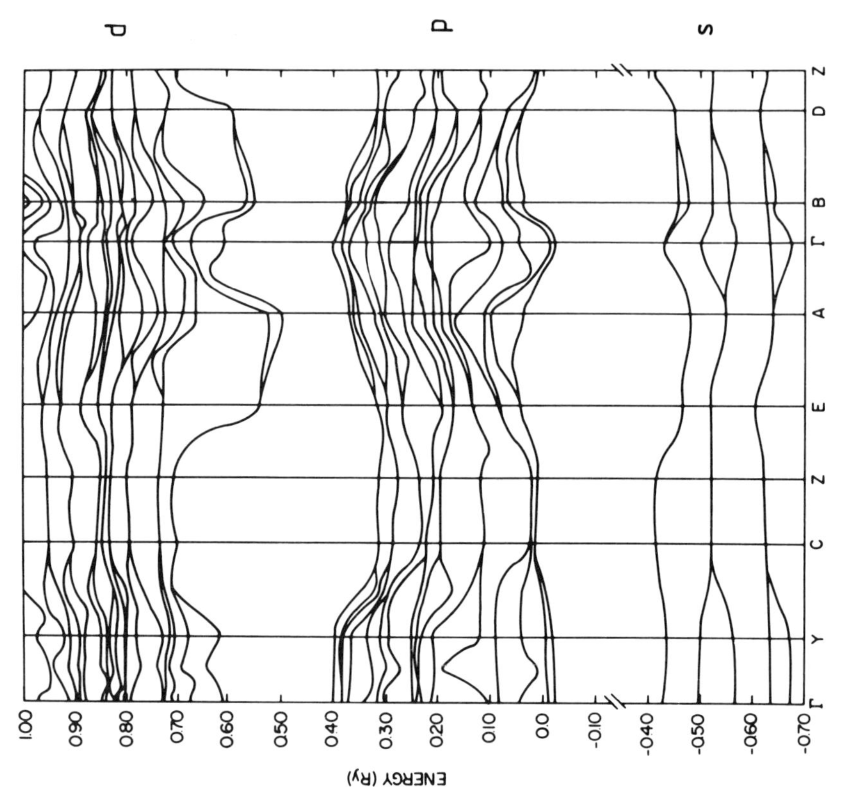

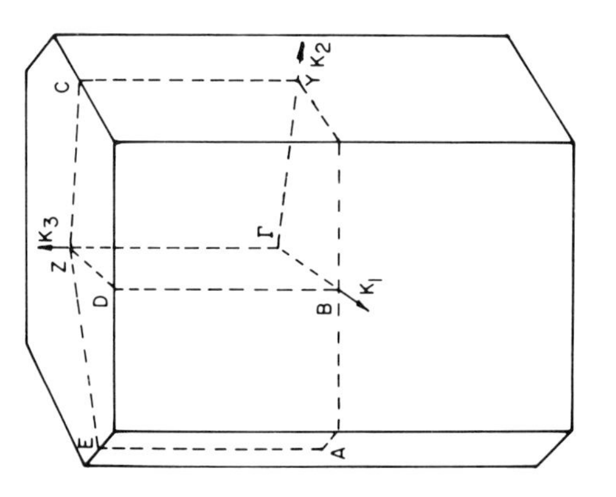

Fig. 11. Electron band structure of $ZrSe_3$ calculated along certain high symmetry lines of the Brillouin zone for the simple monoclinic lattice (Myron *et al.* [79]). The Fermi level lies in the p–d gap at ~0.45 Ry on this energy scale. (© Pergamon Press Ltd.)

more gentle conduction band edge in the KKR density of states. One calculation also reported the sensitivity of the Se *p*-antibonding bands to the $Se_{II}-Se_{III}$ distance: increasing this distance by 5% to 2.48 Å, keeping most other distances constant, caused the indirect band gap to drop from 1.3 to 0.8 eV [79].

The qualitative features of the calculated density of states for $ZrSe_3$ correspond quite well to XPS spectra for this and related MX_3 compounds. Jellinek *et al.* [73] observed a 2.4 eV splitting between bonding (σ_g) and antibonding (σ_u^*) 4*s* XPS peaks for $ZrSe_3$, and 4.0 eV for the corresponding splitting in ZrS_3, and a *p* bandwidth of between 6 and 7 eV in both materials. Similar XPS features have been discussed more recently by Endo and coworkers [74, 75] and by Te-Xiu Zhao *et al.* [77], who carried out a detailed study of the upper valence bands of ZrS_3, $ZrSe_3$, $ZrTe_3$, and $HfSe_3$ using radiation from the Wisconsin synchrotron.

Figure 12 shows the photoelectron energy distribution curves (EDCs) taken at different photon energies on freshly cleaved samples of the four group IV trichalcogenides [77]. For the S and Se compounds five spectral features could be identified, labelled A, B, C, D, and E in order of increasing binding energy. For $ZrTe_3$ only three spectral features were detected; the authors attribute this difference to the longer intrachain M–X distances in this compound. The overall widths *W* of the upper *p*-like valence band for each compound are about 0.5 eV greater than for the corresponding MX_2 layer dichalcogenide. Table IV compares the energy position of EDC features for the MX_3 series and illustrates the agreement between theory and experiment as far as the spectral features of $ZrSe_3$ are concerned.

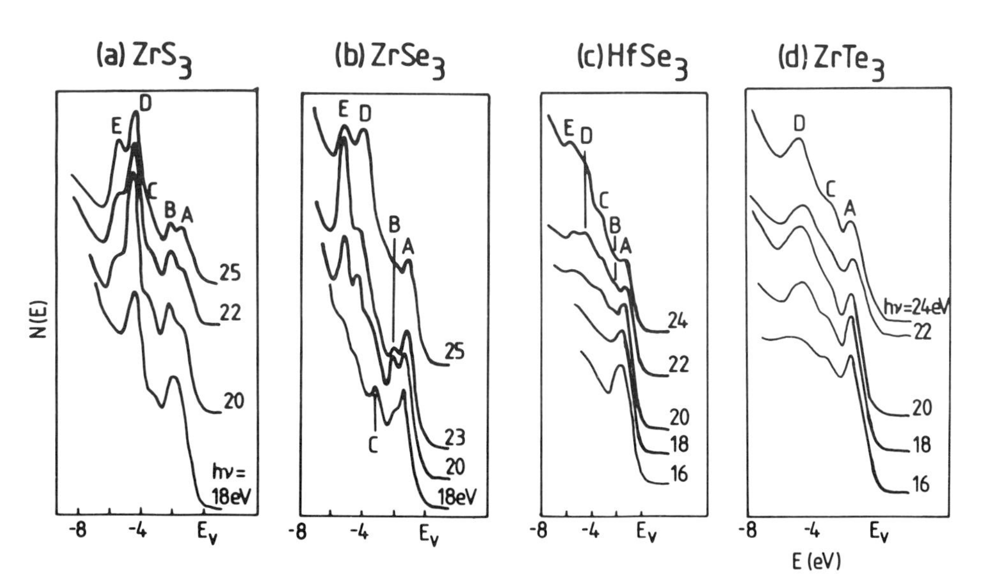

Fig. 12. Photoelectron energy distribution curves (EDCs) recorded at different photon energies in synchrotron radiation photoemission studies of the upper valence band states of (a) ZrS_3, (b) $ZrSe_3$, (c) $HfSe_3$, (d) $ZrTe_3$. (Te-Xiu Zhao *et al.* [77].) (© Società Italiana di Fisica.)

TABLE IV

Energy positions (in eV measured from valence band maximum energy) of EDC features and width W of the upper valence band in group IV MX_3 compounds (Te-Xiu Zhao *et al.* [77]) (© Società Italiano di Fisica.)

EDC peak	A	B	C	D	E	W
$ZrSe_3$, UPS [77]	−1.2	−1.9	−3.1	−4.2	−5.1	5.8
$ZrSe_3$, XPS [73]	−1.5	—	−3.4	—	−5.0	6.5
$ZrSe_3$, theory [79]	−1.2	−2.3	−3.2	−3.8	−5.1	5.8
$ZrSe_3$, theory [62]	−1.2	−2.1	−3.0	−3.8	−5.0	6.0
ZrS_3, UPS [77]	−1.0	−1.9	−3.0	−4.1	−5.2	6.1
ZrS_3, XPS [73]	−1.4	—	—	−4.4	−5.4	6.5
$ZrTe_3$, UPS [77]	−1.5	−2.5	—	—	−4.7	5.9
$HfSe_3$, UPS [77]	−1.2	−2.0	−3.0	−4.4	−5.2	6.0

3.1.4. *Transport Properties*

Electrical transport measurements (d.c. resistivity, Hall coefficient, mobility, and thermoelectric power) have been carried out along the chains (b-axis) of single crystals of ZrS_3 [81], $ZrSe_3$ [82], and TiS_3 [83]. The room temperature resistivity is about 15 Ω cm for ZrS_3 and 900 Ω cm for $ZrSe_3$. For ZrS_3 measurements were made in the temperature range 100–500 K, but for $ZrSe_3$ reliable data could not be obtained below 200 K because of the extremely high resistances of the samples. The activation energies of donor impurities were about 0.20 and 0.25 eV, respectively. At low temperatures the Hall mobility of ZrS_3 is given by $\mu_1 = 6.5 \times 10^{-2} T^{3/2}$ $cm^2 v^{-1} s^{-1}$ and is dominated by ionized impurity scattering; Perluzzo *et al.* [81] estimate the density of acceptors as $4.5 \times 10^{18} cm^{-3}$. At room temperature and above a different mechanism becomes dominant, and the mobility falls with increasing temperature as $T^{-3/2}$ in both compounds. In this regime the dominant scattering is by phonons, and Ikari *et al.* [82] interpret the exponent in terms of carrier scattering by homopolar optical phonons in a two-dimensional system, as is the case for other layer structure semiconductors. Applying the effective mass approximation to carrier conduction in two dimensions, they estimate a value of $0.3m_0$ for the electron density of states mass parallel to the layer.

3.1.5. *Raman and Infrared Spectra of the Group IV Trichalcogenides*

Extensive studies have recently been made of the lattice-dynamical properties of the $ZrSe_3$-type compounds [84–98], in order to provide a better understanding of the relative chain-like and layer-like features of their behaviour. Long-wavelength optical phonon modes have been investigated in TiS_3 [84–86, 88], ZrS_3 [84, 85, 87, 89–92], $ZrSe_3$ [84, 85, 87, 93, 94], $ZrTe_3$ [86, 87], HfS_3 [85, 95, 96], $HfSe_3$ [86, 87], and the ternary phases $HfS_{3-x}Se_x$ [97] and $ZrS_{3-x}Se_x$ [98].

As we have seen, the unit cell (Figure 7) contains two molecular units, i.e eight atoms, and there are thus 8×3 normal modes at the Brillouin zone centre Γ. Zwick and Renucci [87, 93] have carried out the group theoretical analysis to classify these 24 modes according to the irreducible representations of the $2/m-C_{2h}$ point group:

$$\Gamma = 8A_g + 4B_g + 4A_u + 8B_u.$$

Among the 21 long wavelength optical vibrations, the 12 even parity A_g and B_g modes are Raman active while the nine odd parity A_u and B_u are infrared active; because the crystal has inversion symmetry the two activities are mutually exclusive. The modes may also be classified according to the atomic displacements parallel (B_g and A_u) or perpendicular (A_g and B_u) to the chain.

In classifying the crystal modes, it is useful to see how these evolve from the long-wavelength modes of a single $ZrSe_3$ chain. For the single chain the appropriate decomposition into irreducible representations is

$$\Gamma_{\text{chain}} = 4A_1 + 4B_1 + A_2 + 3B_2$$

Figure 13 illustrates the compatibility relationships between chain and crystal modes and the way in which crystal modes are generated through interchain coupling [97]. Each chain mode splits into a $g + u$ pair of crystals modes, and the A and B chain

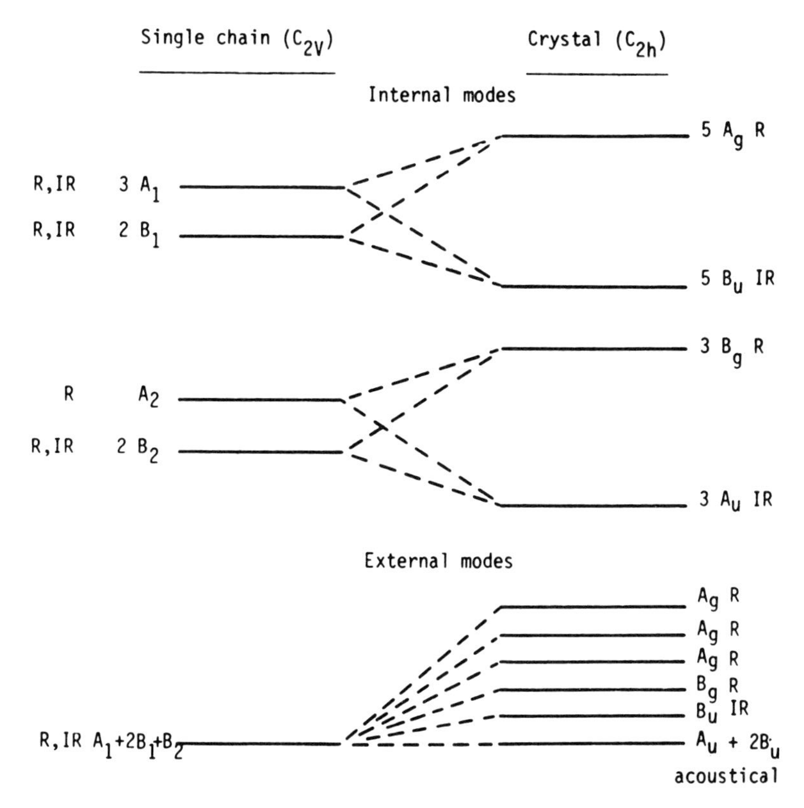

Fig. 13. Correlation diagram relating the long-wavelength chain and crystal modes of the $ZrSe_3$-type structure (R = Raman active, IR = infrared active). (Zwick *et al.* [97].) (© American Physical Society.)

modes are mixed by removal of the twofold axis in the crystal. Four low-frequency Raman modes of the crystal ($3A_g + B_g$) reflect largely rigid-chain motions – three translational modes (A_g, B_g, A_g) along the a, b, and c axes and one librational mode (A_g) about the b-axis.

Observation and classification of the long-wavelength optical phonons can thus provide information about the strength of the interchain binding. Unfortunately not enough of the theoretically predicted $k = 0$ modes have been seen to make mode assignment completely unambiguous in any one compound, but by studying trends in the frequencies of observed peaks throughout this family of compounds, reasonable conclusions can be drawn. The observed spectral lines fall into three groups [97], in order of increasing frequency:

I lines in the low-frequency region originating from the weak coupling between adjacent rigid-chain modes;
II lines largely originating from internal deformations of the chains;
III a single high-frequency line attributed to the $X_{II}-X_{III}$ diatomic stretching mode within the chains.

Longer and longer range interactions are involved as we progress from group III to group I. Wieting, Grisel, and Levy [85, 86] and Sourisseau and Mathey [91] have derived valence force-field parametrizations of the far infrared and Raman frequencies to model the lattice dynamics of these materials. The model (Figure 14)

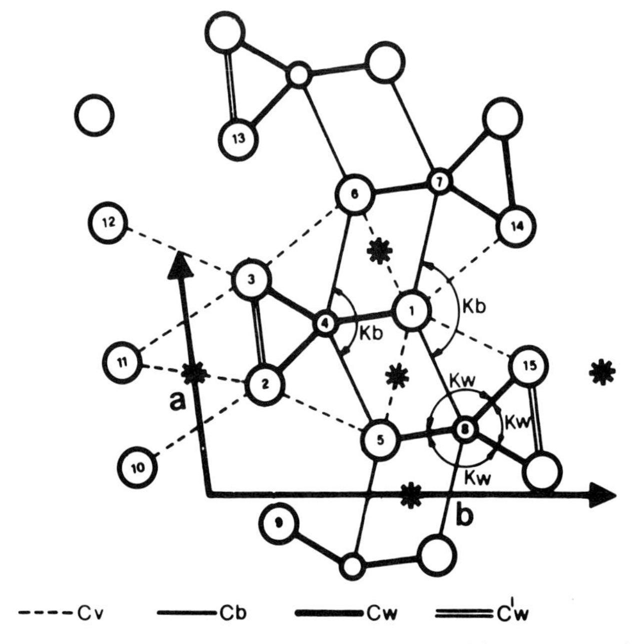

Fig. 14. Projection of the $ZrSe_3$ structure illustrating the four central force constants (C_W, C'_W, C_b, and C_V) and two bond-angle force constants (K_W and K_b) in the valence force model of Grisel *et al.* [85]. (Note that these authors set the monoclinic chain axis along the c direction.) (© North-Holland Publishing Co.)

TABLE V

Force constants derived by Wieting, Grisel, and Lévy [85, 86] for the valence-force lattice dynamical model of group IV transition metal trichalcogenides. (© North-Holland Publishing Co.)

Crystal	Force constants (dyne cm^{-1})					
	C_W	C'_W	C_b	C_V	K_W	K_b
ZrS_3	4.3×10^4	2.3×10^5	7.1×10^4	5.7×10^2	1.6×10^4	1.4×10^4
HfS_3	4.0×10^4	2.4×10^5	5.2×10^4	3.3×10^2	1.1×10^4	1.7×10^4
$ZrSe_3$	3.2×10^4	2.0×10^5	3.9×10^4	7.2×10^2	3.8×10^3	1.7×10^4
TiS_3	2.9×10^4	2.7×10^5	7.1×10^4	5.2×10^2	1.3×10^4	1.8×10^4
$HfSe_3$	3.2×10^4	1.8×10^5	6.2×10^4	4.5×10^2	1.3×10^4	1.5×10^4
$ZrTe_3$	3.4×10^4	1.6×10^5	3.1×10^4	5.3×10^2	3.0×10^3	4.7×10^3

contains four central force constants and two bond-bending force constants. The derived values of these six parameters are reproduced in Table V for the three type *A* compounds (ZrS_3, HfS_3, $HfSe_3$) and three type *B* compounds (TiS_3, $HfSe_3$, $ZrTe_3$). The strongest bond (C'_W) links the $X_{II}-X_{III}$ pairs of chalcogens on one side of the chain. Next in order of strength come C_b (the central M–X forces between chains) and C_W (the central M–X forces between each M atom and the six surrounding X atoms within a chain). The weakest interactions considered are the central van der Waals-like X–X forces between chalcogen atoms on neighbouring chains; all are included within the single force constant C_V. Bond angle force constants are denoted by K_W (representing the X–M–X intrachain bond angle) and K_b (representing the interchain X–M–X angle). The agreement between observed and calculated $k = 0$ optic mode frequencies is reasonably good; results for the six compounds are compared in Table VI.

The optical phonon anisotropies clearly indicate both one- and two-dimensional character in the bonding interactions. The individual chains are strongly coupled into layers by the M–X bonds running perpendicular to the chain axis: C_b and K_b are comparatively large and have a significant influence on the long-wavelength phonons. The approximate equality of C_b and C_W in each compound illustrates that the coordination of the metal atoms is eight rather than six, because of the two interchain M–X bonds. That the interchain forces are highly anisotropic can be seen from the high frequency of the upper A_g acoustic rigid-chain mode (152 cm^{-1} in ZrS_3) compared to the lower A_g acoustic rigid-chain mode (21 cm^{-1} in ZrS_3). However, the interchain coupling does not affect vibrational modes polarized along the chain axis (at least within this valence force model): these modes retain their quasi-one-dimensional character.

3.1.6. *Intercalation Properties*

Another aspect of the low-dimensional character of transition metal trichalcogenides is provided by their intercalation compounds. Like the corresponding dichal-

TABLE VI

Long wavelength phonon frequencies (in cm^{-1}) of group IV trichalcogenides compared with the fitted frequencies of the valence force model [85, 86]. (© North-Holland Publishing Co.)

Vibrational symmetry			ZrS_3		HfS_3		$ZrSe_3$		TiS_3 (cm^{-1})		$HfSe_3$ (cm^{-1})		$ZrTe_3$ (cm^{-1})	
Chain	Crystal or layer	Transf. properties	Raman or IR	Calc.	Raman or IR	Calc.	Raman or IR	Calc.	Raman or IR	Calc.	Raman or IR	Calc.	Raman or IR	Calc.
A_1 (diatomic)	A_g	x^2, y^2, z^2	R 530	529	R 524	530	R 302	303	R 559	561	R 296	299	R 216	216
	B_u	y	IR 515(TO)	529	IR 525(Tr)	530	—	303	—	561	—	299	—	217
			IR 522(LO)											
B_1 (rigid-sublattice)	B_g	yz	—	302	—	235	—	177	—	314	—	180	—	162
	A_u	z	IR 247(TO)	301	IR 215(TO)	234	IR 200(TO)	176	IR 294.5(TO)	313	IR 160(TO)	179	—	162
			IR 306(LO)		IR 226(LO)		IR 223(LO)		IR 328(LO)		IR 181.5(LO)			
A_1	A_g	x^2, y^2, z^2	R 322	333	R 322	329	R 236	240	R 371	385	R 209	215	R 144	161
	B_u	y	—	356	IR 286(TO)	343	—	265	—	411	—	230	—	180
					IR 300(LO)									
B_2	A_g	xy	—	310	—	250	—	176	—	339	—	189	—	146
	B_u	x	IR 305(TO)	315	IR 272(TO)	251	IR 224(TO)	178	IR 333(TO)	347	IR 182(TO)	193	—	147
			IR 334(LO)		IR 280(LO)		IR 248(LO)		IR 348(LO)		IR 193(LO)			
A_1	A_g	x^2, y^2, z^2	R 282	206	R 261	171	R 177	140	R 300	210	R 171	125	R 108	108
	B_u	y	—	207	—	174	—	106	—	197	—	122	—	83
B_2	A_g	xy	—	176	—	164	—	96	—	149	—	98	R 86	76
	B_u	x	IR 247(TO)	186	IR 249(TO)	171	IR 144(TO)	99	IR 221(TO)	159	IR 140(TO)	104	—	78
			IR 254(LO)		IR 252(LO)		IR 146(LO)		IR 227.5(LO)		IR 144.5(LO)			
A_2 (shearing)	B_g	xz	R 152	173	R 140	160	R 77	86	R 102	145	R 69	96	R 64	68
	A_u		IR 180(Tr)	173	IR 160(Tr)	160	—	86	—	145	—	95	—	68
B_1 (shearing)	B_g	yz	—	159	—	150	—	85	—	129	—	85	—	63
	A_u	z	IR 128(Tr)	159	IR 130(Tr)	150	—	84	—	129	—	85	—	63
A_1 (rigid-chain/acous.)	A_g	x^2, y^2, z^2	R 152	135	R 127	107	R 78	71	R 176	134	R 101	82	R 62	52
	B_u	y	—	0	—	0	—	0	—	0	—	0	—	0
B_2 (librational)	A_g	xy	R 110	107	R 73	74	R 50	50	—	112	R 69	67	R 38	39
	B_u	x, R_z	IR 103(Tr)	57	IR 104(Tr)	46	—	29	—	53	—	34	—	21
B_2 (rigid-chain/acous.)	A_g	xy	R 21	21	R 15	15	(16)	16	R 20	20	(12)	12	(11)	11
	B_u	x	—	0	—	0	—	0	—	0	—	0	—	0
B_1 (rigid-chain/acous.)	B_g	yz	—	23	—	15	—	20	—	25	—	14	—	14
	A_u	z	—	0	—	0	—	0	—	0	—	0	—	0

cogenides, several of the MX_3 family can be intercalated with high concentrations of alkali metals such as Li or Na [99–108]. The process has been studied in connection with possible applications as solid-state cathodes.

Each compound can incorporate 3 lithiums per MX_3 unit [99]. In the case of the group V compounds, the intercalation process is completely reversible: the chain structure remains intact and the lattice simply expands in the *ac* plane (Table VII). With the group IV trichalcogenides, the high charge capacity of the first electrochemical discharge is not recovered on subsequent cycles, because of some irreversible change in the atomic geometry. The most likely structural change is a simple twist of adjacent X_3 triangles along individual chains to give the octahedral coordination characteristic of the group IV dichalcogenides MX_2 and some chain compounds such as $BaTaS_3$ [109]. It appears that Li intercalation tends to break down the X–X pair bonds. Chianelli and Dines [99] have shown that in Li_3TiS_3 the $560\,cm^{-1}$ S–S vibration has disappeared in the infrared spectrum, and a strong wide peak (attributed to the Li–S stretch) appears at $425\,cm^{-1}$ which was not present in TiS_3. The breaking of the X–X pairs (along with the lengthening of the two interchain M–X bonds) destabilizes the trigonal prismatic intrachain coordination, and the local coordination of the group IV atom may then relax irreversibly into its preferred octahedral orientation. By contrast the group V atoms tend to prefer trigonal prismatic coordination (e.g. in $NbSe_2$), and for these trichalcogenides no reorientation is necessary and intercalation proceeds reversibly.

TABLE VII

Lattice expansion of Li_3MX_3 (after Chianelli and Dines [99]). (© American Chemical Society.)

	$2a_0$	b_0	c_0	β_0	a	b	c	β	Δa	Δb	Δc	$\Delta\beta$
Li_3TiS_3	9.92	3.40	8.78	97.3	11.1	3.46	9.12	98.0	1.2	+0.06	0.34	0.7
Li_3ZrS_3	10.25	3.62	8.98	97.3	11.2	3.59	9.22	100.0	1.0	−0.03	0.24	2.7
Li_3ZrSe_3	10.82	3.75	9.44	97.5	11.8	3.76	9.68	99.1	1.0	+0.01	0.24	1.6
Li_3HfS_3	10.18	3.60	8.97	97.4	12.1	3.52	9.24	100.0	1.9	−0.08	0.27	2.6
Li_3HfSe_3	10.78	3.72	9.43	97.8	11.5	3.71	9.52	98.1	0.7	−0.01	0.09	0.7
Li_3NbSe_3	15.63	3.48	10.06	109.3	17.0	3.48	11.5	98.1	1.4	0.00	1.4	12.7

Lattice constants in Å, angles in degrees; a_0, b_0, c_0, β_0 are the lattice parameters of the unlithiated trichalcogenides, from Brattås and Kjekshus [58], and from Meerschaut and Rouxel [110].

3.2. TRICHALCOGENIDES OF GROUP V

While the previous family of trichalcogenides are isostructural and their electronic structure and properties could be discussed as a single unit, the group V compounds NbS_3, $NbSe_3$, TaS_3, and $TaSe_3$ (the tritellurides have not yet been produced) all present their own structures and exhibit highly individualistic behaviour. The one-dimensional trigonal prismatic chain still exists as an identifiable structural unit, and the chains are still arranged in layers, but there are subtle differences in the arrangement of chains within the layers. These variations of interchain bonding not only affect the two- and three-dimensional electronic properties, but are also able to influence the relative positions of metal atoms along the chain axis.

As we have seen, the MX_3 compounds of group IV are empty d-band semiconductors (apart from the slight semimetallic p–d overlap in $ZrTe_3$). Their formal configuration can be expressed in the ionic form $M^{4+}(X_2)^{2-}X^{2-}$; the close pairing of two of the chalcogen atoms drives the p-antibonding X_2 state up into the unoccupied conduction bands. In the group V compounds the extra electron per MX_3 unit may be accommodated either by displacing M atoms in alternate directions along the chain axis to establish M–M pair bands, with all M atoms now in a formal d^1 configuration, or by opening some of the X_2 pairs so that M atoms retain their d^0 configuration but the distribution of anion configurations moves towards $M^{5+} \cdot 0.5(X_2)^{2-} \cdot 2X^{2-}$. These two competing mechanisms maintain a delicate balance in each material.

3.2.1. *NbS_3*

The normal structure of NbS_3 [111–113] provides the simplest distortion of the $ZrSe_3$-type structure (Figure 15). Metal atoms are completely paired along the chain axis, to give a doubling of the intrachain lattice parameter analogous to NbI_4.

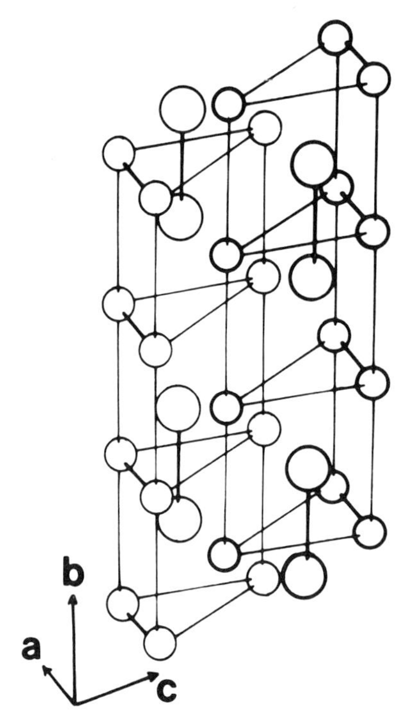

Fig. 15. Schematic picture of the NbS_3 structure. Eight S atoms (small circles) form a bicapped trigonal prism around Nb atoms (large circles). M–M bonds occur parallel to the b-axis. [113, 114]. (© Academic Press, New York.)

Displacements of 0.16 Å from the (pseudo) mirror planes of the coordination polyhedra give Nb–Nb bonding distances of 3.04 Å [113]. The bridging triangle of sulphur atoms is pushed outwards in an attempt to conserve M–X bond lengths; the triangle at the other end of the prism is drawn in, for the same reason. Anion pairs are fully retained in the structure, with the S–S distance equal to 2.05 Å in NbS_3 [113] (compared with 2.09 Å in ZrS_3 [59]). Thus NbS_3 can be formulated as Nb^{4+} $(S_2)^{2-}S^{2-}$ and behaves as a diamagnetic semiconductor [111, 112]. Sulphur atoms on neighbouring chains continue to cap two of the rectangular faces of each prism to provide a weak bonding between chains and overall eightfold coordination of each metal atom.

The density of states for electrons in NbS_3 is reproduced in Figure 16, from calculations in the nonempirical atomic orbital method [114]. The self-consistent *d*-state configuration of the Nb atom in this structure was found to be $d^{3.8}$; the energy levels for isolated neutral atoms in the appropriate configuration are given in Table VIII. Many features of the spectrum are reminiscent of $ZrSe_3$. The density of states separates broadly into three parts which we may loosely label (in order of decreasing binding energy) as chalcogen *s*, chalcogen *p*, and metal *d* bands, respec-

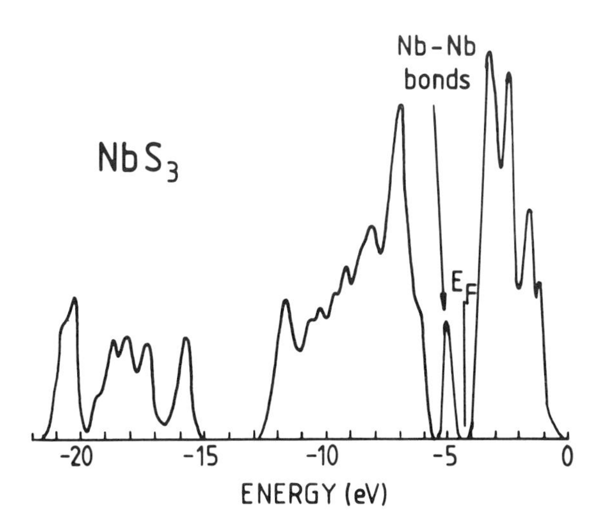

Fig. 16. Calculated valence band density of states for NbS_3, smoothed by a Gaussian of half-width σ = 0.15 eV. (Bullett [114].) (© Academic Press, New York.)

TABLE VIII

Isolated neutral atom energy levels (eV) for NbS_3 [114]. (© Academic Press, New York.)

	Nb ($s^{0.7}p^{0.5}d^{3.8}$)	S (s^2p^4)
s	−5.3	−17.8
p	−3.0	− 8.0
d	−4.4	

tively. These labels denote only the major component of the eigenstates in a particular band. Strong hybridization between *p* and *d* states provides the bonding in these materials and the statement that all '*p* bands' are filled while most '*d* bands' are empty is not intended to mean that these compounds are strongly ionic. If we divide all overlap charges equally between the two contributing atoms, the net charge on the metal atom is about +1 in NbS_3.

The S *s* bands again split roughly into three sub-bands – the bonding and antibonding *s* states of the S_2 pairs, separated by the nonbonding *s* state of the third atom in the S_3 triangle – but because of the lower crystal symmetry the distinction is less precise than in $ZrSe_3$, and the nonbonding and antibonding sub-bands of NbS_3 overlap. Eight *p* bands per formula unit again form the main valence band. Nb–Nb pairing along the *b*-axis is seen in Figure 16 to pull one metal *d* band (per pair of MX_3 units) down below the rest of the *d*-band spectrum. At a nearest neighbour distance of 3.04 Å, the interaction between Nb d_{z^2} orbitals (where *z* is the direction of the *b*-axis) is −0.8 eV (Figure 3). The result is an energy gap of 0.5 eV above the highest occupied band. Thus the calculation accounts naturally for the diamagnetic and semiconducting properties, and black needle-like appearance [111–113] of NbS_3 crystals.

Electron diffraction work by Corne Lissens and coworkers [115] has shown that above room temperature the (h, k, l) reflections with k = odd develop streaks perpendicular to the b^*-axis, suggesting that the correlation between chains gradually disappears. These workers detected two different phases (type I and type II) in their samples. The intense spots form the same arrangement, but for type II the distance between intense spots is subdivided by weaker spots, even at room temperature. This they attribute to chain 'polytypism': different stacking sequences of chains lead to more complicated unit meshes in the plane perpendicular to the chains, and hence superstructure appears in the diffraction pattern. The room temperature diffraction pattern for the type II crystals also exhibits a long-wavelength (incommensurate?) modulation along the chain direction [115, 116]. Superstructure with a period $\sim 2.9b$ is observed. It seems likely that the type II phase contains more than one kind of chain structure, with slightly different numbers of electrons on each kind. The chain with an electron deficit might then contain occasional unpaired Nb atoms in a regular sequence, while the chain with excess electrons might accommodate the excess by breaking a similar sequence of S–S pairs.

At high pressures a third structure, thought to be isomorphous with $NbSe_3$, has been grown [117]. This appears to be the stable high-pressure structure [117–119] for all four compounds (NbS_3, $NbSe_3$, TaS_3, $TaSe_3$), in spite of their individual characteristic structures when prepared under normal conditions (Table IX). Surprisingly, the high-pressure form of NbS_3 is reported to be semiconducting [117], with activation energy 0.18 eV.

3.2.2. *$TaSe_3$*

$TaSe_3$ crystallizes monoclinically [120] with four trigonal prismatic chains per unit cell (Figure 6). The two-dimensional slabs of the $ZrSe_3$ are still identifiable, but can

TABLE IX

Lattice parameters of group V trichalcogenides in their normal and high-pressure[a] phases.

	NbS_3	$NbSe_3$	TaS_3	$TaSe_3$
NORMAL FORMS				
Crystal system	Triclinic	Monoclinic	Orthorhombic	Monoclinic
Space group	$P\bar{1}$	$P2_1/m$	$Cmcm$	$P2_1/m$
Lattice constants	a = 4.963 Å	a = 10.006 Å	a = 36.804 Å	a = 10.402 Å
	b = 6.730 Å	b = 3.478 Å	b = 15.117 Å	b = 3.495 Å
	c = 9.144 Å	c = 15.626 Å	c = 3.340 Å	c = 9.829 Å
	β = 97.17°	β = 109.50°		β = 106.26°
	$\alpha = \gamma$ = 90°			
Reference	Rijnsdorp and Jellinek [113]	Meerschaut and Rouxel [110]	Bjerkelund and Kjekshus [120]	Bjerkelund and Kjekshus [120]
Unit cell volume (Å^3)	303.0	513.3	1865.2	343.0
Formula units/cell	4	6	24	4
Volume/MX_3 unit (Å^3)	75.75	85.55	77.71	85.76
Anion pair lengths:				
intrachain (Å)	2.05	2.374, 2.485 2.909	(Not known)	2.58, 2.90
shorter interchain (Å)	2.91	2.73, 2.93		2.65
HIGH PRESSURE FORMS[a]				
Crystal system	Monoclinic	Monoclinic	Monoclinic	Monoclinic
Lattice constants	a = 9.68 Å	a = 10.02 Å	a = 9.52 Å	a = 10.02 Å
	b = 3.37 Å	b = 3.47 Å	b = 3.35 Å	b = 3.48 Å
	c = 14.83 Å	c = 15.63 Å	c = 14.92 Å	c = 15.65 Å
	β = 109.9°	β = 109.5°	β = 110.0°	β = 109.6°
Unit cell volume (Å^3)	454.3	512.8	446.8	513.4

[a] Prepared at 2GPa and 700 °C by S. Kikkawa *et al.* [117]. For TaS_3, see also A. Meerschaut *et al.* [118] and Kikkawa *et al.* [119].

now be broken down into canted blocks of four columns [62, 64]. Each block contains two structurally distinct types of column: an inner duo of columns in the $ZrSe_3$ setting, with intermediate X–X pairing (2.58 Å), flanked by an outer duo which are reoriented with respect to the $ZrSe_3$ structure and in which X–X pairing is almost completely relaxed (2.90 Å). The weaker Se–Se bonding within the columns is compensated by an intercolumn Se–Se distance of only 2.65 Å.

The calculated electron density of states [62] shows that for this arrangement of columns the Fermi energy E_F lies in a deep, but nonzero, minimum in the density of states. Qualitatively the spectrum of states is similar to those already discussed, but the *s*-band structure is now more complicated (because of the several inequivalent close approaches between Se atoms) and there is an energy overlap between the *p* and *d* bands. The nature of this semimetallic band overlap is illustrated in Figure 10 [62]. The $ZrSe_3$ structure induces a semiconducting gap after 8 valence *p* bands per MX_3 unit; the $TaSe_3$ structure induces a pseudo-gap after 17 *p* bands per pair of MX_3 units. In order to accommodate the extra electrons in the Ta compound,

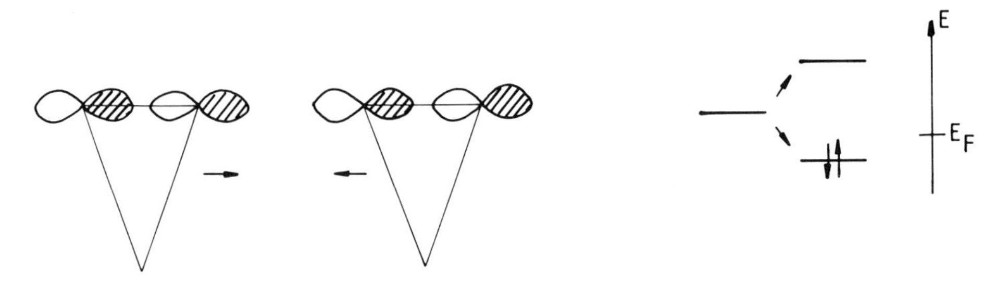

Fig. 17. Alignment of anion pairs in $TaSe_3$, showing how neighbouring intrachain antibonding states interact to produce a valence band interchain bonding state.

separate $TaSe_3$ columns approach and align so that some states which would otherwise have been Se–Se antibonds within a column become Se–Se or M–Se bonding states between columns. Thus rows of four Se atoms (terminated by a metal atom) occur both in this and the $NbSe_3$ structure. It is not hard to deduce (Figure 17) that there remains only one antibonding Se p state per pair of chains in $TaSe_3$, and the electron number is exactly right to fill the density of states to the trough, which is a semimetallic pseudo-gap [121] because the stronger interchain coupling causes stronger dispersion perpendicular to the chain direction within the slabs. There is little reason to expect incipient M–M pairing, nor is there any experimental evidence for a phase transition between 4 and 700 K at atmospheric pressure. The rather complex shape of the Fermi surface and its relation to Shubnikov–de Haas data has been discussed by Wilson [64]. Not surprisingly, strong anisotropies are observed in the compressibility of $TaSe_3$ [122] and in the stress dependence of its superconducting transition temperature [123].

3.2.3. $NbSe_3$

The $NbSe_3$ structure [110, 124] contains six columns per unit cell (Figure 6). Four columns are similar to those of $TaSe_3$, but these blocks are now interconnected through pairs of tight $ZrSe_3$-like columns. Thus there are three kinds of column, containing strong, intermediate and weak X–X pairings [64, 124]. The $ZrSe_3$ part of the structure contributes 8 valence p bands per MX_3 unit, and the $TaSe_3$ part contributes 17 valence bands per pair of MX_3 units. $NbSe_3$ is left with two spare electrons per unit cell (six MX_3 units) to place in the bottom of the conduction d band [6, 62–64, 125]. Superficially the position of the Fermi level in a deep minimum of the density of states (Figure 10) resembles the situation in $TaSe_3$, but the detailed structure of the highest occupied bands is entirely different (Figure 18). A group of six conduction bands in $NbSe_3$ are almost dispersionless in directions normal to the chain axis, but rise very rapidly (because of a strong d_{z^2} component along the b-axis) as **k** moves away from the origin along the chain direction. The result is a highly anisotropic metal, as borne out by experiment: the room temperature anisotropy of electrical conductivity has been determined as $\sigma_b/\sigma_a \sim 500$ [126], while $\sigma_b/\sigma_c \sim 20$ [127].

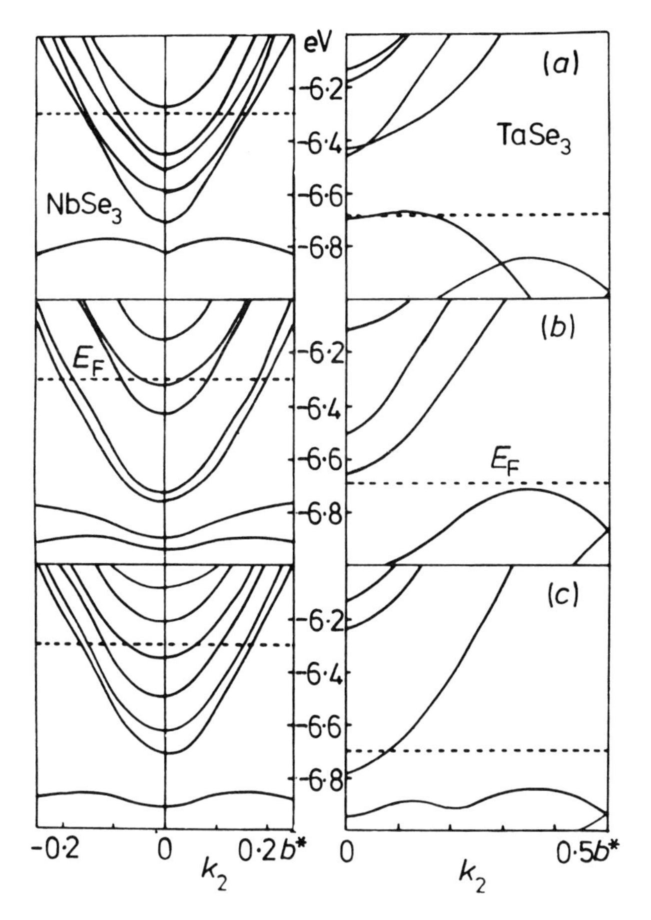

Fig. 18. Comparison of the dispersion parallel to the chain direction for bands near the Fermi energy in $NbSe_3$ and $TaSe_3$, along (a) $(0, k_2, 0)$, (b) $(\pi/a, k_2, 0)$, and (c) $(0, k_2, \pi/c)$. The one-dimensional nature of the bands crossing E_F in $NbSe_3$ leads to Peierls transitions with wavevector $\sim 0.25b^*$. (Bullett [62].) (© The Institute of Physics.)

The extraordinary nonlinear transport properties of $NbSe_3$ have attracted a great deal of recent attention. We only briefly summarize them here since a full discussion is provided in the following article. Two incommensurate phase transitions have been observed [7, 16], at onset temperatures of 144 and 59 K, and diffraction experiments revealed periodic lattice distortions with reduced wavevectors of (0, 0.243, 0) and (0.5, 0.263, 0.5) [8, 11, 18, 128] at the respective transitions. The resistivity increases sharply below each transition, and conductivity and specific heat measurements [129–131] suggest that the upper transition eliminates $\sim 25\%$ of the free carriers and the lower transition removes a further 30–50%. At both transitions the Hall coefficient R_H shows an abrupt increase in magnitude, but no change of sign [129]. The intense interest in $NbSe_3$ was generated because the periodic structural distortions are unusually weakly pinned to the lattice, leading to highly nonohmic behaviour. Application of a small electric field (which can be as low as a few mV per cm for the low-temperature transition and 0.1 V per cm for the higher temperature transition) induces the distortions to slide through the underlying lattice, and the conductivity rises until saturating close to the value expected had no transition occurred [8–10]. The conductivity in the low-temperature state is highly frequency dependent [13–15, 129], rising with frequency to coincide with the high-field zero-frequency value at ~100 MHz. Since the superlattice spots in X-ray diffraction are still visible when the current exceeds the linear threshold, it appears that the current carriers must be charge–density waves sliding through the lattice, as originally envisaged by Fröhlich [132–134]. Any deliberate increase in the number of impurities causes an increase in the threshold electric field by pinning the CDW more strongly to the lattice [5, 135, 136], although the depinning field is only weakly dependent on the concentration of defects introduced by electron irradiation [137–139].

According to the calculated band structure [63], occupation of the 50 fully occupied valence p bands per unit cell puts 16.3 electrons per $NbSe_3$ unit on the strongly X–X paired chains, 16.4 on the intermediately paired chains, and 17.3 on weakly paired chains. The higher electron density on the last chain type comes about because the smaller Se–Se separations in the other types forces Se–Se antibonding weight up into the empty conduction bands. At temperatures above the phase transitions the remaining two electrons per unit cell are distributed with weights of approximately 0.8 electron in pockets concentrated on the strongly paired chains, 1.1 in intermediate chains and 0.1 in the pockets concentrated on weakly X–X paired chains. The natural interpretation of the transitions is then to suppose that as $NbSe_3$ is cooled a periodic distortion develops first on the $ZrSe_3$-like chains, introducing an energy gap at some k_2 close to (but slightly less than) $\pm 0.25\pi/b$, with a similar transition at k_2 just greater in magnitude than $0.25\pi/b$ occurring on the intermediately X–X paired chains at some lower temperature. The actual distortion seems likely to be a metal–metal pairing of slightly less than 50% of metal atoms on the former chains, and slightly more than 50% in the latter [63, 64, 140]. Under a sufficiently large longitudinal electric field these M–M bonds can move freely along the trigonal double chains and the conductivity increases towards its normal value.

The one-dimensional characteristics are rapidly driven out by increases of pressure. The stronger interchain interactions cause the two transition temperatures to fall at 4 K $kbar^{-1}$ [7, 141], and at pressures of a few kilobars $NbSe_3$ can become superconducting [125].

The overall form of the calculated valence bands for $NbSe_3$ and $TaSe_3$ has been verified by the photoelectron energy distribution measurements of Endo and coworkers [142] (Figure 19).

3.2.4. *TaS_3*

The normal crystal structure of TaS_3 is orthorhombic, with a = 36.80 Å, b = 15.17 Å, c = 3.34 Å [143]. There are assumed to be 24 MX_3 units in the unit cell, but

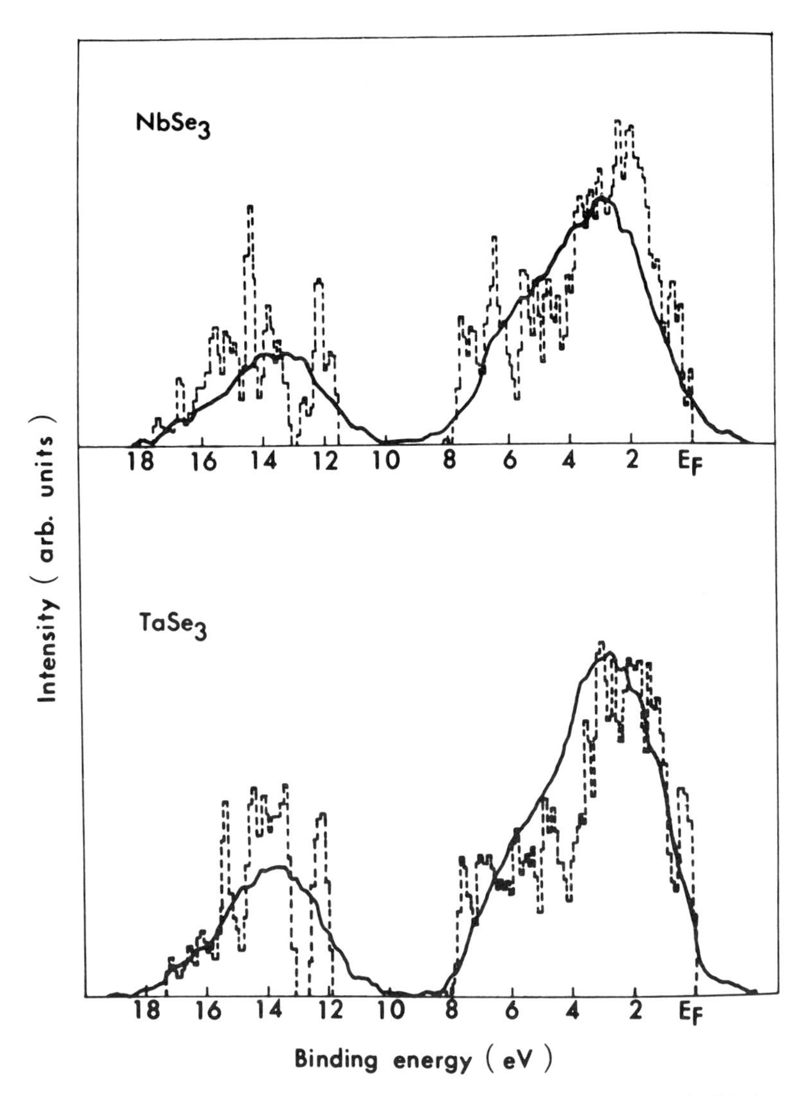

Fig. 19. XPS valence band spectra (solid lines) of $NbSe_3$ and $TaSe_3$, compared with the calculated density of states (dashed lines [62]). (Endo *et al.* [142].) (© North-Holland Publishing Co.)

the atomic coordinates have not been determined. The crystal morphology is similar to those of $NbSe_3$ and $TaSe_3$, and it is widely expected that the structure contains trigonal prismatic columns running parallel to the *c*-axis. Undoubtedly the electron wave functions have many aspects in common with $NbSe_3$ and $TaSe_3$. A monoclinic structure isomorphous with $NbSe_3$ has also been identified [117–119].

Orthorhombic TaS_3 undergoes a metal–semiconductor phase transition at about 215 K [143]. Below this temperature T_{MI} the activation energy for conduction is ~0.15 eV. Diffuse X-ray [144] and electron diffraction [143, 145, 146] studies show a charge–density wave with periodicity 4*c*. The temperature dependence of the order parameter has been established by Raman spectroscopy measurements [147] to be in accordance with the creation of a simple mean-field-theory-like gap at E_F. Figure 20 illustrates the temperature dependence of the low-field ohmic d.c. conductivity σ, together with its temperature derivative [148].

In spite of the structural differences, the nonlinear behaviour of TaS_3 below the transition temperature T_{MI} has strong similarities with that of $NbSe_3$. Above T_{MI} the conductivity is independent of electric field (Figure 21); below T_{MI} σ is strongly field dependent for electric fields above a threshold $E_T \sim 2.2\ \mathrm{V\,cm^{-1}}$, and at high electric fields σ approaches the value appropriate to the metallic state. The dielectric constant is zero above T_{MI}, but extremely high and strongly frequency dependent in the CDW state. The low-field a.c. conductivity is also strongly dependent on frequency below T_{MI} [148].

In both materials the anomalous nonlinear properties appear to arise from a pinned collective mode, which easily becomes mobile under applied electric fields.

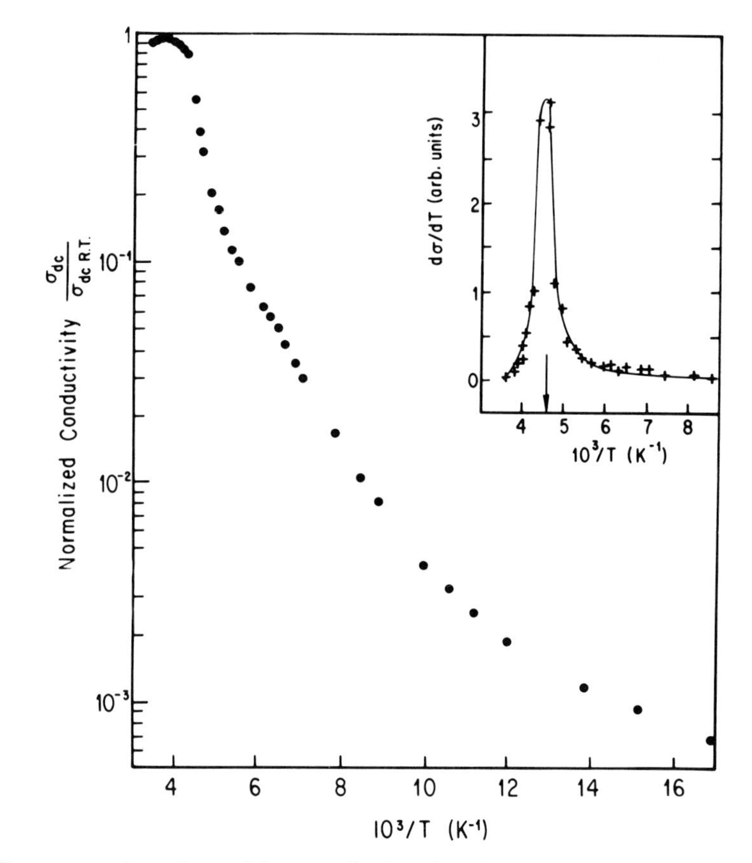

Fig. 20. Temperature dependence of the normalized conductivity of TaS_3 (Thompson *et al.* [148]). Inset shows $d\sigma/dT$ near the transition temperature $T_{MI} = 215\ \mathrm{K}$. The average room temperature conductivity measured on several crystals was $\sigma = (2 \pm 0.5) \times 10^3\ \Omega\,\mathrm{cm}^{-1}$. (© American Physical Society.)

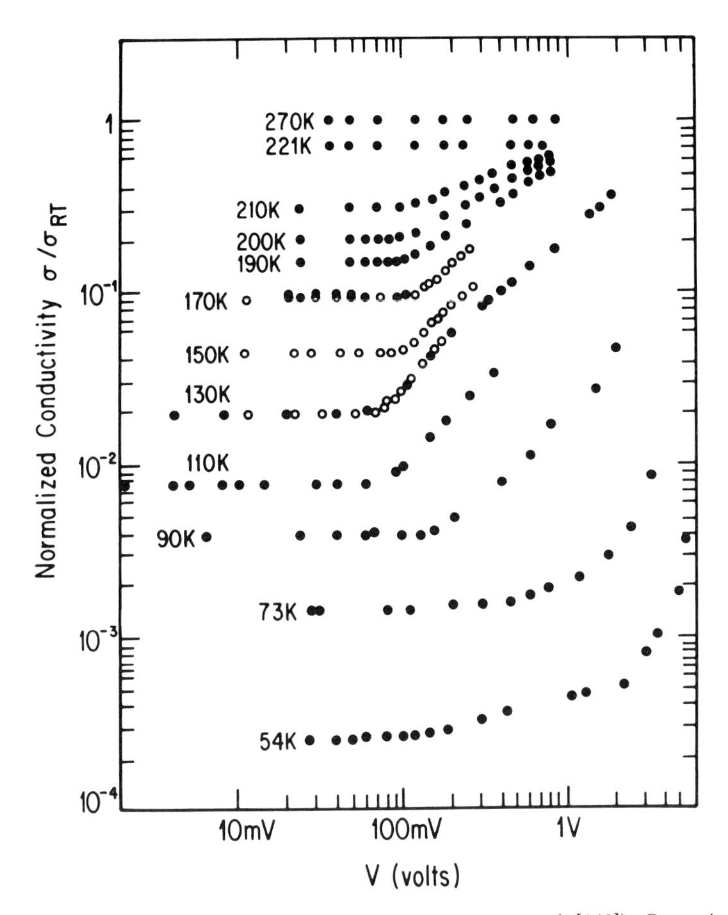

Fig. 21. Field dependence of the conductivity of TaS_3 (Thompson *et al.* [148]). Open circles are d.c. measurements and closed circles are pulse measurements. Sample length was 0.4 mm, indicating a threshold field $E_T \sim 2.2\,\mathrm{V\,cm^{-1}}$ just below the transition. (© American Physical Society.)

Niobium triselenide has substantial two-dimensional character and the CDWs destroy only part of its Fermi surface. The CDWs in $NbSe_3$ are incommensurate and probably pinned by impurities and defects. Tantalum trisulphide is more one dimensional, and the establishing of the CDW removes the whole of the Fermi surface and leads to a semiconducting state below the transition. In TaS_3 the CDW periodicity is locked into the underlying lattice, and it is not yet known whether the dominant pinning of the collective mode is by impurities and defects or by the TaS_3 lattice potential itself.

3.2.5. $FeNb_3Se_{10}$

The normal effect of impurities is to suppress the periodic lattice distortion in $NbSe_3$ and prevent Fröhlich conductivity. However, a compound of Fe and $NbSe_3$ with

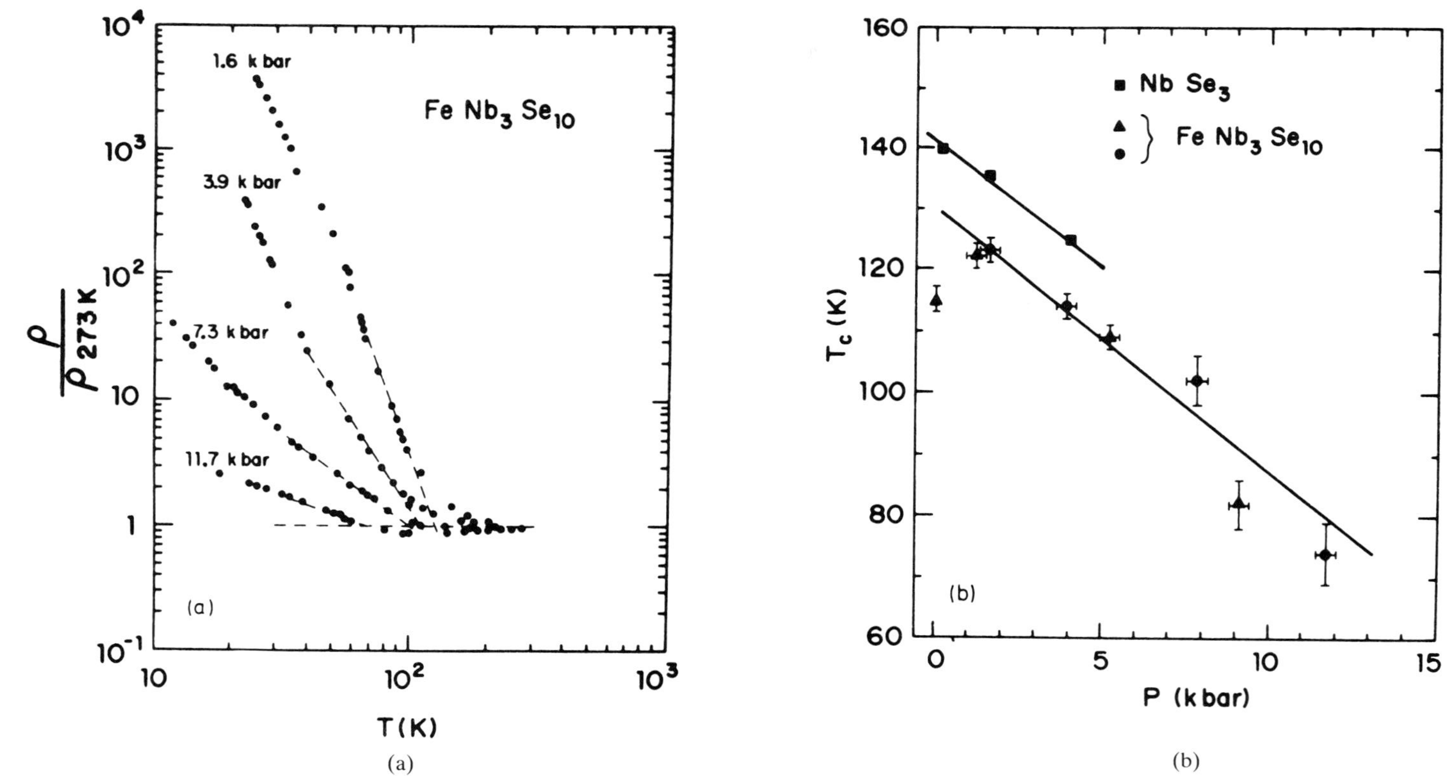

(a)

(b)

Fig. 22. (a) Normalized resistivity of $FeNb_3Se_{10}$ versus temperature for four pressures, showing the rise in resistivity below $T_c \sim 100\,K$. (b) Pressure dependence of T_c for two samples of $FeNb_3Se_{10}$ and of $NbSe_3$. (Hillenius and Coleman [152].) (© American Physical Society.)

nominal composition $FeNb_3Se_{10}$ has recently been discovered to have structural characteristics in common with $NbSe_3$, including the onset of a periodic distortion at 140 K [149–153]. Above this temperature $FeNb_3Se_{10}$ shows a metallic conductivity comparable to that of $NbSe_3$. Below the transition temperature the resistivity rises rapidly and is larger by nine orders of magnitude at 3 K. X-ray scattering reveals an incommensurate distortion [149], with an onset lattice vector of (0.0, 0.27, 0.0). The resistivity rise is greatly reduced by the application of pressure [152] and the onset temperature T_c falls at 5 K kbar^{-1}, as in $NbSe_3$ (Figure 22).

The $FeNb_3Se_{10}$ structure]150, 151] is projected down the monoclinic *b*-axis in Figure 23. Two trigonal chains pass through the unit cell in an orientation almost identical to the two tightly Se–Se paired chains in $NbSe_3$ (Se–Se = 2.34 Å in $FeNb_3Se_{10}$ compared to 2.37 Å in $NbSe_3$). Separating the trigonal columns are double chains of edge-sharing $NbSe_6$ and $FeSe_6$ octahedra. X-ray diffraction data indicate that the octahedral columns are disordered, with Fe and Nb dispersed randomly at each metal atom site. It is thought that the disorder is confined to the octahedral sites, and this imposes a relatively minor perturbation on any incipient instability on the trigonal prismatic columns.

The electron density of states calculated for $FeNb_3Se_{10}$ (Figure 24 [63]) has much in common with that of pure $NbSe_3$. The additional *d* electrons supplied by Fe atoms are exactly compensated by the equal number of Nb atoms that revert to octahedral coordination. The remarkable result is that the Fermi level still lies just above a deep minimum in the state density, and again E_F cuts through a group of quasi-one-dimensional bands. The two lowest of these remain the Nb d_{z^2} bands

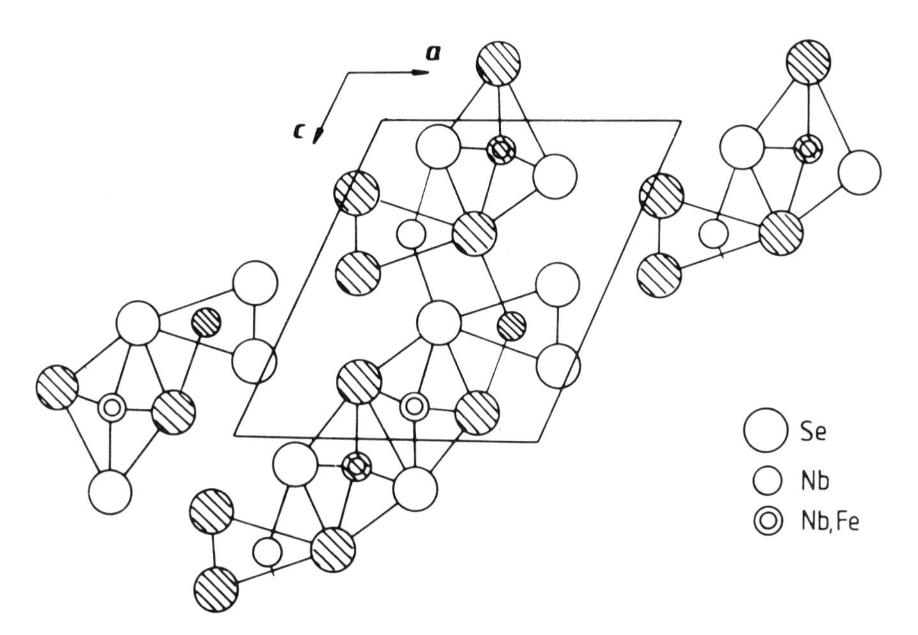

Fig. 23. {010} projection of the $FeNb_3Se_{10}$ structure, showing the trigonal and octahedral double chains. Open circles are at $\frac{1}{4}b$, filled circles at $\frac{3}{4}b$.

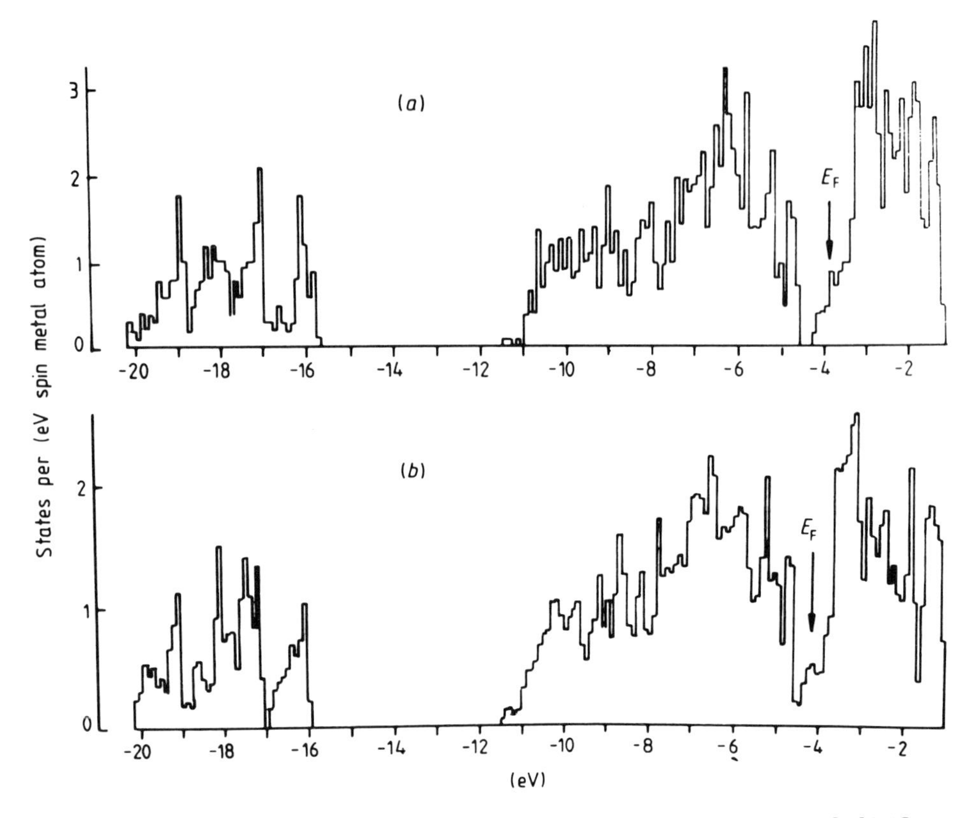

Fig. 24. Calculated density of electron states for (a) $NbSe_3$ and (b) $FeNb_3Se_{10}$. (Bullett [63].) (© The Institute of Physics.)

coupling metal atoms along the axes of the trigonal chains. The calculation suggests there will also be a slight electron occupation of the next higher band, which arises from bonding interactions between transition element atoms in the octahedral chains. The *d*-orbital configuration of Nb(II) octahedrally coordinated atoms is essentially identical to that of the Nb(I) atoms in the trigonal chains (Table X). The total electron count leaves one electron per unit cell to occupy the partially filled band region of Figure 25. Since the $FeNb_3Se_{10}$ structure contains only one crystallographically distinct (albeit disordered) type of trigonal column, we might have predicted that M–M pairing of trigonally coordinated Nb atoms would drive a phase transition to a commensurate structure with period 4*b*. In fact, the reported incommensurate superlattice vector (0, 0.27, 0) deviates slightly from the expected picture. This deviation may indicate that the true composition is richer in iron and therefore contains more electrons per unit cell than would the formula $FeNb_3Se_{10}$. A recent study [153] suggests that this compound is actually nonstoichiometric and exists in the Fe concentration range represented by $Fe_{1+x}Nb_{3-x}Se_{10}$ with $0.25 < x < 0.40$, not $x = 0$ as originally thought. It remains to be seen how the charge–density

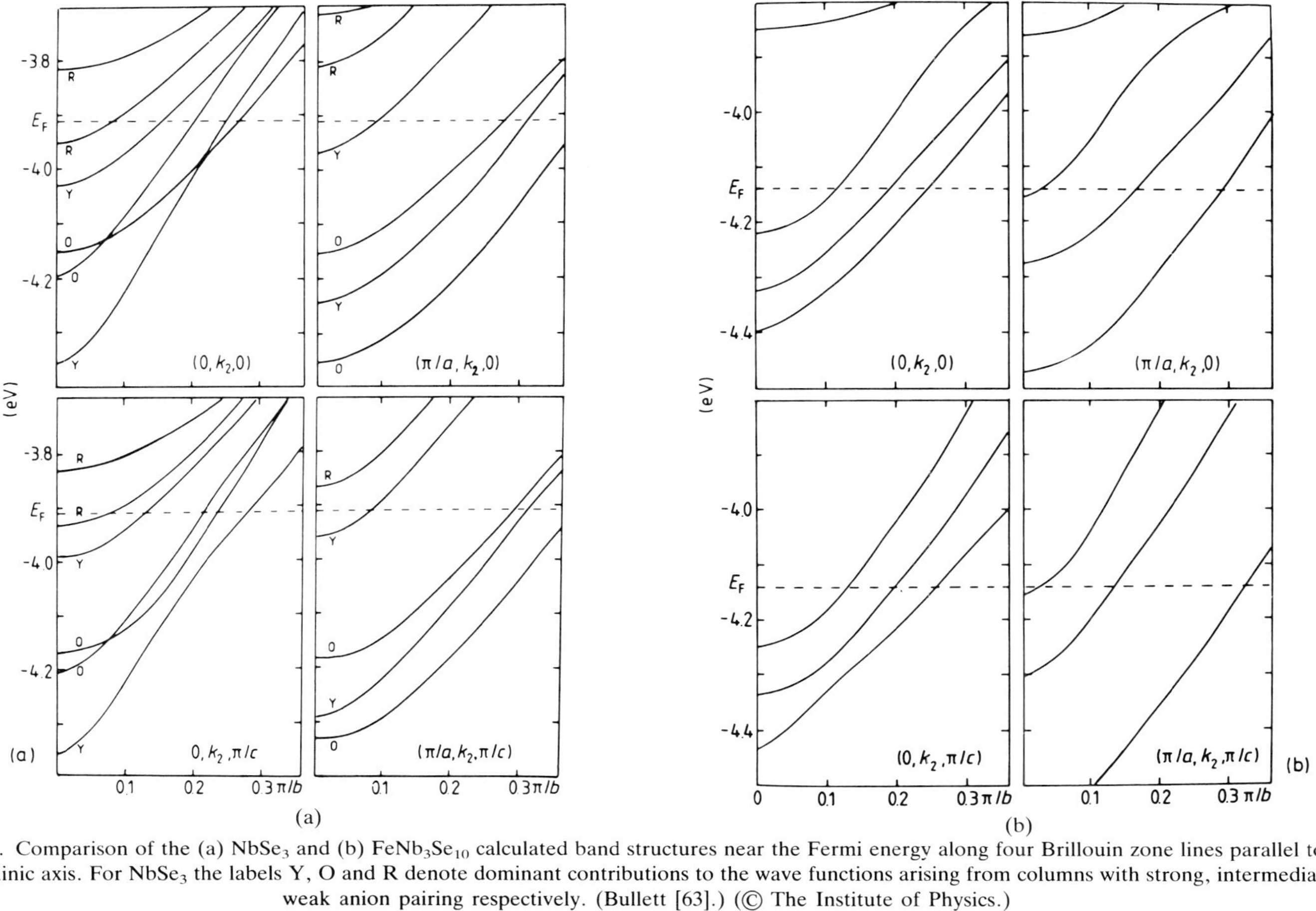

(a) (b)

Fig. 25. Comparison of the (a) $NbSe_3$ and (b) $FeNb_3Se_{10}$ calculated band structures near the Fermi energy along four Brillouin zone lines parallel to the monoclinic axis. For $NbSe_3$ the labels Y, O and R denote dominant contributions to the wave functions arising from columns with strong, intermediate or weak anion pairing respectively. (Bullett [63].) (© The Institute of Physics.)

TABLE X

Energy levels (eV) for isolated neutral atoms and self-consistent configurations for $NbSe_3$ and $FeNb_3Se_{10}$ (Bullett [63]). (© The Institute of Physics.)

	Nb ($d^{3.4}$)	Fe ($d^{7.1}$)	Se
s	−5.1	−5.7	−17.4
p	−2.9	−3.3	−7.4
d	−4.7	−5.5	—

$NbSe_3$:	Nb ($5s^{0.1}5p^{0.1}4d^{3.4}$)
	Se ($4s^{1.9}4p^{4.3}$)
	2 × Se ($4s^{1.9}4p^{4.8}$)
$FeNb_3Se_{10}$:	Fe ($4s^{0.2}4p^{0.1}3d^{7.1}$)
	Nb ($5s^{0.1}5p^{0.1}4d^{3.4}$)
	Se ($4s^{1.9}4p^{4.2-4.9}$)

wave properties, especially the superlattice wavevector, will vary with electron concentration.

Disorder in the arrangement of Nb and Fe atoms on metal (II) octahedral sites will clearly reduce the correlation length of periodic distortions in the trigonal columns, and may repress the more extreme properties of $NbSe_3$. Disorder in $FeNb_3Se_{10}$ prevents the charge–density waves becoming unpinned and sliding through the lattice under modest electric fields. The low-temperature phase builds up more gradually than in pure $NbSe_3$. In $FeNb_3Se_{10}$ there is only a slight resistivity change at 140 K, and it is only below 100 K that the very dramatic rise sets in [150]. Below 19 K the conductivity follows the typical temperature dependence $\rho \propto \exp(T^{-1/4})$ of a hopping mechanism between disorder-localized carrier sites [154–156].

Very recent work [157] has shown that $FeNb_3Se_{10}$ is not the only phase of this structural type in which octahedral chains are interposed between trigonal prismatic $NbSe_3$ chains. Two new compounds with chemical compositions close to (Fe–V) Nb_2Se_{10} and $Cr_2Nb_2Se_{10}$ have now been reported in this family, although attempts to introduce manganese and nontransition elements such as magnesium into the octahedral chains have so far been unsuccessful, as have attempts to synthesize the tantalum derivatives $(MM')Ta_2Se_{10}$ or $(MM')Ta_2S_{10}$.

4. Zirconium and Hafnium Pentatellurides

The transition-metal pentatellurides $ZrTe_5$ and $HfTe_5$ are two further quasi-one-dimensional materials which crystallise in a chain-like structure and show large resistivity peaks at temperatures below room temperature. Transport measurements [158–160] showed an anomalous peak in electrical resistivity near $T_p = 141$ K for $ZrTe_5$ and 83 K for $HfTe_5$ (Figure 26). The Hall coefficient changes sign near T_p [161–163]. The thermoelectric power for both materials is large and positive at high temperatures, but then drops precipitously and crosses zero near T_p [164]. The abrupt change is indicative of a phase transition where the carrier type changes from hole-like to electron-like. At lower temperatures the thermoelectric power reaches a large negative peak and then decreases to zero in a metallic fashion [164].

Although the anomalies in transport properties are reminiscent of those in $NbSe_3$, their effects are much weaker, and the early evidence for their origin in a structural phase transition was not conclusive. In contrast to $NbSe_3$, X-ray reflection anomalies are not seen along the [100] chain axis. DiSalvo *et al.* [159] mounted a single crystal of $ZrTe_5$ so that the scattering plane was (*hk*0), and took long scans at 80 K along [100] and [010] but found no trace of any unexpected diffraction peaks. Scans over restricted sections of the (*hh*0) plane gave hints of a superlattice with a reduced wavevector of $\mathbf{q} = (0.5, 0.5, 0)$, but these extra Bragg peaks were extremely weak and persisted when the crystal was warmed to room temperature, indicating that this small distortion was not associated with the resistivity anomaly.

However Skelton *et al.* [165] have detected anomalous behaviour near T_p in the X-ray line intensities along the *c*-axis, a direction not looked at by the Bell group [159]. Room-temperature scans along [001] revealed the presence of weak odd-order reflections, in violation of the required extinctions of the reported *Cmcm* space group. These 'forbidden' reflections were recorded for every sample of $ZrTe_5$ and $HfTe_5$ on the diffractometer, though they could not be detected in oscillation photographs about the [001] axis. Their most striking property is the temperature dependence of the intensity of the lowest odd-order (001) reflections (Figure 27). In both $ZrTe_5$ and $HfTe_5$ the (001) and (003) peak intensities show a gradual rise (somewhat larger than that predicted by the Debye–Waller effect) on cooling from room temperature. At a lower temperature the intensities increase strongly and sharply up to some maximum value, beyond which the intensities fall with decreasing temperature, contrary to the Debye–Waller factor. These X-ray reflection intensity anomalies were reversible and reproducible. Their close correlation with the temperature dependence of the resistive anomalies provides the strongest evidence that the latter may be associated with a structural transition. Line shifts have also been sought in Raman spectra [166, 167], but a comparison of high- and low-temperature spectra provides no further indication of a phase transition.

The conductivity anomalies in both compounds are strongly frequency dependent [168]. At microwave (9.1 GHz) frequencies the anomalies are almost completely wiped out (Figure 28). Conductivity measurements on the alloy $Hf_{0.5}Zr_{0.5}Te_5$ show that disorder leads to the suppresion of the strong frequency dependence. The resistivity behaviour is rather similar to that of the layer compound $TiSe_2$ [169], in which the hexagonal *a* and *c* lattice parameters double to form a commensurate superlattice below a second-order phase transition at 202 K. A possible explanation of the resistivity anomalies would be the development of a collective mode, which removes part of the Fermi surface when some of the electrons condense at T_p. The frequency-dependent conductivity would then be due to the response of the collective mode to external a.c. driving fields. The collective mode may be pinned by impurities or by the underlying lattice. This pinning would need to be much stronger than in $NbSe_3$ and TaS_3, since no nonlinear d.c. conduction is seen in the pentatel-

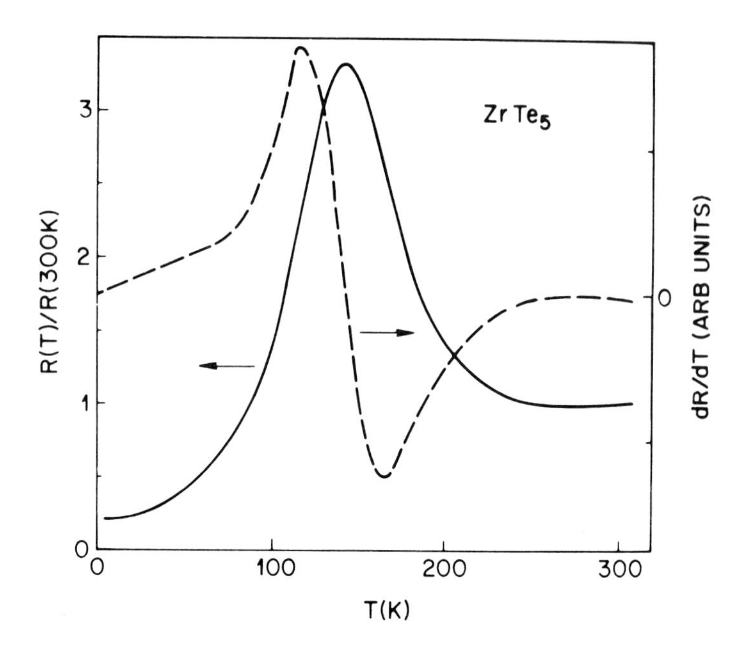

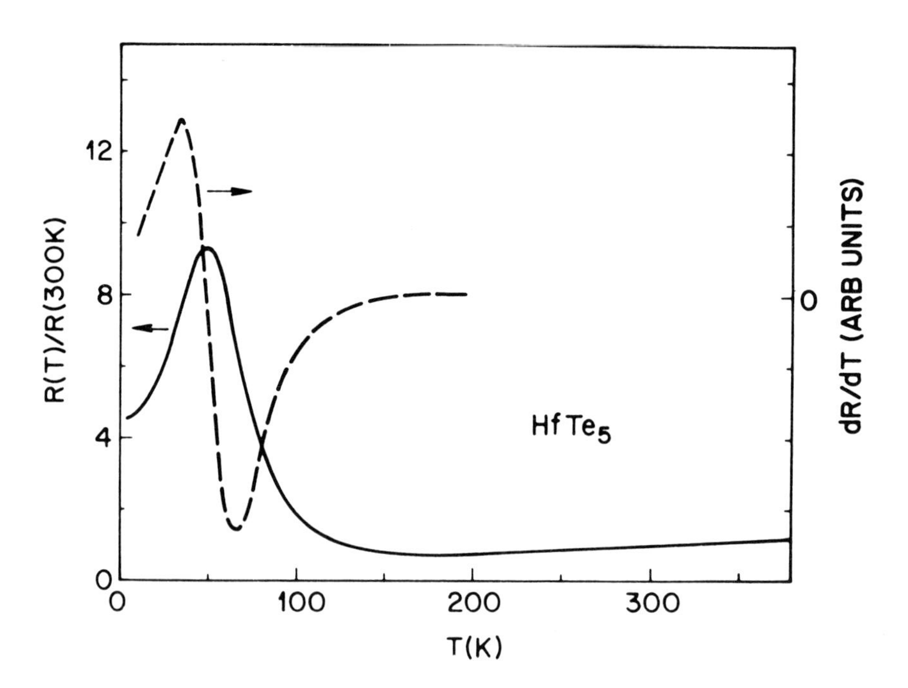

Fig. 26(a) and (b) Electrical resistance (solid line) and its derivative (broken line) of $ZrTe_5$ and $HfTe_5$ from 4.2 to 300 K. The current is parallel to the chain direction. (© American Physical Society.)

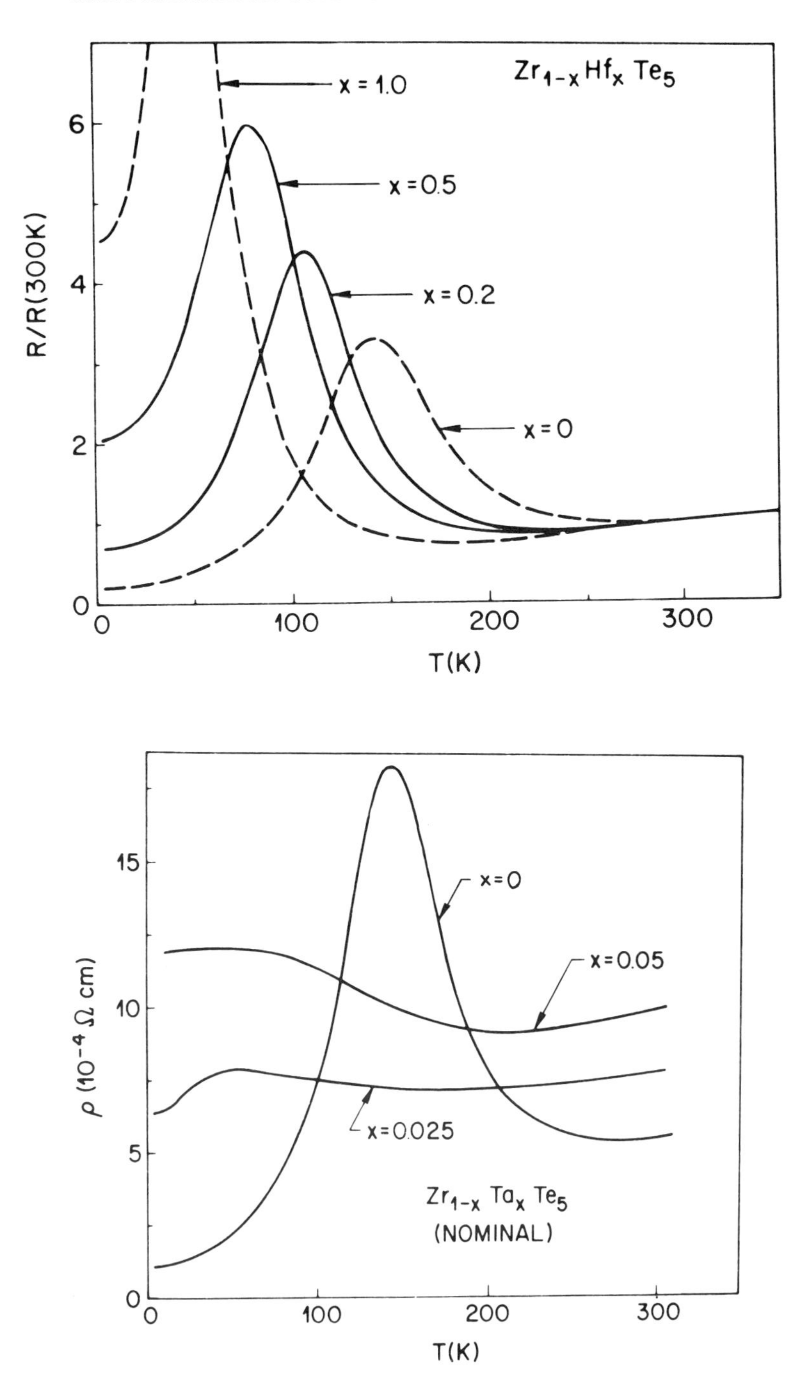

Fig. 26(c) Electrical resistance of $Zr_{1-x}Hf_xTe_5$ (each normalized to R (300 K)) for different compositions. (d) Suppression of the resistivity anomaly by doping with Ta. (DiSalvo *et al.* [160].) (© American Physical Society.)

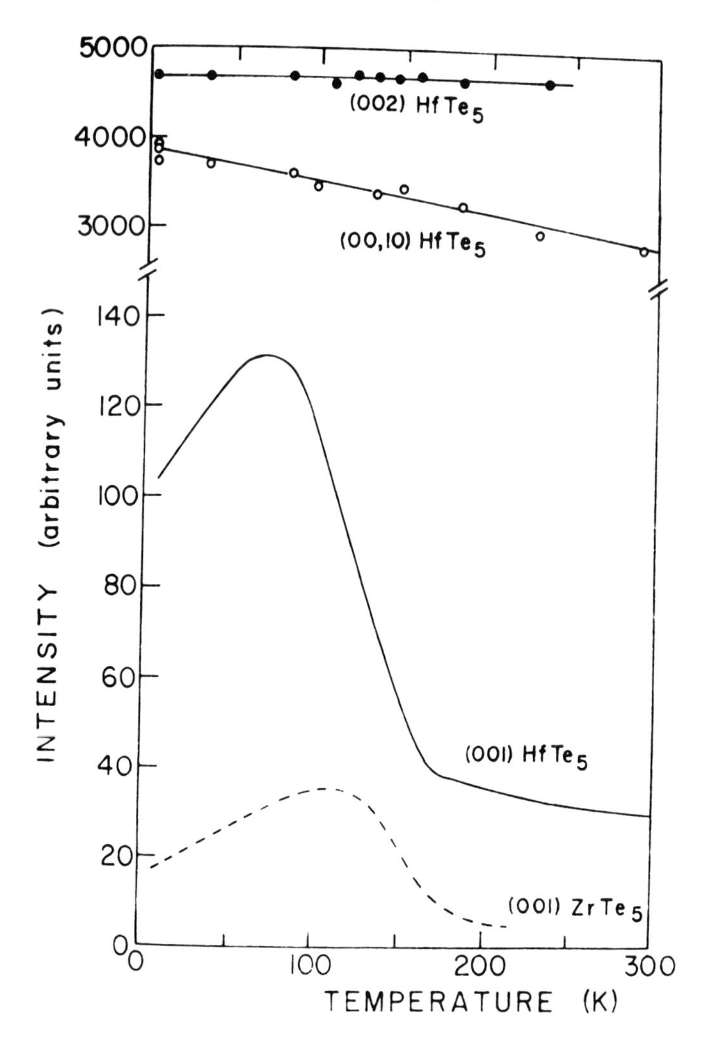

Fig. 27. Temperature dependence of X-ray intensities of the 'forbidden' (001) reflections in $HfTe_5$ and $ZrTe_5$, showing the strong maximum near the temperature where the resistivity anomalies occur. (Skelton *et al.* [165].) (© Pergamon Press Ltd.)

lurides for pulsed fields up to $20\,V\,cm^{-1}$ [158, 165], whereas strongly nonohmic effects are observed in $NbSe_3$ above $10\,mV\,cm^{-1}$ and in TaS_3 above $2\,V\,cm^{-1}$.

The {100} and {010} projections in Figure 29 illustrate the one- and two-dimensional aspects of the structure. The crystals are orthorhombic, with space group $Cmcm(D_{2h}^{17})$. The room-temperature lattice constants determined by Furuseth *et al.* [170, 171] are $a = 3.988$ Å, $b = 14.502$ Å, $c = 13.727$ Å for $ZrTe_5$ and $a = 3.974$ Å, $b = 14.490$ Å, $c = 13.730$ Å for $HfTe_5$. A full structure determination has been made only for the hafnium compound, but, since the lattice constants are so similar, the atomic coordinates are virtually interchangeable between the two materials.

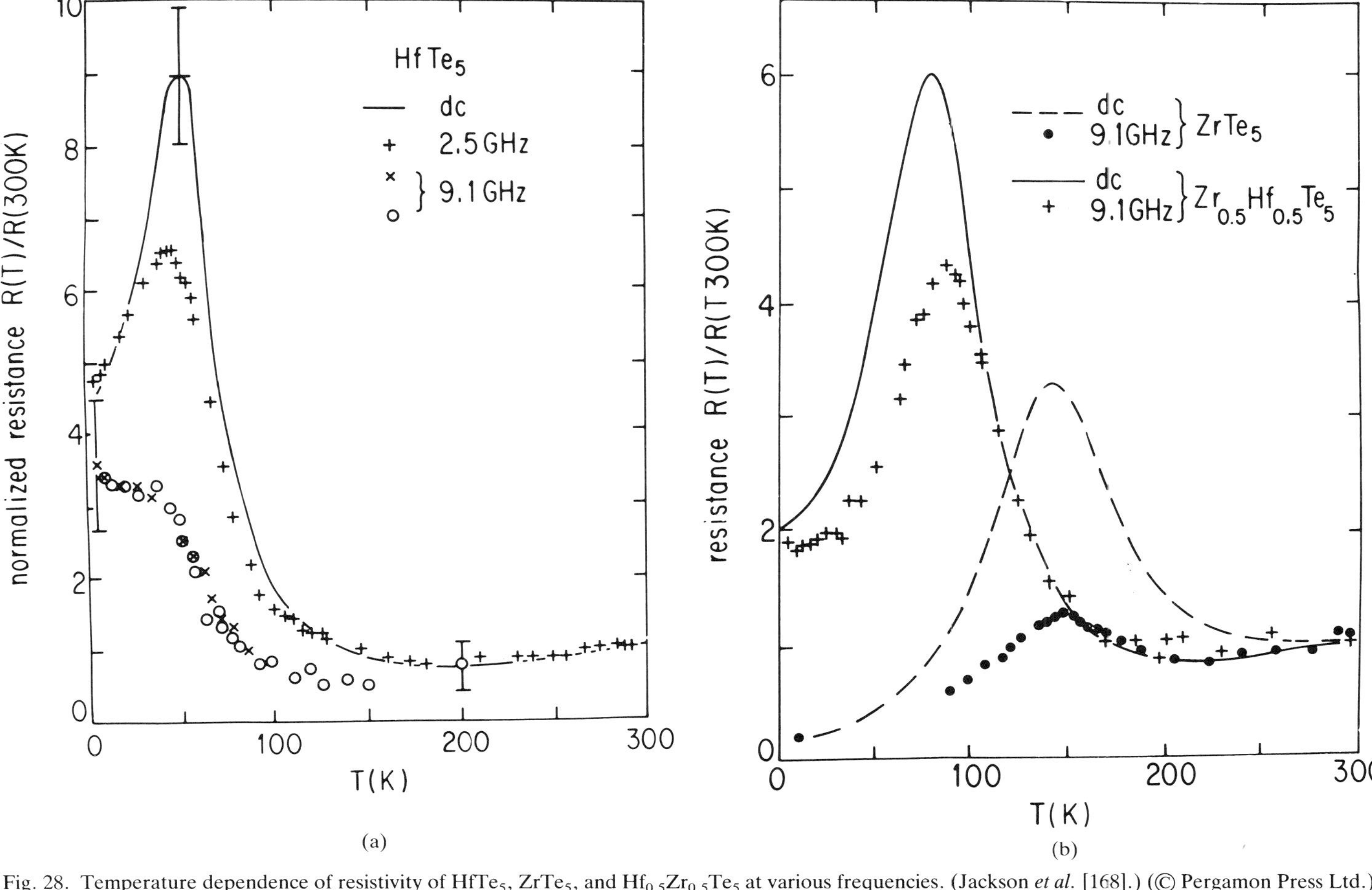

Fig. 28. Temperature dependence of resistivity of $HfTe_5$, $ZrTe_5$, and $Hf_{0.5}Zr_{0.5}Te_5$ at various frequencies. (Jackson *et al.* [168].) (© Pergamon Press Ltd.)

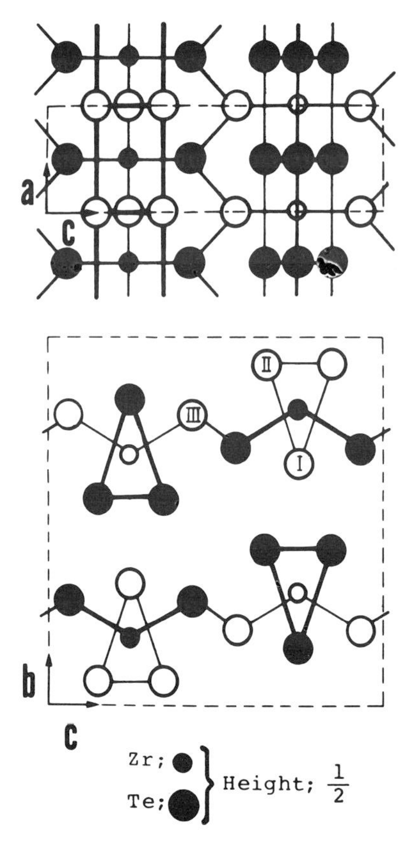

Fig. 29. (01) and (100) projections of the $ZrTe_5$ structure determined by Furuseth *et al.* [171], (from [174], © Physical Society of Japan).

One-dimensional chains of end-sharing MX_6 trigonal prisms run through the structure parallel to the *a*-axis, but instead of being linked directly into pairs, and then layers, of columns as in the trichalcogenides, in the pentachalcogenides the MX_3 chains are linked through intermediate 'Te_2' chains. MTe_3 and Te_2 chains alternate along the *c*-axis in two-dimensional layers, in an arrangement which retains the coordination of eight close Te neighbours around each M atom. Six Te neighbours are at the corners of the triangular prism and two lie outside rectangular faces of the prism. Te–Te pairings are slightly stronger within the prisms (2.763 Å) than in the intervening zigzag chains (2.908 Å). The corresponding single-bond distance in elemental Te falls halfway between these values (2.835 Å). The long interlayer Te–Te contacts (>4.161 Å in $HfTe_5$) illustrate that the mutual interactions between slabs are of only the weak van der Waals type.

Electronic structure calculations have been carried out in the two-centre atomic-orbital-based methods, with the relevant interactions computed either by direct integration [172] or according to the extended Hückel procedure [173]. Figures 30 and 31 show the calculated $ZrTe_5$ energy band structure along some of the edges of

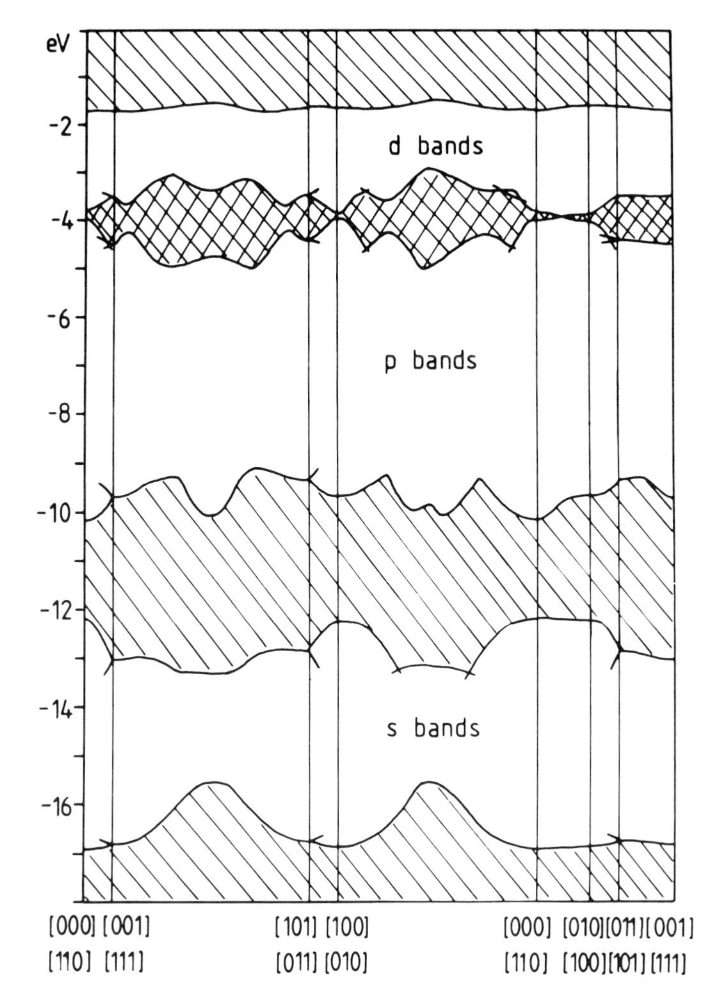

Fig. 30. Calculated electron energy band limits of $ZrTe_5$ along Brillouin zone symmetry lines, showing the semimetallic band crossing along $(0, k_2, 0)$ and $(1, k_2, 0)$. (Bullett [172].) (© Pergamon Press Ltd.)

the irreducible Brillouin zone. States in the energy range −17 to −12 eV are essentially Te *s* states, while the ranges −10 to −4 eV and −4 to −1.5 eV may be conveniently labelled as Te *p* bands and Zr *d* bands, respectively. The Zr 5*s* and 5*p* conduction bands lie several volts higher in energy. Because of the large number (58) of bands involved, only the limiting bands in each group are shown.

The band structure calculations [172, 173] predict quite clearly that $ZrTe_5$ should be a semimetal. Along most of the Brillouin zone edges (and throughout most of the interior) a semiconducting gap ~1 eV separates the highest occupied from the lowest unoccupied band, but coupling between adjacent $ZrTe_5$ layers leads to a semimetallic band crossing along the $(0, k_2, 0)$ and $(1, k_2, 0)$ edges (which are equivalent by symmetry). For a single layer, two bands, both largely derived from Te *p*

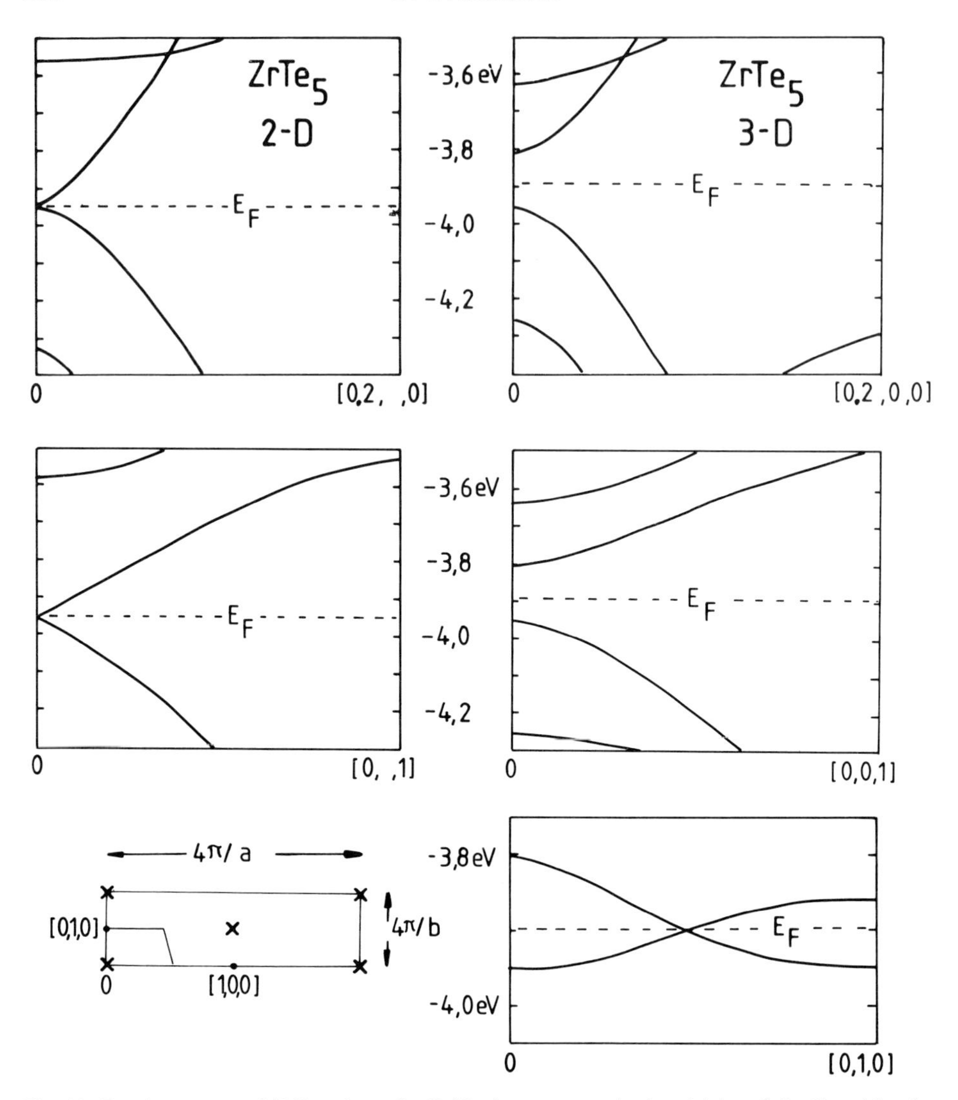

Fig. 31. Band structure of $ZrTe_5$ along the Brillouin zone axes in the vicinity of the Fermi level, as calculated by the nonempirical atomic orbital method (Bullett). Interactions between the 2D layers lead to the semimetallic band crossing near $(0, \pi/b, 0)$ in the 3D structure.

orbitals, come very close together (~0.01 eV [172] or 0.05 eV [173]) at the zone centre Γ (see Figures 31 and 32). The lower (occupied) band is concentrated on Te_I p orbitals oriented parallel to the c-axis; the upper (empty) band consists primarily of p orbitals oriented along the a-axis on the Te_{III} zigzag chains (Figure 29). These two sets of Te p orbitals are orthogonal at $\mathbf{k} = 0$, but mix and repel each other in energy as soon as $\mathbf{k}$ acquires a component along the {100} chain axis. In the three-dimensional crystal, interlayer coupling causes the two relevant bands to cross near $\mathbf{k} = (0, \pi/b, 0)$, and the energy of their crossing point determines the Fermi level.

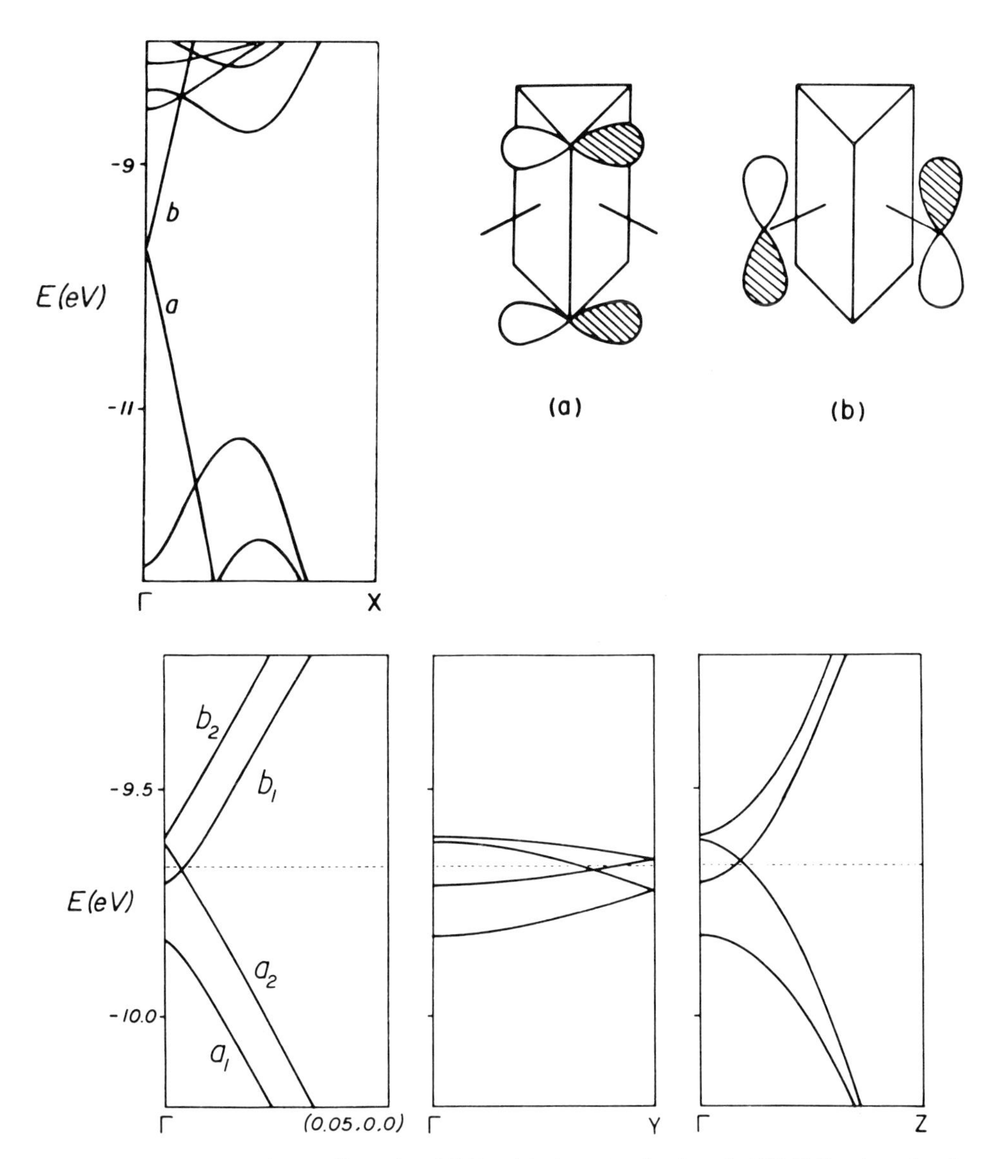

Fig. 32. Band structure of a two-dimensional $ZrTe_5$ slab along the a^*-axis and of 3D $ZrTe_5$ along the a^*, b^*, and c^* axes, as calculated by the extended Hückel method, and a schematic view of the wave function at the top of the valence band and at the bottom of the conduction band in a 2D $ZrTe_5$ slab. (Whangbo *et al.* [173].) (© American Physical Society.)

Formally each Zr atom donates two electrons to fill the Te_I p shell and one electron to each Te_{II} atom; because the Te_{II} atoms are closely spaced the corresponding antibonding p state is forced up into the empty conduction bands. The zigzag chains of Te_{III} atoms (bond angle 86° in $HfTe_5$) retain the formal p^4 configuration of elemental Te, with two p electrons per atom in Te–Te bonding states, two in

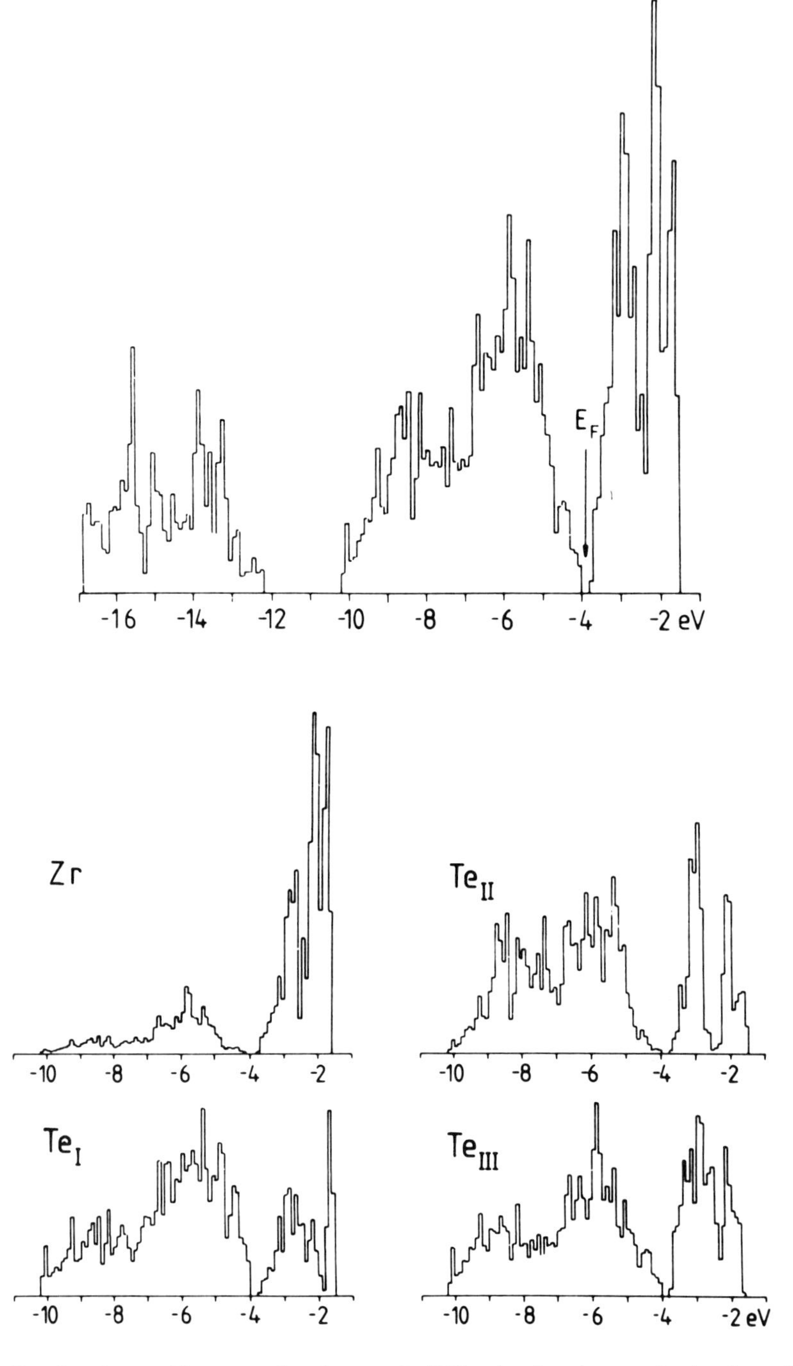

Fig. 33. Density-of-states histograms for electrons in $ZrTe_5$ showing the contribution from different atomic sites (Bullett [172].) (© Pergamon Press Ltd.)

nonbonding states, and two Te–Te antibonding states remaining unoccupied. Considerable hybridization between Zr and Te orbitals actually takes place, as shown in Figure 33, and the calculated orbital occupations of atoms are Zr: $s^{0.1}p^{0.1}d^{2.7}$; Te_I: $s^{1.9}p^{4.7}$; Te_{II}: $s^{1.9}p^{4.4}$; Te_{III}: $s^{1.9}p^{4.0}$.

Obviously this is a very different system from the group IV or group V trichalcogenides. $NbSe_3$ contains excess electrons beyond the number required to fill states up to the incipient semiconducting gap. The excess electrons enter broad, partially occupied, quasi-one-dimensional bands which drive the periodic lattice distortions at 144 and 59 K. NbS_3 is a semiconductor because the electrons left over after filling the sulphur valence *p* bands enter a commensurate system of metal–metal bonds. The group IV MX_3 compounds are semiconductors because there are no electrons left over after filling the chalcogen valence *p* bands (though the tritellurides may be metals by virtue of the indirect valence-conduction gap shrinking to zero). $ZrTe_5$ contains no electron excess: the electron number is exactly right to fill the valence bands, but interchain and interlayer coupling are just sufficient to induce the band-crossing typical of a semimetal. While the physical properties of the pentatellurides will show a pronounced anisotropy by virtue of their crystallographic structure, the bands at the Fermi level are not one-dimensional in the sense of $NbSe_3$ and there is much less reason to expect charge–density wave instabilities. But for the X-ray [165] and frequency-dependent conductivity [168] evidence for a collective mode, one might have attributed the direct current resistivity anomaly to the strong temperature dependence of carrier density and carrier mobility in an ordinary semimetal [160, 172–174]. Since the carrier states at the Fermi level originate primarily from Te atoms, the resistive peak would be retained after alloying with isoelectronic atoms (as in $Zr_{1-x}Hf_xTe_5$ and $Hf_{1-x}Zr_xTe_5$ [160]) but would be suppressed by alloying with electron donating atoms (as in $Zr_{1-x}Ta_xTe_5$ and $Hf_{1-x}Ta_xTe_5$ [160]).

5. Other One-Dimensional Compounds

In this section we very briefly review the electronic structures of various other transition metal compounds that have been investigated because of one-dimensional aspects of their crystal structure and behaviour.

5.1. $NbTe_4$ AND $TaTe_4$

The tetratellurides $NbTe_4$ and $TaTe_4$ provide two further examples of one-dimensional metals with intrinsic charge–density wave instabilities [175]. These materials have a different crystal structure from any previously studied one-dimensional crystals. The atomic arrangements approximate a tetragonal lattice with subcell dimensions $a' = 6.50$ Å, $c' = 6.82$ Å for $NbTe_4$ [175–177] and $a' = 6.51$ Å, $c' = 6.81$ Å for $TaTe_4$ [175, 178, 179], with subcell space groups reported to be either *P4/mcc* or *P4cc* or, for $NbTe_4$ perhaps *P422*. Superimposed on the X-ray reflections of this sublattice are faint superlattice spots indicating that the *a*-axis is doubled and the c-axis is trebled. For $TaTe_4$ the c-axis superlattice is commensurate, for $NbTe_4$ the superlattice periodicity deviates from $3c$ by a few per cent.

The atomic arrangement of the subcells consists of columns along the c-axis of distorted square antiprisms of Te atoms with the metal atoms near their centres (Figure 34). For the M atom in its ideal central position, each Nb atom in $NbTe_4$ would have eight Te neighbours at 2.90 Å and two Nb atoms at 3.42 Å along the c-

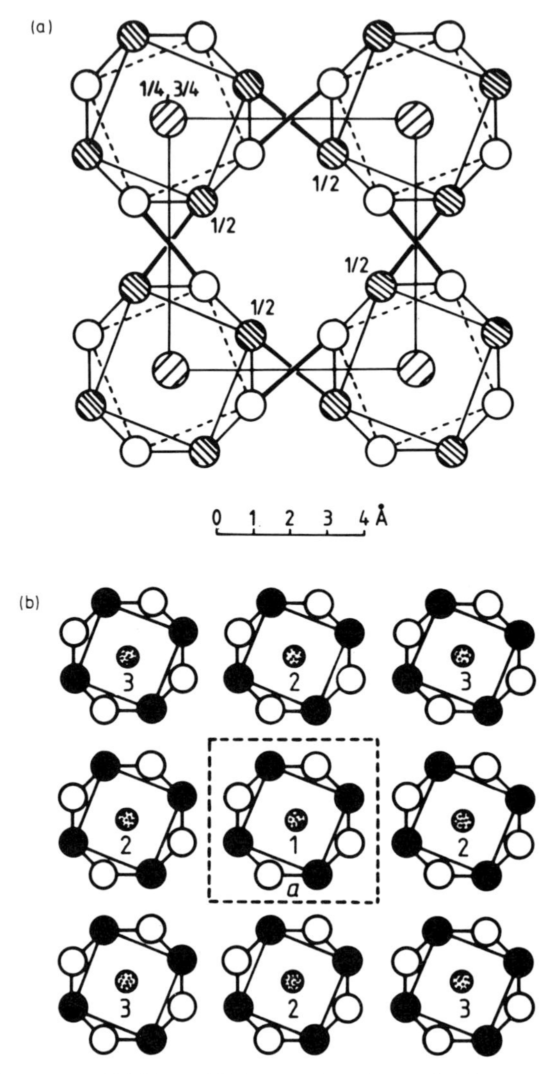

Fig. 34. (a) Subcell structure of $NbTe_4$ and $TaTe_4$, projected along [001], emphasizing the linear chains of metal atoms and the Te–Te pairing between the distorted-square-antiprismatic MTe_4 columns. Metal atoms are at height $\frac{1}{4}$, $\frac{3}{4}$; Te atoms are at height 0 (open circles) and $\frac{1}{2}$ (shaded circles). (b) Superlattice structure of $NbTe_4$ and $TaTe_4$, showing the three types of column in the $2a \times 2a \times 3c$ ($TaTe_4$) or $2a \times 2a \times \sim 3c$ ($NbTe_4$) unit cell. (Boswell *et al.* [175].) (© The Institute of Physics.)

axis. Each Te has two Te neighbours at 3.32 Å in the same square face of the antiprism, but is bonded more strongly to a single Te atom at 2.87 Å in a neighbouring $NbTe_4$ column (2.92 Å in $TaTe_4$), a distance only slightly longer than the 2.84 Å interatomic distance in elemental Te. One way to look at the structure is thus as a collection of Te_2 singly bonded diatomic molecules interspersed between linear chains of transition metal atoms. The formal charge of the diatomic ligands is $(Te_2)^{2-}$, since in the simplest analysis the p σ-antibonding state is unoccupied. One electron per group V transition-element atom remains in the metal chain and the situation would appear ripe for a commensurate M–M pairing Peierls transition analogous to NbS_3. As there are already two metal atoms along the c-axis of the sublattice unit cell, M–M pairing need not alter the c parameter.

However the MTe_4 structure is not one-dimensional to the same extent as NbS_3 or NbI_4. Electronic structure calculations reveal that a simple M–M pairing displacement within the sublattice unit cell is insufficient to eliminate the Fermi surface completely. Figure 35 compares the energy distribution of electron states in the valence and conduction band region, calculated (a) for Nb in the centre of the $NbTe_8$ antiprisms and (b) for alternate axial displacements of 0.25 Å, so that Nb atoms are paired with a bond length of 2.9 Å. Although one would obviously expect small displacements of the Te_2 pairs to accompany the Nb–Nb bonding, the calculation does suggest that in the tetratellurides the three-dimensional coupling between MTe_4 columns is too strong to allow a semiconducting gap in this simple arrangement. A full explanation of the real crystal structures remains an unsolved problem. The superlattice spots observed by X-ray and electron diffraction in these

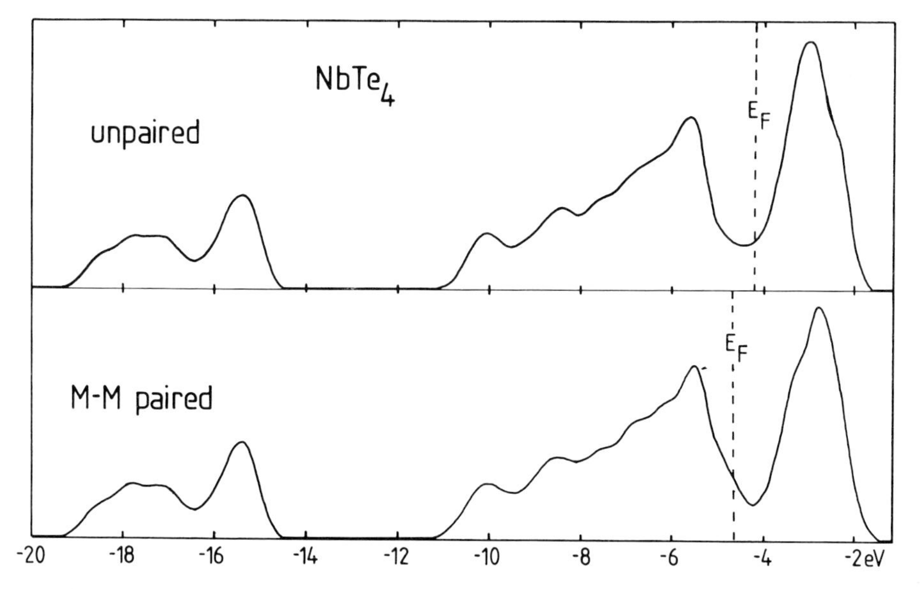

Fig. 35. Calculated electronic structure for the $NbTe_4$ subcell and for a M–M paired model, showing that in this structure a simple M–M pairing displacement is insufficient to eliminate the Fermi surface. (D. W. Bullett, *J. Phys* **C17**, 253 (1984). © The Institute of Physics.)

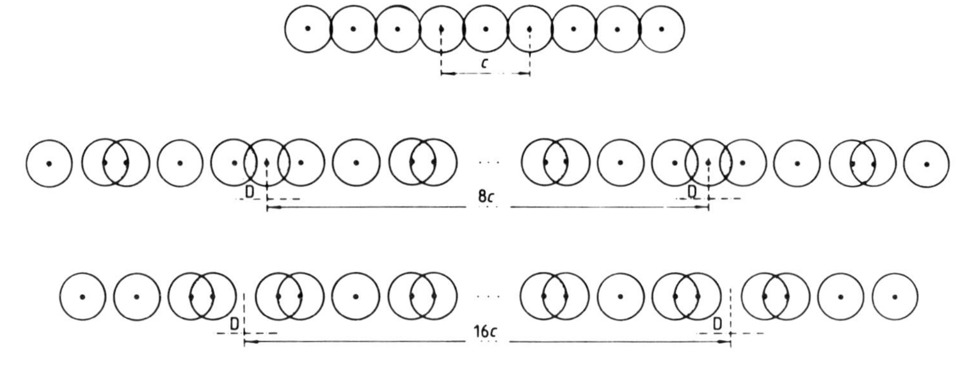

Fig. 36. Diagrammatic model of possible discommensurations in $NbTe_4$. Upper diagram shows undistorted chain with delocalized electrons. Middle diagram shows a discommensurate distortion in type 1 chains and lower diagram a discommensurate distortion in type 2 chains. D indicates the discommensuration. (Boswell *et al.* [175].) (© The Institute of Physics.)

quasi-one-dimensional metals are interpreted in terms of charge–density waves, which are unusual in existing even at room temperature. It appears that three slightly different types of column occur simultaneously. In $NbTe_4$ two types of chain have incommensurate distortions with wavevectors $\mathbf{q}_1 = (0, 0, 0.311c^*)$ and $\mathbf{q}_2 = (0.5a^*, 0.5b^*, 0.344c^*)$. The third type has a very weak distortion with $\mathbf{q}_3 = (0.5a^*, 0, \frac{1}{3}c^*)$. For $TaTe_4$ the wavevectors are similar except that the component along the metal chain is exactly $\frac{1}{3}c^*$ in each case. The superlattice periodicities $\mathbf{q}_1$ and $\mathbf{q}_2$ are independent of temperature but can be varied by substitution of Ta for Nb [175, 180], such that $q_{1c} = (0.333 - \delta)c^*$ and $q_{2c} = (0.333 + \delta)c^*$. On cooling to 50 K the superlattice row types 1 and 2 are unchanged but the spots of the third superlattice sharpen and strengthen indicating an increasing correlation distance in both transverse and longitudinal directions. Boswell *et al.* [175] have proposed a model akin to that for $NbSe_3$ in which type 3 columns have exactly two-thirds electron per cation available to form M–M pairs. Alternating M–M pairs and single M atoms superimposed on the sublattice symmetry would give periodicity $3c$ along the chain axis (Figure 35). Type 1 columns then have, on average, slightly fewer electrons per cation (10 electrons per 16 cations would give the required discommensuration frequency $q_{1c} = \frac{5}{16}c^* = 0.312c^*$), while type 2 columns would have on average slightly more; 22 electrons per 32 cations in either an incommensurate charge–density wave or a discommensurate lattice distortion would lead to the observed $q_{2c} = \frac{11}{32}c^* = 0.344c^*$. It remains to be seen whether nonlinear conductivity is possible in this system.

5.2. Nb_3X_4

Like so many of the transition-metal chalcogenides discussed in this article, the Nb_3X_4 (X = S, Se or Te) compounds form as needle-shaped crystals when the elements in the correct proportion are heated to ~1000 °C using iodine vapour

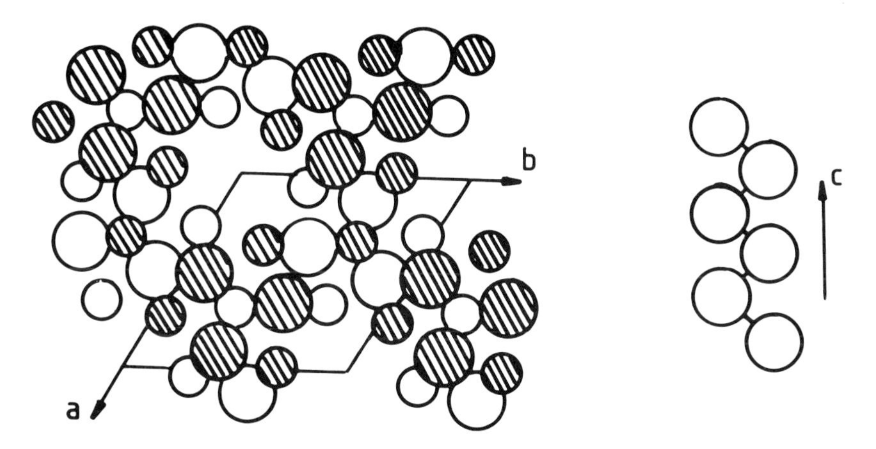

Fig. 37. Crystal structure of Nb_3S_4 projected down the c-axis. Nb atoms are denoted by large circles, S by small circles. Open circles are at $z = \frac{1}{4}$; hatched circles at $z = \frac{3}{4}$. The structure contains zigzag Nb chains along the z-axis. [114, 182]. (© Academic Press, New York.)

transport in a sealed tube [181–183]. The crystal lattice is hexagonal (space group $P6_3/m$) and the structure is built up from NbS_6 octahedra, linked by shared faces and edges (Figure 37). There are some similarities to the structures already discussed: each metal atom is displaced towards one of the faces of the coordinating octahedron, so that zigzag Nb–Nb–Nb chains are formed, running along the c-direction (which is also the needle axis). The Nb–Nb distances within these chains are comparable to those in elemental Nb. There are no close X–X approaches.

The quasi-one-dimensional aspect of the crystal structure is revealed in the highly anisotropic metallic resistivity [183, 184]. Nevertheless, Nb_3S_4, Nb_3Se_4, and Nb_3Te_4 eventually become superconducting, at temperatures of 4.0 K, 2.0 K, and 1.8 K respectively [183–185], without passing through a metal-insulator transition or strong resistivity anomaly characteristic of more strongly one-dimensional conductors. Ishihara and Nakada [186] have reported a very weak anomaly at 80 K in Nb_3Te_4. Below 50 K the electrical resistivity along the chain axis of Nb_3S_4 and Nb_3Se_4 varies as T^3 [184], rather than the usual T^2 form of conventional metals.

The electronic properties have been investigated by LCAO methods, both in the two-centred approximation [114] and to full self-consistency in an iterative process based on the set of numerical Bloch functions and using Diophantine integration techniques for the multicentre integrals [187, 188]. The results are very similar by either method. The energy range near the Fermi level divides roughly into three regions, corresponding to Nb–X bonding states, Nb–Nb states, and Nb–X antibonds (Figure 38). In particular, in both calculations the close approach of Nb atoms in the zigzag chains leads to three quasi-one-dimensional bands each of width ~1.5–2.5 eV. The conduction electron number is sufficient to occupy only one-third of this group of bands. The three conduction bands have mainly Nb d_{zx} and d_{yz} orbital character directed along the zigzag chains (where z denotes the chain axis). Along directions perpendicular to the z-axis these three bands are rather flat; ratios

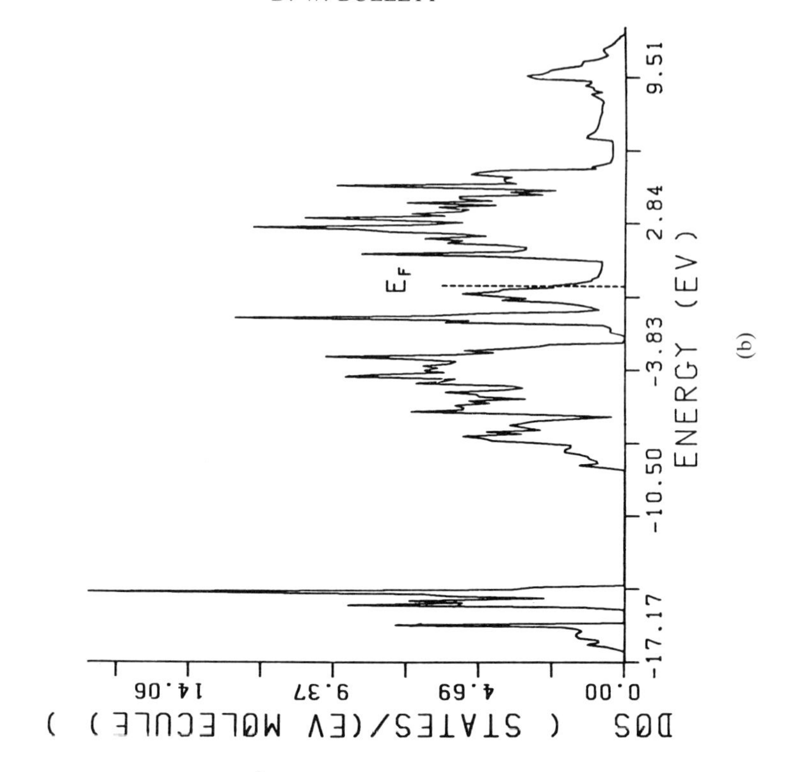

DOS (STATES/(EV MOLECULE))
0.00
4.69
9.37
14.06
ENERGY (EV)
-17.17
-10.50
-3.83
2.84
9.51
E_F

(b)

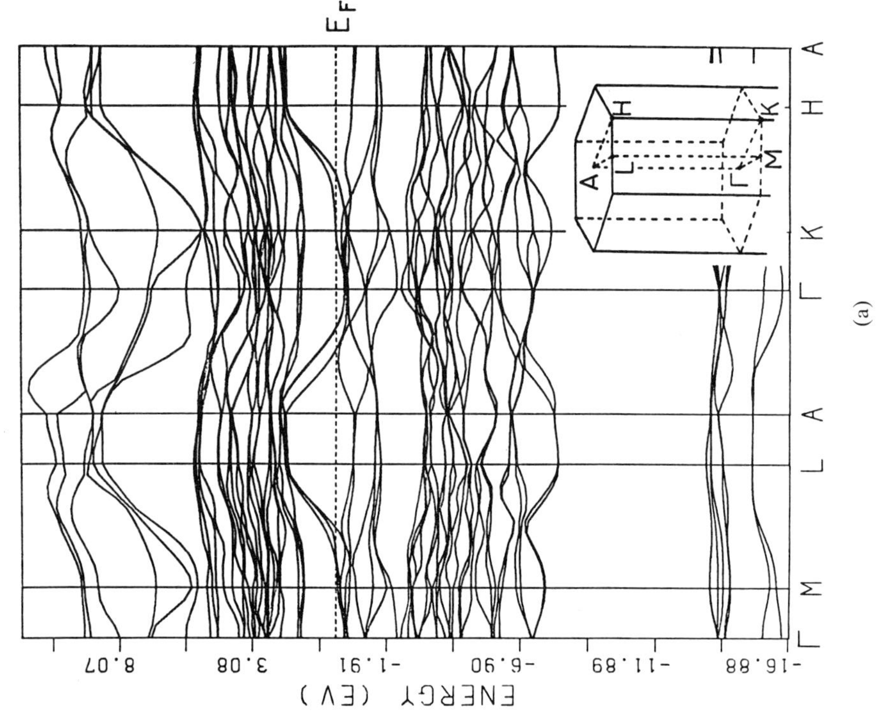

ENERGY (EV)
8.07
3.08
-1.91
-6.90
-11.89
-16.88
E_F
Γ
M
L
A
Γ
K
H
A
A
L
H
Γ
M
K

(a)

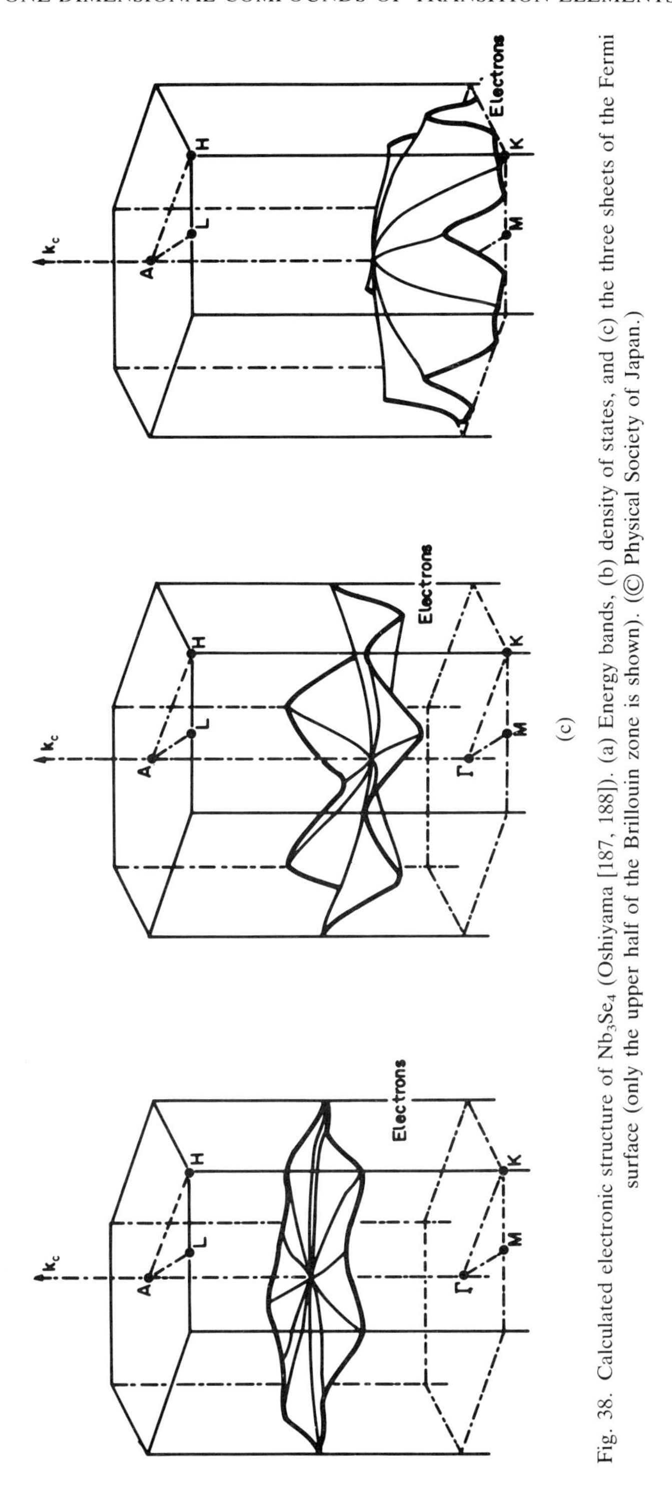

(c)

Fig. 38. Calculated electronic structure of Nb_3Se_4 (Oshiyama [187, 188]). (a) Energy bands, (b) density of states, and (c) the three sheets of the Fermi surface (only the upper half of the Brillouin zone is shown). (© Physical Society of Japan.)

of conduction bandwidths parallel and perpendicular to the chain axis are about 15 for Nb_3S_4, 18 for Nb_3Se_4, and 19 for Nb_3Te_4 [187, 188]. Thus one is justified in calling these compounds one-dimensional metals. The Fermi surfaces consist of undulating plane-like sheets [187].

Oshiyama [188] has shown that the calculated density of states at the Fermi level is reasonably consistent with the experimental value obtained from specific heat measurements for Nb_3S_4, and that the characteristic features of the optical reflectivity spectrum can be explained qualitatively by the calculated joint density of states. Electron–electron Umklapp scattering across the calculated quasi-one-dimensional Fermi surfaces could explain the observed T^3 dependence of the electrical resistivity along the chain axis of Nb_3S_4 and Nb_3Se_4.

5.3. Other Nb Chalcogenides, Halides, and Chalcogenide Halides

A large number of other compounds of Nb with chalcogens and halogens exhibit one-dimensional anisotropies to various degrees. Atomic orbital studies have been presented [189] for the electronic structures of NbS_2Cl_2 and $NbSe_2Cl_2$ [190–192], $Nb_3Se_5Cl_7$ [193], Nb_2Se_9 [194], and $I_{0.33}NbSe_4$ [195]. Each of these crystal structures is built up from Nb_2X_4 clusters (X = S, Se) containing a strong metal–metal bond.

These cage-like units contain two X_2 pairs and may be nominally described as $(Nb_2)^{4+} \cdot 2(X_2^{2-})$. The different geometrical arrangements of these units and the

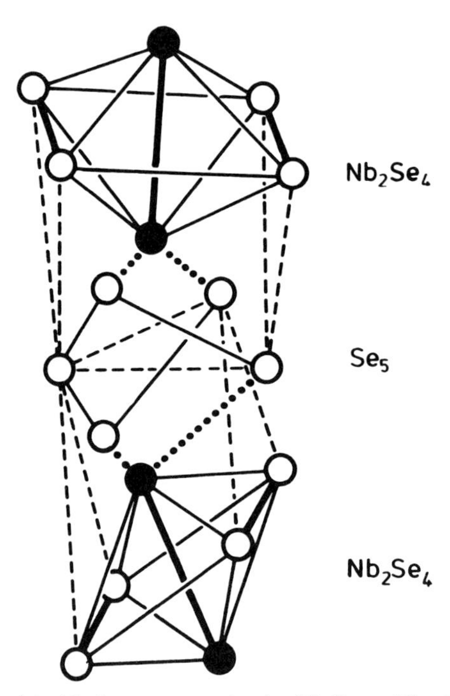

Fig. 39. Schematic picture of the Nb_2Se_9 structure, showing Nb_2Se_4 and Se_5 clusters alternating along the chain direction (*c*-axis). [189, 194]. (© Physical Society of Japan.)

surrounding halogen ions and chalcogen groups stabilize the various crystal structures by inducing a semiconducting gap in each compound. Metal–metal pairing contributes strongly to the narrow bands at the edges of the gap.

As two examples we include here Nb_2Se_9 and $I_{0.33}NbSe_4$ (Figures 39 to 41). In Nb_2Se_9 [194] the Nb_2Se_4 cages are separated by Se_5 units in such a way that each Nb atom is coordinated by Se in a bicapped trigonal prism. The Nb atoms form non-linear chains in which long (3.76 Å) and short (2.89 Å) intermetallic distances alternate along the *c*-axis. Narrow *d*-like bands are formed on either side of a ~1 eV semiconducting gap [189].

The expected one-dimensional compound $NbSe_4$ does not appear to exist in the pure form, but oxidized single crystals of composition $I_{0.33}NbSe_4$ have been grown [195]. These display a semiconducting gap ~0.4 eV and marked one-dimensional properties. There is a 45° rotation (but not always in the same sense) of Nb_2Se_4 cages from one unit to the next along the *c*-axis, giving chains of rectangular $NbSe_8$ antiprisms. Iodine atoms occupy electron acceptor sites in the channels between the chains, and the Se–Se pairing is within the cage units of each chain rather than between chains as in $NbTe_4$. There are six $NbSe_4$ units in the repeat distance along the chain, and metal–metal pairing splits the d_{z^2} band into occupied and unoccupied sub-bands: each short (3.06 Å) Nb–Nb bond is followed by two longer (3.25 Å) Nb–Nb distances. Thus every third Nb atom is unpaired and compensated by an interchain iodide ion. There would appear to be a relation with the 3*c* superlattice in $NbTe_4$ and $TaTe_4$. Nominally the unit cell contains two M–M bonds:

$$I_2Nb_6Se_{24} = (Se_2^{2-})_{12}(I^-)_2(Nb^{5+})_2(Nb^{4+}-Nb^{4+})_2$$

although the real Nb–Se charge transfers reached in the full calculation [189] are very small.

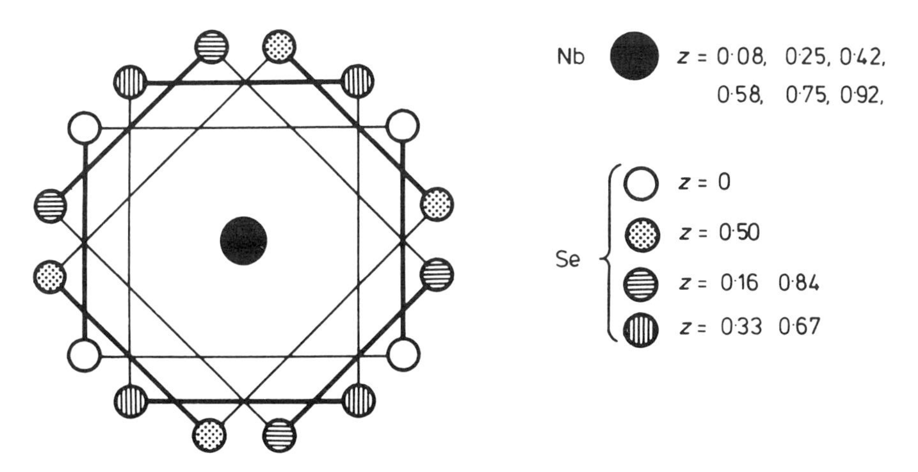

Fig. 40. Structure of the linear chains in $I_{0.33}NbSe_4$. Successive Nb_2Se_4 clusters share a common atom and rotate so that each Nb is coordinated by 8Se in a rectangular antiprism. Each short [3.06 Å) Nb–Nb bond is followed by two longer (3.25 Å) Nb–Nb distances [189, 195]. (© Physical Society of Japan.)

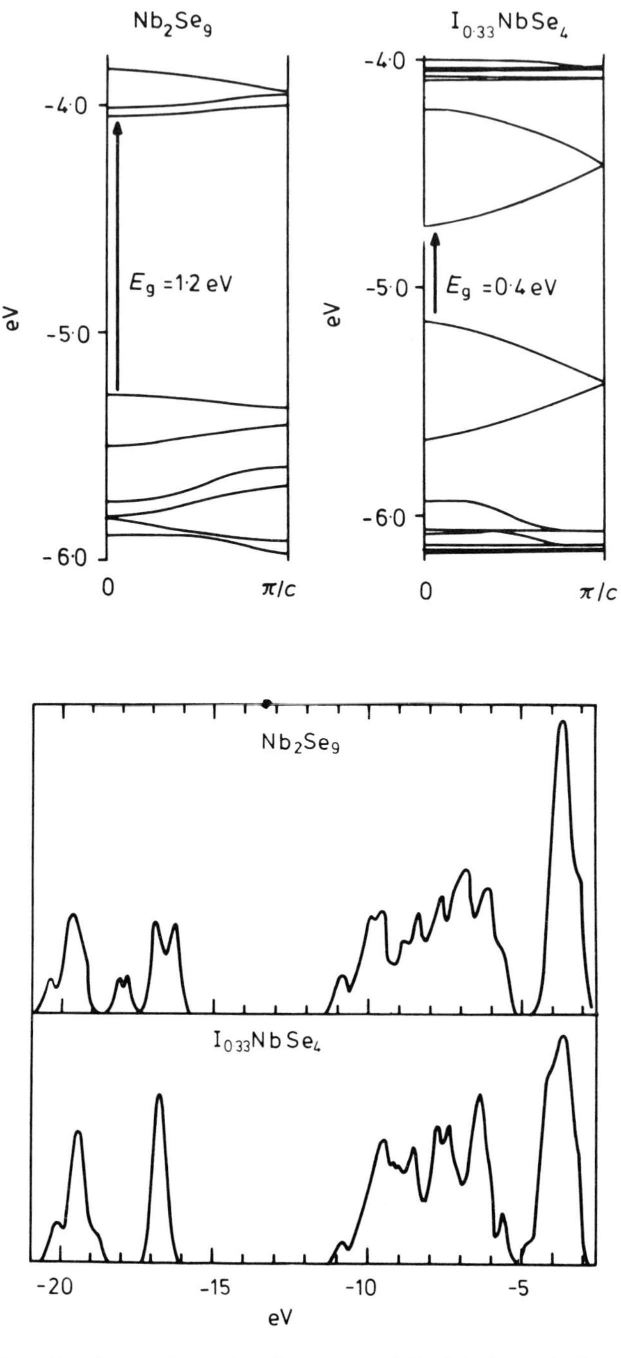

Fig. 41. Dispersion of bands near the semiconducting gap E_g for **k** in the chain direction of (a) Nb_2Se_9 and (b) $I_{0.33}NbSe_4$, and the densities of electron states calculated for the two compounds. (Bullett [189].) (© Physical Society of Japan.)

5.4. $K_{0.30}MoO_3$

Nonlinear transport properties are no longer unique to the two trichalcogenides $NbSe_3$ and TaS_3. Very recently nonlinear electrical conductivity, associated as in the two trichalcogenides wth a large noise voltage, has been observed in the charge–density wave state of the blue bronze $K_{0.30}MoO_3$ [196, 197]. At room temperature this material behaves as a quasi-one-dimensional metal [198], accounted for by the presence of infinite chains of MoO_6 octahedra parallel to the axis of highest conductivity [199]. At $T_c = 180\,K$ there is a Peierls transition to a semiconducting state [200] and weak satellite peaks appear in X-ray scattering at (0, 0.74b^*, 0.5c^*), corresponding to an incommensurate period along b [201]. Under electric fields greater than some threshold field E_t in the semiconducting temperature regime, nonohmic conduction sets in (Figure 42) [196, 197], and is assumed to be associated with the collective sliding of the charge–density wave through the lattice. In this system E_t is very dependent on temperature, reaching a maximum $\sim 0.25\,V\,cm^{-1}$ near 100 K.

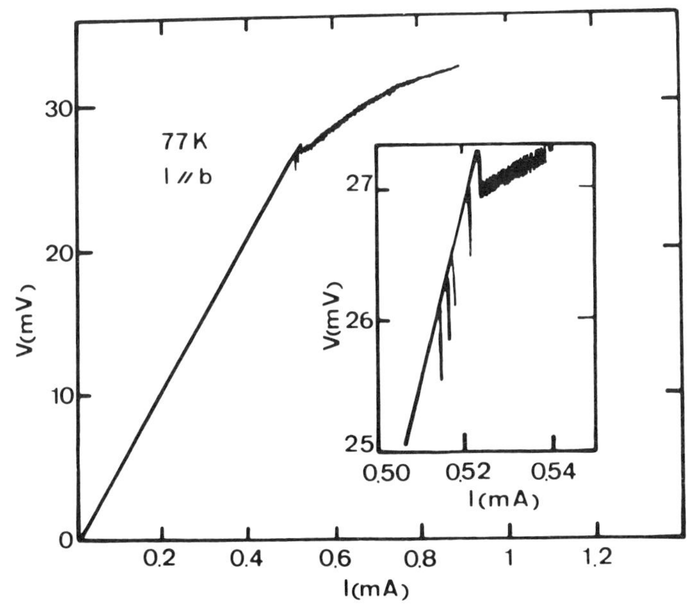

Fig. 42. D.c. voltage–current curve for the quasi-one-dimensional blue bronze $K_{0.30}MoO_3$, showing a sharp threshold field E_t and switching from the ohmic regime to the nonohmic one (distance between voltage contacts = 2 mm; $E_t = 0.136\,V\,cm^{-1}$). Inset shows the onset of pulses for fields just less than E_t and the noise for $E > E_t$ (measured over 5 min). (Dumas *et al.* [196].) (© American Physical Society.)

5.5. $TlMo_3Se_3$

We close with an example of infinite chain polymers derived by condensation of larger M_xX_y clusters. Several ternary Mo chalcogenides with chemical composition ZMo_3X_3 (Z = In, Tl, Na, K, Rb, Cs; X = S, Se, Te) have been reported [202–204].

Unlike the related (quasi-zero-dimensional) Chevrel phases [45, 205, 206] these compounds are stoichiometric. Klepp and Boller [207, 208] had previously synthesized a compound of formula $Tl_2Fe_6Te_6$ with the same structure. Figure 43 illustrates the projection of the hexagonal structure on the {001} plane. One-dimensional $(Mo_3X_3)_\infty$ chains parallel to the hexagonal c-axis characterize the structure, separated by parallel chains of Z atoms. Each Mo chain is best regarded as a condensation of slightly distorted Mo_6X_8 clusters [24], consisting of an octahedron of Mo

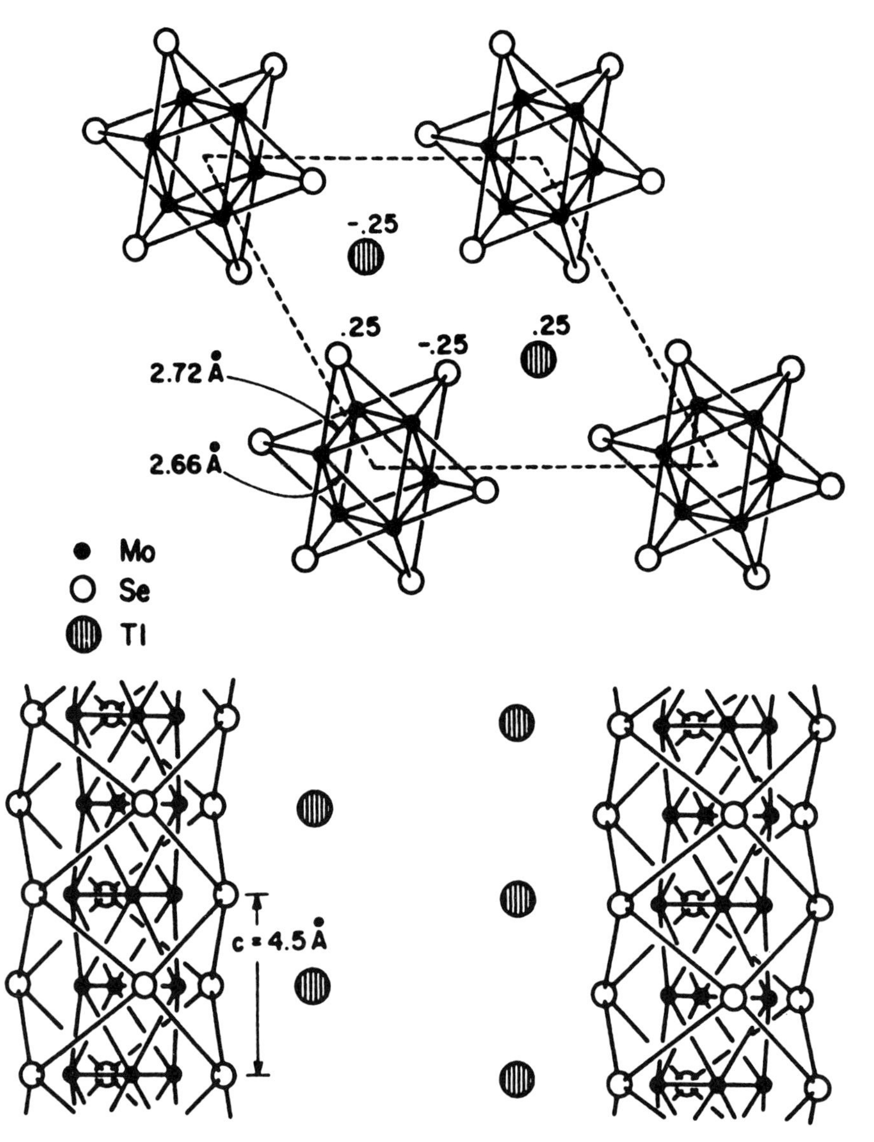

Fig. 43. Crystal structure of the linear chain compound $Tl_2Mo_6Se_6$. All atoms lie in planes perpendicular to the c-axis and separated by $c/2$. (Potel *et al.* [202].) (from [212], © American Chemical Society.)

atoms with X atoms above each triangular Mo_3 face (the basic building block of the Chevrel phases). Two X atoms from opposite faces of each Mo_6X_8 unit are eliminated when the Mo_6 octahedra condense into a linear chain by sharing opposite faces (Figure 43). The shortest intrachain Mo–Mo distance of about 2.6 Å is close to that in metallic *bcc* molybdenum, and less than half of the shortest interchain Mo–Mo distance. Since there are no strong Mo–X–Mo interchain links through chalcogen atoms, a large anisotropy in the physical properties is to be expected. For $TlMo_3Se_3$ Armici *et al.* [209] found a resistivity ratio $\rho_\perp/\rho_\| \sim 10^3$ and an anisotropy ~26 in the critical field for the superconducting ($T_c \sim 3$ K [210]) state.

Kelly and Andersen [211] have used the LMTO band structure method in the atomic sphere approximation to investigate the electronic structures of the In and Tl containing compounds. Hughbanks and Hoffmann [212] have presented a similar discussion for the $(Mo_3Se_3)_\infty$ chain in the extended Hückel model. Figure 44 shows the $TlMo_3Se_3$ for **k** along the *c*-axis, where the twofold screw axis has been used to unfold the bands corresponding to a single $TlMo_3Se_3$ unit cell. The threefold

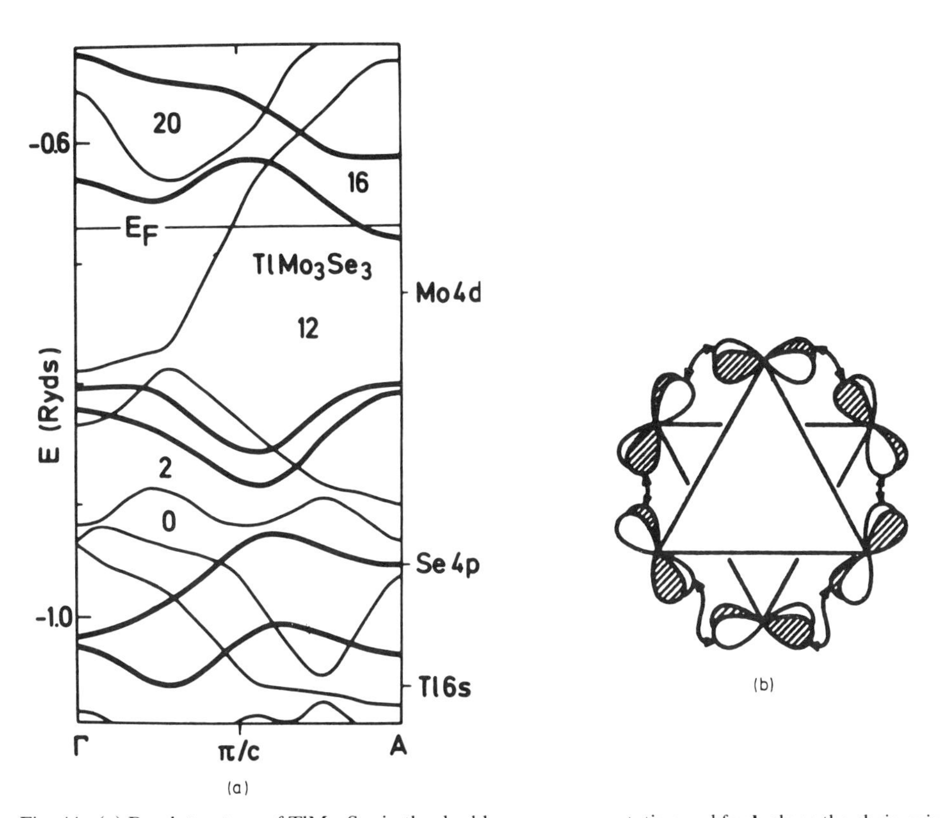

Fig. 44. (a) Band structure of $TlMo_3Se_3$ in the double zone representation and for **k** along the chain axis. Numbers between the bands give the Mo *d*-electron count. Doubly and singly degenerate bands are indicated by strong and weak lines respectively (Kelly and Andersen [211]). (b) Wave function of the dispersive helical band which provides the Fermi surface as it winds its way around the $(Mo_3)_\infty$ chains [211, 212]. (© American Chemical Society.)

symmetry axis gives rise to doubly and singly degenerate representations. The nine Se p-like bands occupy the approximate energy range -11 to -15 eV. The 15 Mo d-like bands are about 7 eV wide and overlap the top of the Se p bands. The Tl 6s state forms a filled band at -12 eV, whereas the Tl 6p states and the Mo s- and p-like states form empty bands above the Mo d states. Thus Tl is nominally Tl^+ and 13 electrons enter the Mo d bands. The Mo chain structure is stabilized by the division of the d bands into six bonding and eight antibonding states, separated by the gap $\sim$1.3 eV around -10 eV in Figure 44.

The remaining singly degenerate band is highly dispersive along the chain axis, with a width $\sim$4 eV, and crosses this energy gap. Kelly and Andersen [211] describe this as a 'helix' band because of its composition from two Bloch states which wind along the column in a left-handed and right-handed helix. In their calculation E_F lies $\sim$0.1 eV above the bottom 'octahedron' band of the Mo antibonding states, and the Fermi surface consists of two extremely one-dimensional sheets at $k_F = \pm(\pi/c - \delta)$ arising from the helix band, and two small electron pockets from the octahedron band. The calculated undulations in the sheets are in reasonable accord with the observed anisotropy in physical properties. Probably it is essential to have some electrons in the pockets of the octahedron band, in order to stabilize the structure against a Peierls distortion in the one-dimensional direction. Indeed it is thought [212] that the equivalent potassium compound $K_2Mo_6Se_6$ may well exhibit a pairing distortion which leads to semiconducting properties. Kelly and Andersen [211] find that the occupancy of the pockets increases with increasing c/a ratio, but at the same time the carrier mass decreases. Thus it may be that the particular c/a ratio of $TlMo_3Se_3$ provides in the Fermi level density of states a maximum which favours superconductivity at this particular point in the sequence of materials (in order of increasing c/a) $TlMo_3Te_3$, $InMo_3Te_3$, $TlMo_3Se_3$, $InMo_3Se_3$. The properties of these electron pockets have yet to be investigated experimentally.

References

1. S. Barišić, A. Bjeliš, J. R. Cooper, and B. A. Leontić (eds), *Proc. Int. Conf. on Quasi-One-Dimensional Conductors*, Springer-Verlag, Berlin (1979).
2. P. Monceau, *Physica B + C,* **109**, 1890 (1982).
3. J. Bardeen, *Phys. Rev. Lett.* **45**, 1978 (1980).
4. A. Briggs, P. Monceau, M. Nunez-Regueiro, J. Peyrard, M. Ribault, and J. Richard, *J. Phys.* **C13**, 2117 (1980).
5. J. W. Brill, N. P. Ong, J. C. Eckert, J. W. Savage, S. K. Khanna, and R. B. Somoana, *Phys. Rev.* **B23**, 1517 (1981).
6. D. W. Bullett, *Sol. St. Commun.* **26**, 563 (1978).
7. J. Chaussy, P. Haen, J. C. Lasjaunias, P. Monceau, G. Waysand, A. Waintal, A. Meerschaut, P. Molinié, and J. Rouxel, *Sol. St. Commun.* **20**, 759 (1976).
8. R. M. Fleming, D. E. Moncton and D. B. McWhan, *Phys. Rev.* **B18**, 5560 (1978).
9. R. M. Fleming and C. C. Grimes, *Phys. Rev. Lett.* **42**, 1423 (1979).
10. R. M. Fleming, *Phys. Rev.* **B22**, 5606 (1981).
11. K. K. Fung and J. W. Steeds, *Phys. Rev. Lett.* **45**, 1696 (1980).
12. J. C. Gill, *Sol. St. Commun.* **39**, 1203 (1981).

13. G. Grüner, C. C. Tippie, J. Sanny, W. G. Clark, and N. P. Ong, *Phys. Rev. Lett.* **45**, 935 (1980).
14. G. Grüner, W. G. Clark, and A. M. Portis, *Phys. Rev.* **B24**, 3641 (1981).
15. G. Grüner, A. Zawadowski, and P. M. Chaikin, *Phys. Rev. Lett.* **46**, 511 (1981).
16. P. Monceau, N. P. Ong, A. M. Portis, A. Meerschaut, and J. Rouxel, *Phys. Rev. Lett.* **37**, 602 (1976).
17. N. Shima, *J. Phys. Soc. Japan* **51**, 11 (1982).
18. K. Tsutsumi, T. Takagaki, M. Yamamoto, Y. Shiozaki, M. Ido, T. Sambongi, K. Yamaya, and Y. Abe, *Phys. Rev. Lett.* **39**, 1675 (1977).
19. K. Krogmann, *Angew. Chem. Int. Ed. Engl.* **8**, 35 (1969).
20. J. S. Miller and A. J. Epstein, *Prog. Inorg. Chem.* **20**, 1 (1976).
21. L. V. Interrante and R. P. Messmer, *Chem. Phys. Lett.* **26**, 225 (1974).
22. D. W. Bullett, *Sol. St. Commun.* **27**, 467 (1978).
23. M.-H. Whangbo and R. Hoffmann, *J. Am. Chem. Soc.* **100**, 6093 (1978).
24. A. Simon, *Angew. Chem. Int. Ed. Engl.* **20**, 1 (1981).
25. F. Hulliger, *Structure and Bonding* **4**, 83 (1968).
26. A. F. Wells, *Structural Inorganic Chemistry* (4th edn), Oxford U.P. (1975).
27. C. N. R. Rao and K. P. R. Pisharody, *Prog. Sol. St. Chem.* **10**, 207 (1976).
28. H. F. Franzen, *Prog. Sol. St. Chem.* **12**, 1 (1978).
29. *Physics and Chemistry of Materials with Low-Dimensional Structures*, Series A: *Layered Structures* (general ed. E. Mooser), D. Reidel, Dordrecht, Vols 1–6 (1976–79).
30. D. W. Bullett, *Solid State Physics*: Advances in Research and Applications (eds H. Ehrenreich, F. Seitz, and D. Turnbull), Academic Press, New York, Vol. 35, p. 129 (1980).
31. W. A. Harrison, *Electronic Structure and the Properties of Solids*, Freeman, San Francisco (1980).
32. R. Hoffmann, *J. Chem. Phys.* **39**, 1397 (1963).
33. M.-H. Whangbo, R. Hoffmann, and R. B. Woodward, *Proc. R. Soc. London* **A366**, 23 (1979).
34. A. Zunger and A. J. Freeman, *Phys. Rev.* **B15**, 4716 (1977).
35. A. Zunger and A. J. Freeman, *Phys. Rev.* **B16**, 906 (1977).
36. A. Oshiyama and H. Kamimura, *J. Phys.* **C14**, 5091 (1981).
37. J. C. Slater, *Quantum Theory of Molecules and Solids*, McGraw-Hill, New York, Vol. 4, (1974).
38. O. K. Andersen, *Europhysics News* **12**, 5, 1 (1981).
39. O. K. Andersen, *The Electronic Structure of Complex Systems*, (ed. W. Temmerman and P. Phariseau, NATO Advanced Study Institute, Gent, Plenum, New York (1982).
40. O. Jepsen, J. Madsen, and O. K. Andersen, *Phys. Rev.* **B18**, 605 (1978).
41. O. Jepsen, J. Madsen, and O. K. Andersen, *Phys. Rev.* **B26**, 2790 (1982).
42. H. L. Skriver, *The LMTO-method*, Springer-Verlag, Berlin (1983).
43. O. K. Andersen, *Sol. St. Commun.* **13**, 133 (1973).
44. O. K. Andersen, *Phys. Rev.* **B12**, 3060 (1975).
45. O. K. Andersen, W. Klose, and H. Nohl, *Phys. Rev.* **B17**, 1209 (1978).
46. O. K. Andersen, H. L. Skriver, H. Nohl, and B. Johansson, *Pure and Appl. Chem.* **52**, 93 (1979).
47. J. D. Corbett and P. Seabaugh, *J. Inorg. Nucl. Chem.* **6**, 207 (1958).
48. L. F. Dahl and D. L. Wampler, *J. Am. Chem. Soc.* **81**, 3150 (1962).
49. L. F. Dahl and D. L. Wampler, *Acta Cryst.* **15**, 903 (1962).
50. D. L. Kepert and R. E. Marshall, *J. Less-Common Metals* **34**, 153 (1974).
51. D. W. Bullett, *Inorg. Chem.* **19**, 1780 (1980). See also M.-H. Whangbo and M. J. Foshee, *Inorg. Chem.* **20**, 118 (1981); and S. S. Shaik and R. Bar, *Inorg. Chem.* **22**, 735 (1983).
52. H. Kawamura, I. Shirotani, and K. Tachikawa, *Phys. Lett.* **A65**, 335 (1978).
53. H. Kawamura, I. Shirotani, and K. Tachikawa, *J. Sol. St. Chem.* **27**, 223 (1979).
54. H. G. von Schnering and H. Wöhrle, *Angew. Chem. Int. Ed. Engl.* **2**, 558 (1963).
55. H. Schäfer and H. G. von Schnering, *Angew. Chem.* **76**, 833 (1964).
56. H. Schäfer and K.-D. Dohmann, *Z. Anorg. Allg. Chem.* **311**, 134 (1961).
57. V. W. Krönert and K. Plieth, *Z. Anorg. Allg. Chem.* **336**, 207 (1965).
58. L. Brattås and A. Kjekshus, *Acta Chem. Scand.* **26**, 3441 (1972).
59. S. Furuseth, L. Brattås, and A. Kjekshus, *Acta Chem. Scand.* **A29**, 623 (1975).
60. A. P. Chernov, N. I. Zbykovskaya, O. E. Panchuk, and S. A. Dembovskii, *Inorg. Mater.* **14**, 1545

(1978); translation of *Izv. Akad, Nauk SSSR Neorg. Mater.* **14**, 1986 (1978).
61. N. I. Zbykovskaya, O. E. Panchuk, A. P. Chernov, and S. A. Dembovskii, *Inorg. Mater.* **14**, 1548 (1978); translation of *Izv. Akad. Nauk SSSR Neorg. Mater.* **14**, 1990 (1978).
62. D. W. Bullett, *J. Phys.* **C12**, 277 (1979).
63. D. W. Bullett, *J. Phys.* **C15**, 3069 (1982).
64. J. A. Wilson, *Phys. Rev.* **B19**, 6456 (1979).
65. H. G. Grimmeiss, A. Rabenau, H. Hahn, and P. Neiss, *Z. Elektrochem.* **65**, 776 (1961).
66. W. Schairer and M. W. Shafer, *Phys. Status Solidi* **A17**, 181 (1973).
67. F. S. Khumalo and H. P. Hughes, *Phys. Rev.* **B4**, 2078 (1980).
68. F. S. Khumalo, C. G. Olson, and D. W. Lynch, *Physica B + C*, **105**, 163 (1981).
69. G. Perluzzo, S. Jandl, and P. E. Girard, *Can. J. Phys.* **58**, 143 (1980).
70. S. Kurita, J. L. Staehli, M. Guzzi, and F. Levy, *Physica B + C*, **105**, 169 (1981).
71. S. C. Bayliss and W. Y. Liang, *J. Phys.* **C14**, L803 (1981).
72. S. F. Nee, T. W. Nee, S. F. Fan, and D. W. Lynch, *Phys. Status Solidi* **B113**, K5 (1982).
73. F. Jellinek, R. A. Pollack, and M. W. Shafer, *Mat. Res. Bull.* **9**, 845 (1974).
74. K. Endo, H. Ihara, K. Watanabe, and S. Gonda, *J. Sol. St. Chem.* **39**, 215 (1981).
75. K. Endo, H. Ihara, K. Watanabe, and S. Gonda, *J. Sol. St. Chem.* **44**, 268 (1982).
76. G. Margaritondo, A. D. Katnani, and N. G. Stoffel, *J. Electron Spectrosc. Relat. Phenom. (Netherlands)* **20**, 69 (1980).
77. Te-Xiu Zhao, A. D. Katnani, P. Perfetti, and G. Margaritondo, *N. Cimento* **1D**, 549 (1982).
78. F. K. McTaggart, *Austral. J. Chem.* **11**, 471 (1958).
79. H. W. Myron, B. N. Harmon, and F. S. Khumalo, *J. Phys. Chem. Solids* **42**, 263 (1981).
80. D. W. Bullett, *J. Phys.* **C11**, 4501 (1978).
81. G. Perluzzo, A. A. Lakhani, and S. Jandl: *Sol. St. Commun.* **35**, 301 (1980).
82. T. Ikari, R. Provencher, S. Jandl, and M. Aubin, *Sol. St. Commun.* **45**, 113 (1983).
83. S. Kikkawa, M. Koizumi, S. Yamanaka, Y. Onuki, and S. Tanuma, *Phys. Status Solidi* **A61**, K55 (1980).
84. T. J. Wieting, A. Grisel, F. Levy, and P. Schmid, p. 354 of Ref. [1].
85. A. Grisel, F. Levy, and T. J. Wieting, *Physica B + C,* **99**, 365 (1980).
86. T. J. Wieting, A. Grisel, and F. Levy, *Physica B + C*, **105**, 366 (1981).
87. A. Zwick, M. A. Renucci, and A. Kjekshus, *J. Phys.* **C13**, 5603 (1980).
88. D. W. Galliardt, W. R. Nieveen, and R. D. Kirby, *Sol. St. Commun.* **34**, 37 (1980).
89. J.-Y. Harbec, C. Deville-Cavellin, and S. Jandl, *Phys. Status Solidi* **B96**, K117 (1979).
90. S. Jandl, M. Banville, and J.-Y. Harbec, *Phys. Rev.* **B22**, 5697 (1980).
91. C. Sourisseau and Y. Mathey, *Chem. Phys.* **63**, 143 (1981).
92. A. Zwick, M. A. Renucci, R. Carles, N. Saint-Cricq, and J. B. Renucci, *Physica B + C*, **105**, 361 (1981).
93. A. Zwick and M. A. Renucci, *Phys. Status Solidi* **B96**, 757 (1979).
94. S. Jandl and J. Deslandes, *Can. J. Phys.* **59**, 936 (1981).
95. C. Deville-Cavellin and S. Jandl, *Sol. St. Commun.* **33**, 813 (1980).
96. S. Jandl and J. Deslandes, *Phys. Rev.* **B24**, 1040 (1981).
97. A. Zwick, G. Landa, M. A. Renucci, R. Carles, and A. Kjekshus, *Phys. Rev.* **B26**, 5694 (1982).
98. A. Zwick, G. Landa, R. Carles, M. A. Renucci, and A. Kjekshus, *Sol. St. Commun.* **45**, 889 (1983).
99. R. R. Chianelli and M. B. Dines, *Inorg. Chem.* **14**, 2417 (1975).
100. R. R. Chianelli, *J. Cryst. Growth* **34**, 239 (1976).
101. A. J. Jacobson, *Sol. St. Ionics* **5**, 65 (1981).
102. M. Zanini, J. L. Shaw, and G. J. Tennenhouse, *J. Electrochem. Soc.* **128**, 1647 (1981).
103. D. W. Murphy and F. A. Trumbore, *J. Electrochem Soc.* **123**, 960 (1976).
104. F. A. Trumbore, *Pure Appl. Chem.* **52**, 119 (1980).
105. J. Broadhead, F. A. Trumbore, and S. Basu, *J. Electroanal. Chem. Interfacial Electrochem.* **118**, 241 (1981).
106. M. S. Whittingham, *J. Electrochem. Soc.* **123**, 315 (1976).

107. M. S. Whittingham, R. R. Chianelli, and A. J. Jacobson, *Proc. Electrochem. Soc.* **80–7**, 206 (1980).
108. M. S. Whittingham, *Prog. Sol. St. Chem.* **12**, 41 (1978).
109. P. A. Gardner, M. Vlasse, and A. Wold, *Inorg. Chem.* **8**, 2784 (1969).
110. A. Meerschaut and J. Rouxel, *J. Less-Common Metals* **39**, 197 (1975).
111. L. A. Grigoryan and A. V. Novoselova, *Dokl. Akad. Nauk. SSSR* **144**, 795 (1962).
112. F. Kadijk and F. Jellinek, *J. Less-Common Metals* **19**, 421 (1969).
113. J. Rijnsdorp and F. Jellinek, *J. Sol. St. Chem.* **25**, 325 (1978).
114. D. W. Bullett, *J. Sol. St. Chem.* **33**, 13 (1980).
115. T. Corne Lissens, G. van Tendeloo, J. van Landuyt, and S. Amelinckx, *Phys. Status Solidi* **A48**, K5 (1978).
116. F. W. Boswell and A. Prodan, *Physica B + C*, **99**, 361 (1980).
117. S. Kikkawa, N. Ogawa, M. Koizumi, and Y. Onuki, *J. Sol. St. Chem.* **41**, 315 (1982).
118. A. Meerschaut, L. Guémas, and J. Rouxel, *J. Sol. St. Chem.* **36**, 118 (1981).
119. S. Kikkawa, M. Koizumi, S. Yamanaka, Y. Onuki, R. Inada, and S. Tanuma, *J. Sol. St. Chem.* **40**, 28 (1981).
120. E. Bjerkelund and A. Kjekshus, *Acta Chem. Scand.* **19**, 701 (1965).
121. E. Bjerkelund, J. H. Fermor, and A. Kjekshus, *Acta Chem. Scand.* **20**, 1836 (1966).
122. K. Yamaya and G. Oomi, *J. Phys. Soc. Japan* **51**, 3512 (1982).
123. K. Yamaya and Y. Abe, *Physica* **108B**, 1235 (1981).
124. J. L. Hodeau, M. Marezio, C. Roucau, R. Ayroles, A. Meerschaut, J. Rouxel, and P. Monceau, *J. Phys.* **C11**, 4117 (1978).
125. R. Hoffmann, S. Shaik, J. C. Scott, M.-H. Whangbo, and M. J. Foshee, *J. Sol. St. Chem.* **34**, 263 (1980).
126. P. Monceau, J. Peyrard, J. Richard, and P. Molinié, *Phys. Rev. Lett.* **39**, 161 (1977).
127. N. P. Ong and J. W. Brill, *Phys. Rev.* **B18**, 5265 (1978).
128. K. Tsutsumi and T. Sambongi, *J. Phys. Soc. Japan* **49**, 1859 (1980).
129. N. P. Ong and P. Monceau, *Phys. Rev.* **B16**, 3443 (1977).
130. S. Tomić, K. Biljaković, D. Djurek, J. R. Cooper, P. Monceau, and A. Meerschaut, *Sol. St. Commun.* **38**, 109 (1981).
131. S. Tomić, *Sol. St. Commun.* **40**, 321 (1981).
132. H. Fröhlich, *Proc. R. Soc. London* **A223**, 296 (1954).
133. P. A. Lee, T. M. Rice, and P. W. Anderson, *Sol. St. Commun.* **14**, 703 (1974).
134. P. A. Lee and T. M. Rice, *Phys. Rev.* **B19**, 3970 (1979).
135. N. P. Ong, J. W. Brill, J. C. Eckert, J. W. Savage, S. K. Khanna, and R. B. Somoana, *Phys. Rev. Lett.* **42**, 811 (1979).
136. W. W. Fuller, P. M. Chaikin, and N. P. Ong, *Phys. Rev.* **B24**, 1333 (1981).
137. W. W. Fuller, P. M. Chaikin, and N. P. Ong, *Sol. St. Commun.* **39** (1981), 547.
138. W. W. Fuller, G. Grüner, P. M. Chaikin, and N. P. Ong, *Phys. Rev.* **B23**, 6259 (1981).
139. P. Monceau, J. Richard, and R. Lagnier, *J. Phys.* **C14**, 2995 (1981).
140. J. A. Wilson, *J. Phys.* **F12**, 2469 (1982).
141. J. Richard and P. Monceau, *Sol. St. Commun.* **33**, 635 (1980).
142. K. Endo, H. Ihara, S. Gonda, and K. Watanabe *Physica B + C*, **105**, 159 (1981).
143. T. Sambongi, K. Tsutsumi, Y. Shiozaki, M. Yamamoto, K. Yamaya, and Y. Abe, *Sol. St. Commun.* **22**, 729 (1977). Two measurements of the temperature dependence of the thermoelectric power have recently been reported. While one group (R. Allgeyer, B. H. Suits, and F. C. Brown, *Sol. St. Commun.* **43**, 207 (1982)) finds a negative thermopower, as for electrons, over the entire temperature range, the other (B. Fisher, *Sol. St. Commun.* **46**, 227 (1983)) reports a positive Seebeck coefficient indicating transport by positive carriers above and below the phase transition. A sharp rise in the magnitude of the thermopower below the transition temperature was found in both experiments.
144. K. Tsutsumi, T. Sambongi, S. Kagoshima, and T. Ishiguno, *J. Phys. Soc. Japan* **44**, 1735 (1978).
145. C. Roucau, R. Ayroles, P. Monceau, L. Guémas, A. Meerschaut, and J. Rouxel, *Phys. Status*

Solidi **A62**, 483 (1980).
146. G. van Tendeloo, J. van Lauduyt, and S. Amelinckx, *Phys. Status Solidi* **A43**, K137 (1977).
147. J. C. Tsang, C. Hermann, and M. W. Shafer, *Phys. Rev. Lett.* **40**, 1528 (1978).
148. A. H. Thompson, A. Zettl, and G. Grüner, *Phys. Rev. Lett.* **47**, 64 (1981). As in $NbSe_3$, low-frequency broad band noise in the current carrying CDW state of orthorhombic TaS_3 has recently been observed by A. Zettl and G. Grüner, *Sol. St. Commun.* **46**, 29 (1983).
149. S. J. Hillenius, R. V. Coleman, R. M. Fleming, and R. J. Cava, *Phys. Rev.* **B23**, 1567 (1981).
150. A. M. Meerschaut, P. Gressier, L. Guémas, and J. Rouxel, *Mat. Res. Bull.* **16**, 1035 (1981).
151. R. J. Cava, V. C. Himes, A. D. Mighell, and R. S. Roth, *Phys. Rev.* **B24**, 3634 (1981).
152. S. J. Hillenius and R. V. Coleman, *Phys. Rev.* **B25**, 2191 (1982).
153. M.-H. Whangbo, R. J. Cava, F. J. DiSalvo, and R. M. Fleming, *Sol. St. Commun.* **43**, 277 (1982).
154. P. W. Anderson, *Phys. Rev.* **109**, 1492 (1958).
155. F. J. DiSalvo, J. A. Wilson, and J. V. Waszczak, *Phys. Rev. Lett.* **36**, 885 (1976).
156. F. J. DiSalvo and J. E. Graebner, *Sol. St. Commun.* **23**, 825 (1977).
157. A. Ben Salem, A. Meerschaut, L. Guémas, and J. Rouxel, *Mat. Res. Bull.* **17**, 1071 (1982).
158. S. Okada, T. Sambongi, and M. Ido, *J. Phys. Soc. Japan* **49**, 839 (1980).
159. M. Izumi, K. Uchinokura, and E. Matsuura, *Sol. St. Commun.* **37**, 641 (1981).
160. F. J. DiSalvo, R. M. Fleming, and J. V. Waszczak, *Phys. Rev.* **B24**, 2935 (1981).
161. M. Izumi, K. Uchinokura, S. Harada, R. Yoshizaki, and E. Matsuura, *Proc. Int. Conf. Low-Dimensional Conductors, Boulder* (1981).
162. M. Izumi, K. Uchinokura, S. Harada, R. Yoshizaki, and E. Matsuura, *Mol. Cryst. Liq. Cryst.* **81**, 141 (1982).
163. M. Izumi, K. Uchinokura, E. Matsuura, and S. Harada, *Sol. St. Commun.* **42**, 773 (1982).
164. T. E. Jones, W. W. Fuller, T. J. Wieting, and F. Levy, *Sol. St. Commun.* **42**, 793 (1982).
165. E. F. Skelton, T. J. Wieting, S. A. Wolf, W. W. Fuller, D. V. Gubser, T. L. Francavilla, and F. Levy, *Sol. St. Commun.* **42**, 1 (1982).
166. A. Zwick, G. Landa, R. Carles, M. A. Renucci, and A. Kjekshus, *Sol. St. Commun.* **44**, 89 (1982).
167. I. Taguchi, A. Grisel, and F. Levy, *Sol. St. Commun.* **45**, 541 (1983).
168. C. M. Jackson, A. Zettl, G. Grüner, and F. J. DiSalvo, *Sol. St. Commun.* **45**, 247 (1983).
169. I. Taguchi and M. Asai, *Sol. St. Commun.* **40**, 187 (1981).
170. L. Brattås and A. Kjekshus, *Acta Chem. Scand.* **25**, 2783 (1971).
171. S. Furuseth, L. Brattås, and A. Kjekshus, *Acta Chem. Scand.* **27**, 2367 (1973).
172. D. W. Bullett, *Sol. St. Commun.* **42**, 691 (1982).
173. M.-H. Whangbo, F. J. DiSalvo, and R. M. Fleming, *Phys. Rev.* **B26**, 687 (1982).
174. S. Okada, T. Sambongi, M. Ido, Y. Tazuke, R. Aoki, and O. Fujita, *J. Phys. Soc. Japan* **51**, 460 (1982).
175. F. W. Boswell, A. Prodan, and J. K. Brandon, *J. Phys.* **C16**, 1067 (1983).
176. K. Selte and A. Kjekshus, *Acta Chem. Scand.* **18**, 690 (1964).
177. E. Bjerkelund and A. Kjekshus, *Acta Chem. Scand.* **22**, 3336 (1968).
178. E. Bjerkelund and A. Kjekshus, *J. Less-Common Metals* **7**, 231 (1964).
179. J. K. Brandon and G. Lessard (to be published, 1983).
180. F. W. Boswell and A. Prodan *Mat. Res. Bull.* **19**, 93 (1984).
181. K. Selte and A. Kjekshus, *Acta Cryst.* **17**, 1568 (1964).
182. A. F. J. Ruysink, F. Kadijk, A. J. Wagner, and F. Jellinek, *Acta Cryst.* **B24**, 1614 (1968).
183. E. Amberger, K. Polborn, and P. Grimm, *Sol. St. Commun.* **26**, 943 (1978).
184. Y. Ishihara and I. Nakada, *Sol. St. Commun.* **42**, 579 (1982).
185. W. Biberacher and H. Schwenk, *Sol. St. Commun.* **33**, 385 (1980).
186. Y. Ishihara and I. Nakada, *Sol. St. Commun.* **45**, 129 (1983).
187. A. Oshiyama, *Sol. St. Commun.* **43**, 607 (1982).
188. A. Oshiyama, *J. Phys. Soc. Japan* **52**, 587 (1983).
189. D. W. Bullett, *J. Phys.* **C13**, 1267 (1980).
190. H. Schäfer and W. Beckmann, *Z. Anorg. Allg. Chem.* **347**, 225 (1966).
191. H. G. von Schnering and W. Beckmann, *Z. Anorg, Allg. Chem.* **347**, 231 (1966).
192. J. Rijnsdorp, G. J. de Lange, and G. A. Wiegers, *J. Sol. St. Chem.* **30**, 365 (1979).

193. J. Rijnsdorp and F. Jellinek, *J. Sol. St. Chem.* **25**, 325 (1978).
194. A. Meerschaut, L. Guémas, R. Berger, and J. Rouxel, *Acta Cryst.* **B35**, 1747 (1979).
195. A. Meerschaut, P. Palvadeau, and J. Rouxel, *J. Sol. St. Chem.* **20**, 21 (1977).
196. J. Dumas, C. Schlenker, J. Marcus, and R. Buder, *Phys. Rev. Lett.* **50**, 757 (1983).
197. J. Dumas and C. Schlenker, *Sol. St. Commun.* **45**, 885 (1983).
198. G. Travaglini, P. Wachter, J. Marcus, and C. Schlenker, *Sol. St. Commun.* **37**, 599 (1981).
199. J. Graham and A. D. Wadsley, *Acta Cryst.* **20**, 93 (1966).
200. W. Fogle and J. H. Perlstein, *Phys. Rev.* **B6**, 1402 (1972).
201. J. P. Pouget, S. Kagoshima, J. Marcus, and C. Schlenker, *J. Phys.* (*Paris*), *Lett.* **44**, L113 (1983).
202. M. Potel, R. Chevrel, and M. Sergent, *Acta Cryst.* **B36**, 1545 (1980).
203. M. Potel, R. Chevrel, M. Sergent, J. C. Armici, M. Decroux, and Ø. Fischer, *J. Sol. St. Chem.* **35**, 286 (1980).
204. W. Hönle, H. G. von Schnering, A. Lipka, and K. Yvon, *J. Less-Common Metals* **71**, 135 (1980).
205. *Superconductivity in Ternary Compounds* (eds Ø. Fischer and M. B. Maple), Springer-Verlag (1982).
206. H. Nohl, W. Klose, and O. K. Andersen in Ref. [205].
207. K. Klepp and H. Boller, *Acta Cryst.* **A34**, 5160 (1978).
208. K. Klepp and H. Boller, *Monatsh. Chemie* **110**, 677 (1979).
209. J. C. Armici, M. Decroux, Ø. Fischer, M. Potel, R. Chevrel, and M. Sergent, *Sol. St. Commun.* **33**, 607 (1980).
210. S. Z. Huang, J. J. Mayerle, R. L. Greene, M. K. Wu, and C. W. Chu, *Sol. St. Commun.* **45**, 749 (1983).
211. P. J. Kelly and O. K. Andersen, *Superconductivity in d- and f-Band metals* (eds W. Buckel and W. Weber), Kernforschungszentrum Karlsruhe, p. 137 (1982).
212. T. Hughbanks and R. Hoffmann, *Inorg. Chem.* **21** 3578 (1982); *J. Am. Chem. Soc.* **105**, 1150 (1983).

Note Added in Proof

Studies of the properties of quasi-one-dimensional chalcogenides and halides of transition elements have continued apace since this manuscript was completed in March 1983. Some of the more recent work on specific compounds may be found in the references below, or in the proceedings of four recent symposia: *J. Phys. Colloq.* (France) **44**, C-3 (1983); Proc. Int. Symp. Nonlinear Transport and Related Phenomena in Inorganic Quasi-One-Dimensional Conductors, Hokkaido Univ., Sapporo, Japan (1983); 'Charge Density Waves in Solids' Budapest Conference 1984; eds. G. Hutiray and J. Sólyom. Lecture Notes in Physics 217 Springer-Verlag Berlin 1985; *Phil. Trans. R. Soc.* **A314**, 1–198 (1985): see especially the review by J. A. Wilson on page 159 of these proceedings.

$ZrTe_3$: D. J. Eaglesham, J. W. Steeds, and J. A. Wilson, *J. Phys.* **C17**, L697 (1984); S. Takahishi, T. Sambongi, J. W. Brill, and W. Roark, *Sol. St. Commun.* **49**, 1031 (1984).

TaS_3: A. W. Higgs and J. C. Gill, *Sol. St. Commun.* **47**, 737 (1983); A. Maeda, M. Naito, and S. Tanaka, *Sol. St. Commun.* **47**, 1001 (1983); A. Zettl and G. Grüner, *Phys. Rev.* **B28**, 2091 (1983); C. H. Chen and R. M. Fleming, *Sol. St. Commun.* **48**, 777 (1983); G. Mihaly, L. Mihaly, and H. Mutka, *Sol. St. Commun.* **49**, 1009 (1984); J. P. Stokes, A. N. Bloch, A. Janossy, and G. Grüner, *Phys. Rev. Lett.* **52**, 372 (1984).

$ZrTe_5$: J. W. Brill and T. Sambongi, *J. Phys. Soc. Japan* **53**, 20 (1984).

$NbTe_4$: D. J. Eaglesham, D. Bird, R. L. Withers and J. W. Steeds, *J. Phys.* **C18**, 1 (1985).

Nb_3X_4: Y. Ishihara, I. Nakada, K. Suzuki, and M. Ichihara, *Sol. St. Commun.* **50**, 657 (1984).

$(MSe_4)_nI$: M. Maki, M. Kaiser, A. Zettl and G. Grüner, *Sol. St. Commun.* **46**, 497 (1983); H. Fujishita, M. Sato, and S. Hoshino, *Sol. St. Commun.* **49**, 313 (1984); P. Gressier, A. Meerschaut, L. Guémas, J. Rouxel, and P. Monceau, *J. Sol. St. Chem.* **51**, 141 (1984); A. Meerschaut, P. Gressier, L. Guémas, and J. Rouxel, *J. Sol. St. Chem.* **51**, 307 (1984).

ELECTRONIC STRUCTURE OF $NbSe_3$

NOBUYUKI SHIMA and HIROSHI KAMIMURA

Department of Physics, Faculty of Science,
University of Tokyo, Bunkyo-ku, Tokyo 113, Japan

1. Low-Dimensional Material $NbSe_3$

1.1. Introduction

In this article we survey theoretical investigations for the electronic structure of $NbSe_3$ and its characteristics. During the last decade much interest has been aroused for MX_3 compounds, especially $NbSe_3$ [1–5] because of the existence of its two CDW phase transitions and accompanied Fröhlich sliding mode [6, 7]. Many experimental [8–82] and theoretical [83–108] works have been carried out on $NbSe_3$, and many anomalous properties have been reported. But there has been little understanding of the microscopic mechanisms of characteristic phenomena up to now, because of a lack of precise knowledge of its electronic structure.

In this section, we summarize the general characters of MX_3 compounds, paying particular attention to the case of $NbSe_3$. In Section 2, the methodology of the relativistic LCAO-Xα-SCF band calculation which is extensively used in the following sections is presented. In Section 3 the results of the calculation are shown and compared with some of experimental results. In Section 4 we discuss the relation between the Fermi surfaces of $NbSe_3$ and the two CDW transitions, and calculate some transport properties. We then discuss some unsolved problems.

1.2. General characters of MX_3 compounds

1.2.1. *Structures and Electronic Properties of MX_3 Compounds*

Before studying $NbSe_3$, it is better to summarize the general characters of MX_3 compounds. A more detailed description of MX_3 is presented in the preceding article by Bullett. MX_3 has been obtained in a process of synthesizing a layered-structure compound MX_2 [109]. The method of synthesizing MX_3 – where M represents a transition metal atom which belongs to group IV (Ti, Zr, Hf) or group V (Nb, Ta), and X is the chalcogenide atom (S, Se, Te) – is described in Reference [14]. There are also other types of MX_3 ($RuBr_3$, $PuBr_3$, $TiCl_3$, $ThTe_3$, ...).

While a unit cell of MX_2 consists of some layers, a unit cell of MX_3 consists of some trigonal chains. A crystal structure of $ZrSe_3$ is shown in Figure 1, for example. A trigonal chain is made by a linear stacking of irregular prismatic cages, each of which consists of six chalcogenide atoms at the corners and one transition metal at the center. On the other hand, a layer in a unit cell of 2H–MX_2, for example, is formed by a planar spreading of regular prismatic cages. From a comparison

H. Kamimura (ed.), Theoretical Aspects of Band Structures and Electronic Properties of Pseudo-One-Dimensional Solids, 231–274.

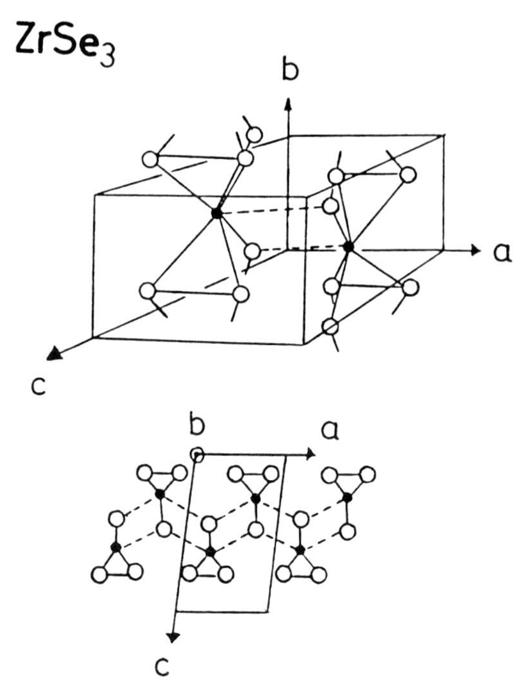

Fig. 1. Crystal structure of $ZrSe_3$. ● represents the Zr atom and ○ the Se atom. Three Se atoms of a chain and Zr atoms of adjacent chains are in the same plane which is parallel to the **a**–**c** plane. In a chain, all Zr atoms are in a row along the chain direction, and all Se atoms of each vertex sites of a trigonal chain are also in a row along the chain axis.

between the structures of MX_2 and MX_3, it may be supposed that the triangular chain structure of MX_3 is constructed by cutting some of the covalent bonds in the layers of MX_2 and recombining suitable bonds so as to produce irregular trigonal chains of MX_3.

A cross-section of a trigonal chain of MX_3 is a nearly isosceles triangle, and two sides of the triangle are rather longer than the remaining side. Concerning the origin of the deformation of an equilateral triangle in MX_2 to an isosceles one in MX_3, Hoffmann *et al.* [105] suggested in the case of $NbSe_3$ the deformation of a single chain due to a kind of Jahn–Teller distortion, corresponding to a breaking of the two-fold degeneracy of the $(e'')^3$ configuration, where e'' represents the highest orbital set of p orbitals of Se atoms and has two-fold degeneracy. In Figure 2 the schematic electronic structures of a regular prismatic MX_3 cluster and of an irregular prismatic MX_3 cluster obtained by Hoffmann *et al.* [105] are shown. In this figure arrows represent the electrons which occupy energy levels.

In the case of the group IV MX_3 compound, there are two electrons in the e'' level (see the upper part of the figure). After distortion which makes the distance between two X atoms of a triangle shorter, the doubly degenerate e'' level splits into a bonding level σ and an antibonding level σ^*. The two electrons occupy the bonding σ level forming a covalent bond between two X atoms, which stabilizes the

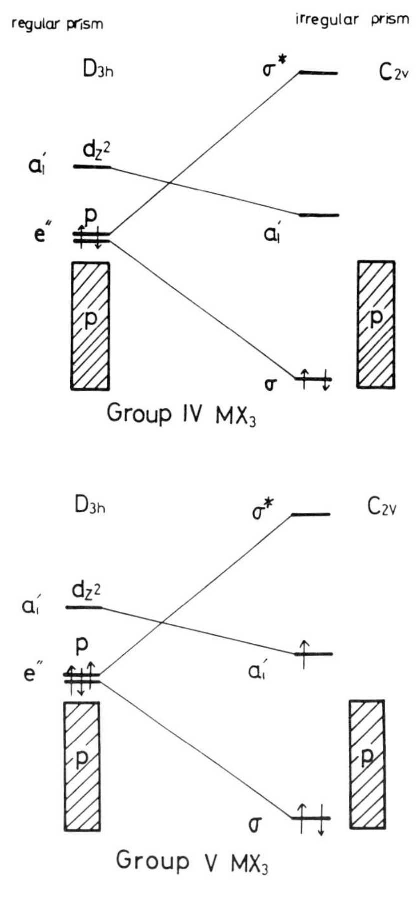

Fig. 2. Schematic electronic structures of regular and irregular prisms of group IV and V MX_3 cluster (after [105]). The hatched part represents occupied levels. Arrows stand for electrons.

irregular prismatic structure and makes the distance between two X atoms short enough. Two X ligands are transformed into a coordinated X_2 ligand as Wilson [101] and Hoffmann *et al.* [105] have discussed.

In the case of the group V MX_3 compounds, there are three electrons in the e'' level (see the lower part of the figure). After distortion two of them occupy the σ level of an irregular prism. The remaining electron occupies the a_1' level rather than the σ^* level, because the a_1' level is lower than the σ^* level in energy, where the a_1' level corresponds to the lowest energy level of d orbitals of an M atom. In this case the σ bond between two X atoms also stabilizes the irregular prismatic structure. However, the energy gain in this case depends on the magnitude of the energy shift of the a_1' level and on the degree of lowering of the σ level by distortion. Because the location of the a_1' level is expected to be dependent on a crystal field, to which a transition metal atom is subject, the degree of distortion of a chain – or, in other words, the distance between two X atoms – may take a different value according to the species of M and X atoms and to the position and the number of chains in a unit cell. (We shall see that in the group V MX_3 compounds there are two types of chains: A and B.)

After all, in the case of the group IV MX_3 compounds, the distortion of a prismatic chain is stabilized by a strong covalent σ bond between two X atoms. On the other hand, in the case of the group V MX_3 compounds, distortion does not always occur in the same way as the case of the group IV MX_3 compounds.

The group IV MX_3 compounds are insulators or semiconductors. On the other hand, there are different types in the group V MX_3 compounds. Some of characteristics of typical MX_3 are listed in Table I. $TaSe_3$ is semimetallic from room temperature to low temperatures and becomes superconducting at $T = 2.3\,K$. $NbSe_3$ is metallic in all region of temperature and exhibits two CDW transitions. We shall describe $NbSe_3$ in more in detail later. TaS_3, which has two different structures, is metallic at room temperature and becomes semiconducting after exhibiting CDW transition. NbS_3 is a semiconductor below $T = 700\,K$, and is considered to be in a non-metallic state caused by the Peierls transition whose transition temperature is assumed to be higher than $T = 700\,K$. Phase transitions in these materials occur very clearly and sharply.

As seen below, whether these materials become an insulator, a semiconductor or a metal can be understood qualitatively by a simple energy band scheme, as Bullett discussed [104]. (For MX_2 materials, Wilson and Yoffe [109] have discussed their electronic properties by a similar energy band scheme.) We shall describe this in the following section.

1.2.2. *Relation between Crystal Structures and Electronic Properties of MX_3*

First, we discuss the qualitative electronic structure of a single chain of MX_3, assuming ionic bonds between the M atom and the neighbouring X atoms. For simplicity, only the *p* orbitals of X atoms and the *d* orbitals of an M atom are taken into account.

Generally there are nine *p* orbital levels of three X atoms for an MX_3 cluster. Then we assume that there is a covalent *p–p* bonding orbital between two X atoms on the shortest side of an isosceles triangle which is formed by three X atoms. In this case the energy level of an antibonding orbital is considered to be rather high compared with that of a bonding orbital if the distance between these two X atoms is short enough. Thus only the strong covalent bond and the remaining seven *p* orbital levels, which we call 'bond levels', may be considered [Case A].

On the other hand, if the distance between two X atoms is not short enough (compared with the distance between two X atoms of solid X, which is for example 4.34 a.u. in the case of Se), the energy level of an antibonding orbital is close to the level of a bonding orbital. Then in this case the bonding and antibonding orbitals and the remaining seven *p* orbitals must all be considered at the same time. As there is one more level here than in Case A, the MX_3 cluster is expected to be more ionic [Case B].

Thus there are eight or nine energy levels in Case A or B, respectively. This

TABLE I

Some properties of typical MX_3 compounds

MX_3	Sym.	Chain	ρ	T_{CDW}	q_{CDW}	T_{super}	Other properties
$TaSe_3$	$P2_1/m$	4	M. $5 \times 10^{-4}\,\Omega$ cm	—	—	2.3 K	
$NbSe_3$	$P2_1/m$	6	M. $6 \times 10^{-4}\,\Omega$ cm	142 K, I	(0, 0.243, 0)	<50 mK	Nonlinear
				58 K, I	(0.5, 0.263, 0.5)	$\frac{dT_c}{dp} = 0.6$ K/kbar	$E_{T_1} = 0.05$ V/cm $E_{T_2} = 0.005$ V/cm
TaS_3	*Cmcm* Orthorhomb.	[24]	S.C. ← M. $[3 \times 10^{-4}]$	218 K, C	(0.5, 0.125, 0.25)	—	
	Monoclinic	[6]	S.C.	240 K, I 160 K, I		—	Nonlinear $E_{T_1} = 0.5$ V/cm $E_{T_2} < E_{T_1}$
NbS_3	$P\bar{1}$	2	S.C.	700 K <, C	(0, 0.5, 0)	—	
$ZrSe_3$	$P2_1/m$	2	S.C.			—	

The column of 'Sym.' shows a space group of each MX_3 compound. The column of 'chain' presents the number of chains in a unit cell. In the column of 'ρ', electronic properties of MX_3 compounds, M. (metallic) or S.C. (semiconducting), and resistivities at room temperature and shown. The column of 'T_{CDW}' shows transition temperatures of 'I' (incommensurate) or 'C' (commensurate) CDW transitions. In the following column, nesting vectors of CDW, q_{CDW}, are shown. The column of 'T_{super}' shows superconducting transition temperatures. In the last column, other properties are shown. 'E_T' represents the threshold electric field of non-Ohmic conduction which is explained in Section 1.3.

picture, on which the bonding and antibonding orbitals between two X atoms play an important role, has been proposed by Bullett [104].

As the M atom is surrounded by six X atoms, it feels approximately a hexagonal ligand field. In this field, an energy level of five d orbitals splits into a three-fold degenerate energy level of t_{2g} orbitals (d_{z^2}, d_{xy}, $d_{x^2-y^2}$) and a two-fold degenerate energy level of e_g orbitals (d_{xz}, d_{yz}), where the z-axis is parallel to the chain direction of MX_3. Because e_g orbitals spread along the direction to X atoms, the mixing of e_g and p orbitals of X atoms is expected to be large. Thus the ionic bonds between the M atom and the X atoms have few covalent characters, but the number of levels does not change. In fact, the mixing of e_g and p orbitals can be seen in the partial

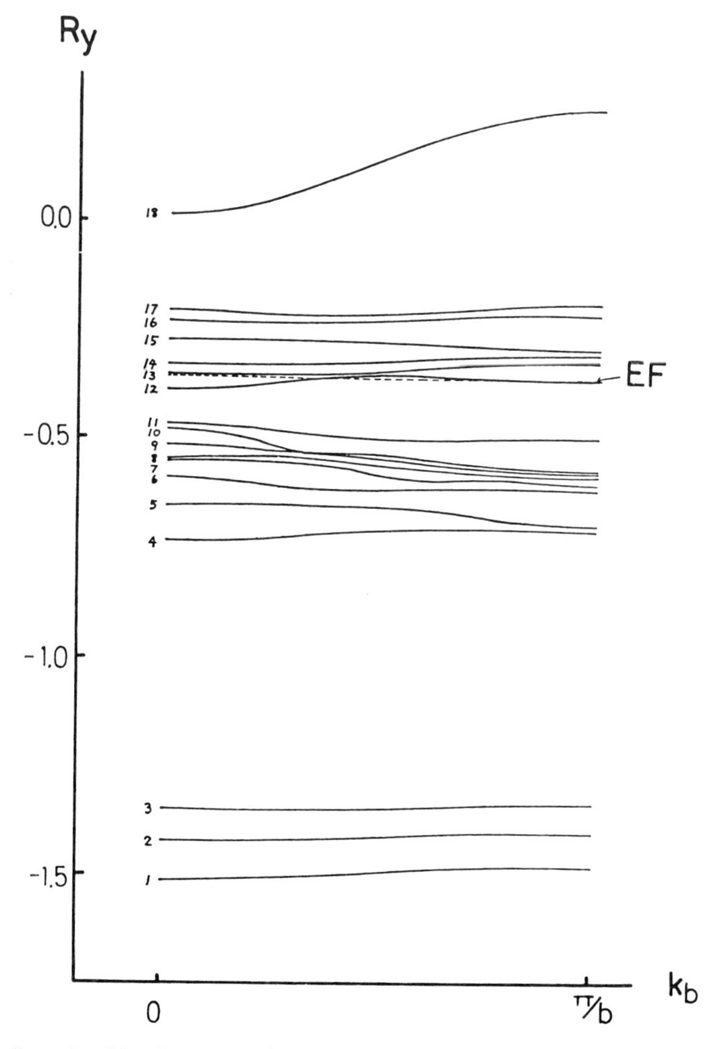

Fig. 3. One-dimensional band structure of a type A chain of $NbSe_3$. Bands 1–3 are 4s bands of three Se atoms. A conduction band is the 12th band which has mainly, a d_{z^2} character.

density of states of $NbSe_3$ in Figures 18 and 19. This will be pointed out in Section 3.3. The three-fold degenerate t_{2g} energy level splits into two parts: a d_{z^2} level (lower part) and a d_{xy}, $d_{x^2-y^2}$ level (upper part) in a trigonal prismatic field, as in the case of MX_2 [110]. Consequently, the d_{z^2} level locates just above the levels corresponding to the eight Case A (or nine Case B) energy levels.

In order to ascertain the above simple picture in a real system where the energy levels of a cluster change into the energy bands of a periodic system, a preliminary band calculation of a single chain of $NbSe_3$ was performed previously Case A by one of the present authors [111]. The resulting one-dimensional band structure is shown in Figure 3. The vertical axis represents energy in Rydberg (=13.6 eV) units; and the horizontal axis represents crystal momentum k_b along the chain axis in a Brillouin zone. There are eight bond-bands (4th to 11th) and a partially filled d_{z^2} band (12th) which is a conduction band in this figure. These bands correspond to the previous bond picture of Case A.

Let us consider the electronic structure of a typical group IV MX_3 compound, $ZrSe_3$. As explained before, a group IV MX_3 compound is expected to have a strong covalent σ bond between two X atoms of a chain. Since such a chain belongs to Case A, we call it a 'type A' chain and the covalent bond a 'σ_A' bond. On the other hand, if the chain belongs to Case B, we shall call it a 'type B' chain and the covalent bond a 'σ_B' bond in the following.

$ZrSe_3$ has two trigonal chains belonging to 'type A' in its unit cell [110, 114].

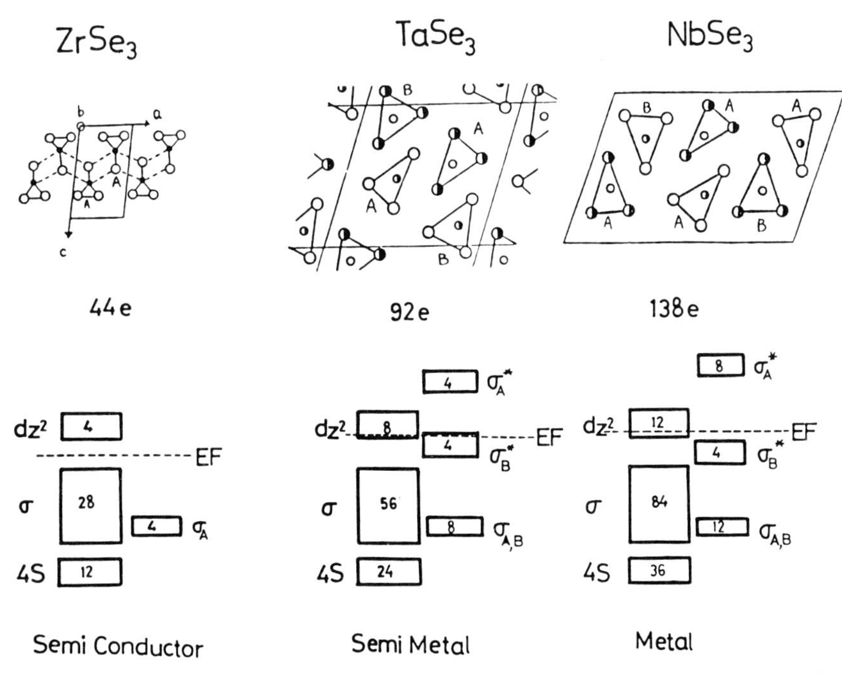

Fig. 4. Chain structures of some MX_3 compounds and their schematic bond-bands. Each square represents a band. The numeral in the square stands for the acceptable numbers of electrons of the band in a unit cell.

Figure 4 shows the schematic bond-bands and their acceptable numbers of electrons in a unit cell. As there are six Se atoms in a unit cell, the number of Se $4s$ bands is six, and the $4s$ bands can accept twelve electrons. Because there are two 'type A' chains in a unit cell, there are 14 $[=7 \times 2]$ σ bond-bands, 2 $[=1 \times 2]$ σ_A bond-bands, and 2 $[=1 \times 2]$ d_{z^2} bands, whose acceptable numbers of electrons are 28, 4, and 4, respectively. The antibonding band σ_A^* is located at a much higher energy level. The electronic configuration of Zr is $4d^2 5s^2$ and that of Se is $4s^2 4p^4$. The total number of electrons accommodated in the $4s$ level of Se and in higher levels in a unit cell is 44 $[=(4+6 \times 3) \times 2]$. Thus a Fermi level locates between the σ and d_{z^2} bands. Then $ZrSe_3$ becomes a semiconductor.

Next we consider the electronic structures of group V MX_3 compounds. As described above, a group V MX_3 compound is assumed to have chains which may be distorted differently according to their crystal structure – namely, to be able to have both type A and type B chains.

$TaSe_3$, which has four chains in a unit cell, has two type A and two type B chains [115, 116]. The antibonding band σ_B^* associated with the type B chain appears between the σ band and the σ_A^* band associated with the type A chain which locates at much higher energy. If the σ_B^* band locates below the d_{z^2} band, a Fermi level lies between the σ_B^* and d_{z^2} bands. Thus, $TaSe_3$ may become a semiconductor or a semimetal according to the relative location of the σ_B^* band to that of the d_{z^2} band.

$NbSe_3$ has four type A and two type B chains in a unit cell [9, 14]. The σ_B^* band locates between the σ and σ_A^* bands, as in the case of $TaSe_3$. However, even if the σ_B^* band locates below the d_{z^2} band, there are still two electrons in the d_{z^2} band which can accept six electrons. Thus, $NbSe_3$ becomes metallic.

NbS_3 has two type A chains in a unit cell [112]. The band structure of NbS_3 is the same as that of the group IV MX_3 compounds. In this case, however, there are two more electrons in a unit cell which partially occupy the d_{z^2} band. The d_{z^2} band becomes half-filled, and NbS_3 is metallic above the Peierls transition temperature, which is higher than $T = 700$ K.

1.2.3. *Electronic Structure and CDW Transition*

Second, we investigate the relation between the electronic structures of group V MX_3 compounds and their CDW transitions.

As described above, $TaSe_3$ may become a semiconductor or semimetallic. Here, in order to explain a semimetallic properties of $TaSe_3$, the top of the σ_B^* band and the bottom of the d_{z^2} band are considered to overlap each other. A Fermi level crosses the top of the σ_B^* band and the bottom of the d_{z^2} band. Because of the location of a Fermi level, a Fermi surface of $TaSe_3$ is expected to be very complicated. As this is not favourable to the nesting mechanism, the CDW transition does not occur in $TaSe_3$; it remains semimetallic from room temperature to low temperatures. $NbSe_3$ has two electrons in the d_{z^2} bands at least. As the existence of six overlapping conduction bands is expected from the crystal structure, it is likely to have some nesting vectors, i.e. CDW transitions may occur. Because NbS_3 has a half-filled conduction band, a Peierls transition occurs in this system. At room temperature NbS_3 is a semiconductor.

Next we consider the relation between interchain interaction and the nesting vectors of NbS_3 and $NbSe_3$. As the d_{z^2} band is not expected to be dispersive along a direction perpendicular to the chain axis, the constant energy surface becomes flat, especially around the Fermi energy level.

In the case of NbS_3, if we do not take account of an interchain interaction, two d_{z^2} bands are degenerate. Then a nesting vector is (0.0, 0.5**b***, 0.0) at which a Peierls transition occurs (see Figure 5). Because of an interchain interaction, however, the two degenerate d_{z^2} bands in NbS_3 split into two bands. As the d_{z^2} band has a large dispersion along the chain direction, the shape of the band around the nesting region is nearly linear in the **b***-direction. Therefore, if the interchain interaction is not very large, a nesting vector is also expected to remain at (0.0, 0.5**b***, 0.0), resulting in a Peierls distortion.

In the case of $NbSe_3$, there is a sixfold degenerate d_{z^2} conduction band if we

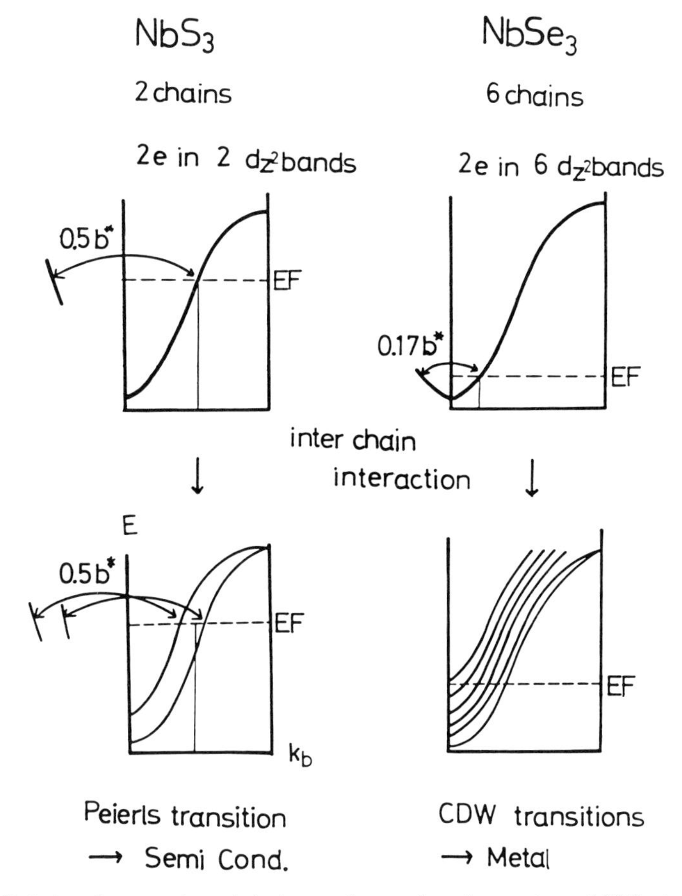

Fig. 5. Relations between interchain interactions and nesting vectors of NbS_3 and $NbSe_3$.

neglect interchain interactions. Putting two electrons into this conduction band, only one nesting vector of (0.0, 0.17**b***, 0.0) appears, where 0.17**b*** corresponds to one-sixth of **b***, the edge of the first Brillouin zone along the chain axis. Because of interchain interactions, this degenerate band splits into six bands. The Fermi level may generally cross six bands. For the lowest conduction band, the Fermi level crosses the band at the largest k value among the six bands. For the highest conduction band, on the other hand, the Fermi level crosses the band at a smaller k value. Thus there are certain differences in the nesting vectors for the three bands, and their lengths are about 0.2**b***. Then several incommensurate CDW transitions are expected for $NbSe_3$. As the innermost Fermi surface is near the point **G** in a Brillouin zone, where $\mathbf{k} = (0.0, 0.0, 0.0)$, this Fermi surface may have a warped and partially closed shape, which may still survive even after some incommensurate CDW transitions, causing metallic conduction.

1.2.4. *Important Factors Determining the Electronic Properties of MX_3 Compounds*

In the picture mentioned above, we have seen that the relation between the number of electrons and the number of type A and B chains in a unit cell determines whether MX_3 becomes metallic or semiconducting. In particular, in the case of group V MX_3 compound which may have both type A and B chains at the same time, the relative location of the σ_B^* band to that of the d_{z^2} band is essential in determining its electronic properties. Further, in the case of the metallic state, CDW transition may occur. In that case the number of chains and the strength of the interchain interactions determine the number and lengths of nesting vectors and the number of CDW transitions. In a real system the effects of such important factors as the location of the σ_B^* band and the strength of the interchain interaction, must be determined from the first principle band calculation for the whole system of MX_3.

1.3. ELECTRONIC PROPERTIES OF $NbSe_3$

1.3.1. *Structure of $NbSe_3$*

Synthesized $NbSe_3$ was first reported ten years ago [8, 9]. Owing to a quasi-one-dimensional chain structure, a crystal of $NbSe_3$ has a form of ribbon-like fibrous shape whose typical size in the directions of the **b**, **c**, and **a** axes is $5 \times 0.02 \times 0.005\ \text{mm}^3$. The Bravais lattice of $NbSe_3$ is monoclinic. Its space group is $P2_1/m$ [14]. That is, it has screw axes which are parallel to the chain axis **b** of a unit cell, mirror planes which are parallel to the **a**–**c** plane, and space inversion points. A unit cell is shown in Figure 6, where $a = 19$ a.u., $b = 6.6$ a.u., $c = 30$ a.u., and $\beta = 109°$. Six Se atoms form a prismatic cage with a Nb atom at its center. A trigonal chain is made by a linear stacking of these cages along the **b**-axis. The unit cell of $NbSe_3$ consists of these six trigonal chains. Each chain is displaced from the adjacent one by $b/2$ along the **b**-axis. This $b/2$ displacement between adjacent chains is characteristic of MX_3 compounds. While in 2H–$NbSe_2$ a Nb atom has six Se atoms as its neighbors, in

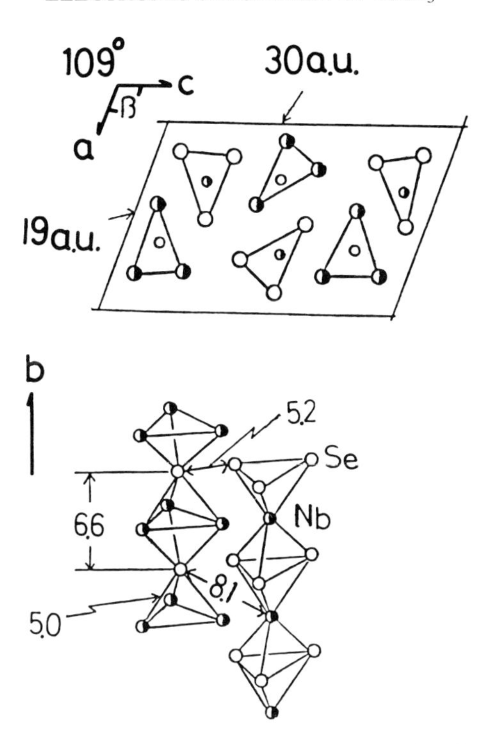

Fig. 6. Crystal structure of $NbSe_3$. In a unit cell, the atoms represented by ○ are in the same **a**–**c** plane, whose height along a chain direction is zero. The atoms represented by ◑ are in another **a**–**c** plane, whose height along the chain direction is $b/2$.

$NbSe_3$ a Nb atom has eight Se atoms in its neighborhood. Because of the existence of the screw axes, there are three pairs of two trigonal chains which have the same structures and the same sizes but are connected to each other by inversion or screw symmetric operation. The shortest distance between Nb atoms of a chain is 6.6 a.u. The shortest distance between a Nb atom of a chain and that of an adjacent one is about 8 a.u. Distances between a Nb atom in a chain and Se atoms in the same or adjacent chain are about 5 a.u. Thus interactions between a Nb atom and Se atoms in different chains cannot be neglected.

The electronic configuration of a Se atom is $4s^2 4p^4$ and that of a Nb atom is $4d^4 5s$. It is expected that in $NbSe_3$ some of the electrons of an Nb atom are transferred to Se atoms, and conduction bands are expected to be made of $4d$ orbitals of Nb atoms as discussed in Section 1.2.

1.3.2. *Experimental Results of* $NbSe_3$

The anomalous properties of $NbSe_3$ were reported by P. Haen *et al.* in 1975 [8]. Since then, many experimental [8–82] and theoretical [83–108] studies have been carried out with interest in its CDW transitions and nonlinear conductivities [3–6] which are explained below.

As shown in Figure 7, the resistivity ρ of $NbSe_3$ is metallic at room temperatures [11]. Its magnitude is about $6 \times 10^{-4}\,\Omega\,cm$ at room temperature, and decreases linearly with temperatures. At $T_1 = 142\,K$, ρ increases abruptly and has a peak around $T = 125\,K$. It then decreases linearly again. At $T_2 = 58\,K$, ρ increases again, has a peak around $T = 49\,K$, and then decreases to lower temperatures, as a metal. It is assumed that a phase transition occurs at T_1 and T_2. By electron diffraction [12, 13] and X-ray [14, 15, 16] experiments, two independent wavevectors representing a superstructure, $\mathbf{q}_1 = (0.0, 0.243\mathbf{b}^*, 0.0)$ and $\mathbf{q}_2 = (0.5\mathbf{a}^*, 0.263\mathbf{b}^*, 0.5\mathbf{c}^*)$ were observed at T_1 and T_2, respectively. These facts indicate that these two independent structural transitions correspond to two independent CDW phase transitions and some of the conducting carriers are frozen at each transition temperature.

A Hall constant was measured [33–37] to get some information about carriers of $NbSe_3$. The temperature dependence of the Hall constant is shown in Figure 8. It is negative at room temperature, which means that majority carriers are electron-like. It changes its sign from negative to positive at $T = 15\,K$. At T_1 and T_2 the Hall constant changes abruptly, indicating drastic changes of the number of carriers and Fermi surfaces. Other properties such as specific heat [25–27], magnetic susceptibility [28, 29], thermoelectric power [38, 39], NMR [41, 42], Seebeck effect [43],

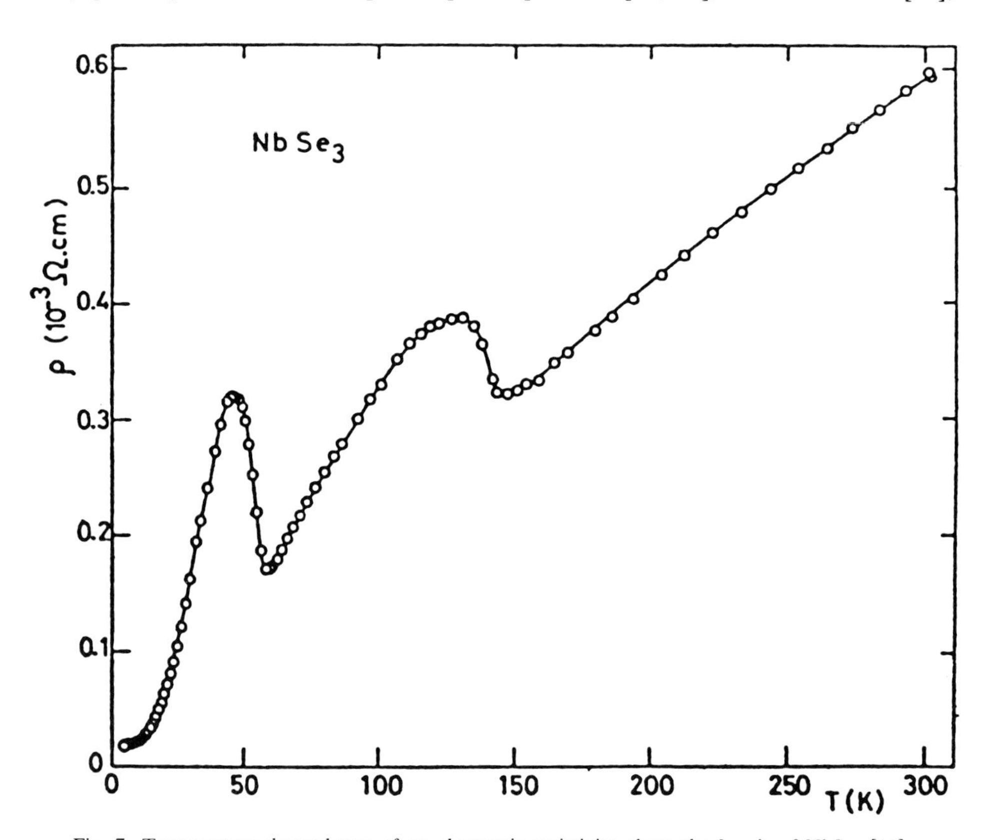

Fig. 7. Temperature dependence of an electronic resistivity along the **b**-axis of $NbSe_3$ [11].

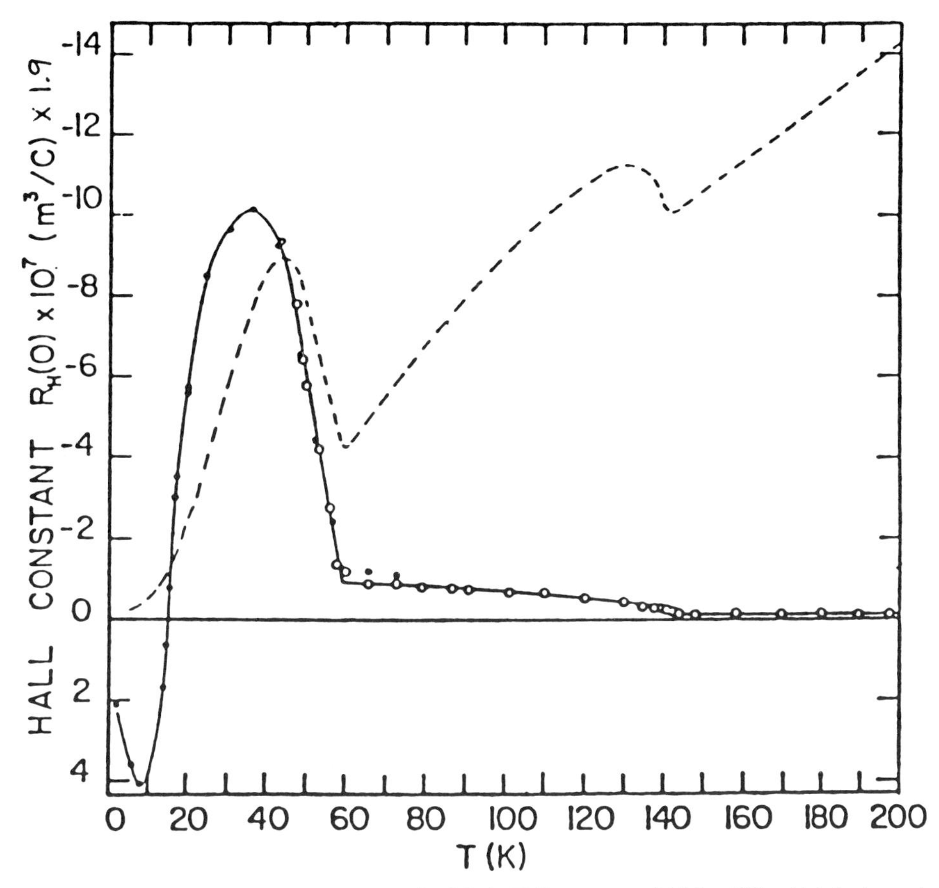

Fig. 8. Temperature dependence of a zero-field-limit Hall constant of $NbSe_3$ [33] which is shown by circles. Electric and magnetic fields are applied along the **b**-axis and perpendicular to the **a**–**c** plane, respectively. Resistivity is also shown by a dashed line for comparison.

Young modulus [52], thermal conductivity [53], dielectric constant [78], etc., also behave anomalously at the transition temperatures. All these large anomalous behaviors around the transition temperatures also indicate drastic changes of electronic structure due to CDW transitions.

At very low temperatures such as $T = 1.5$ K, quantum oscillations were observed in a magnetoresistance [30–32]. The result is shown in Figure 9. This oscillatory behavior suggests the existence of a rather simple and partially closed Fermi surface below two transition temperatures.

With the application of pressure, the transition temperatures shift to lower temperatures and the heights of the peaks in ρ become smaller [11, 24]. Furthermore, $NbSe_3$ becomes superconducting at $T = 2.4$ K under the pressure of 7 kbar [44–51].

Nonohmic behavior was observed below transition temperatures [17–19, 55–62]. The result is shown in Figure 10. The heights of the two peaks of the resistivity curve become smaller with the application of electric fields whose typical magnitudes are

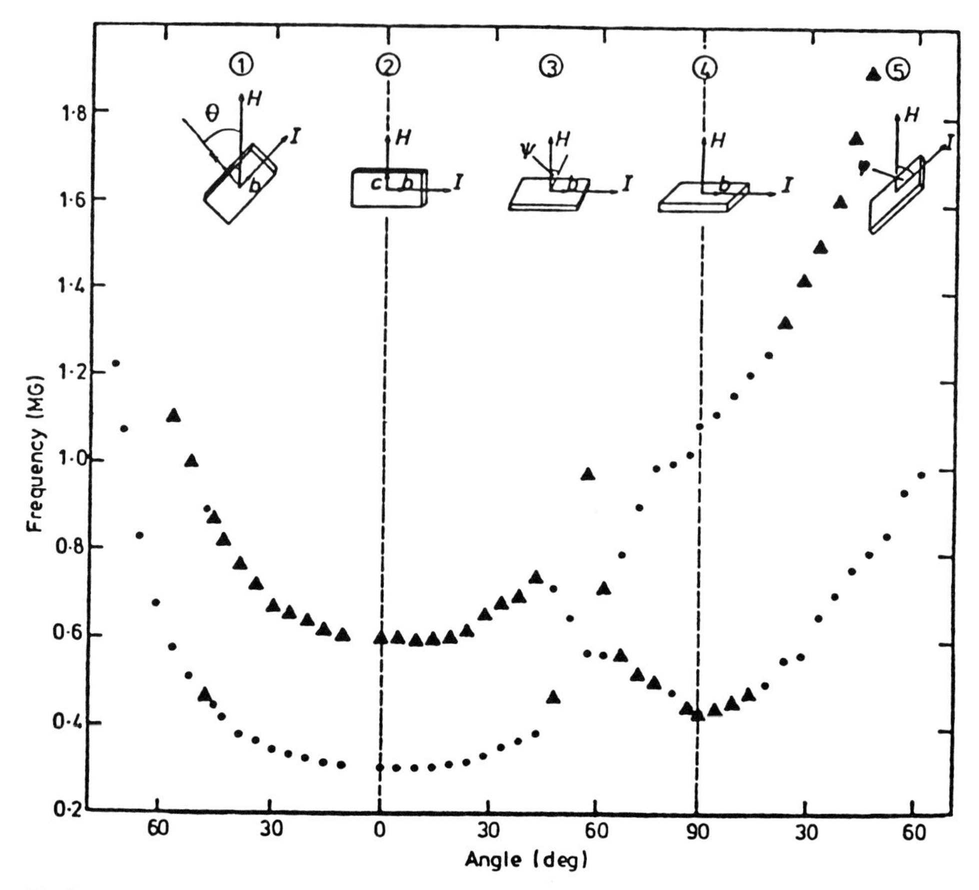

Fig. 9. Angular variation of the Shubnikov–de Haas frequencies of $NbSe_3$ [31]. H represents a magnetic field and I an electric current.

$1\,V\,cm^{-1}$ for the first transition and $0.1\,V\,cm^{-1}$ for the second. By increasing the applied electric field, the anomalous peaks disappear and the curve of ρ approaches the line of normal metallic resistivity, as if the carriers which were lost because of the CDW transitions had recovered again. The observed temperature and electric field dependences of conductivity in $NbSe_3$ are reproduced by the following empirical relation [22–24].

$$\sigma(T, E) = \begin{cases} \sigma_a(T) & (E < E_{\mathrm{th}}(T)) \\ \sigma_a(T) + \sigma_b(T)\exp[-E_0(T)/(E - E_{\mathrm{th}}(T))] & (E \geqslant E_{\mathrm{th}}(T)) \end{cases}$$

where

	E_0	E_{th} ($V\,cm^{-1}$)
$T \simeq 130\,K$	1	0.05
$T \simeq 50\,K$	0.1	0.005

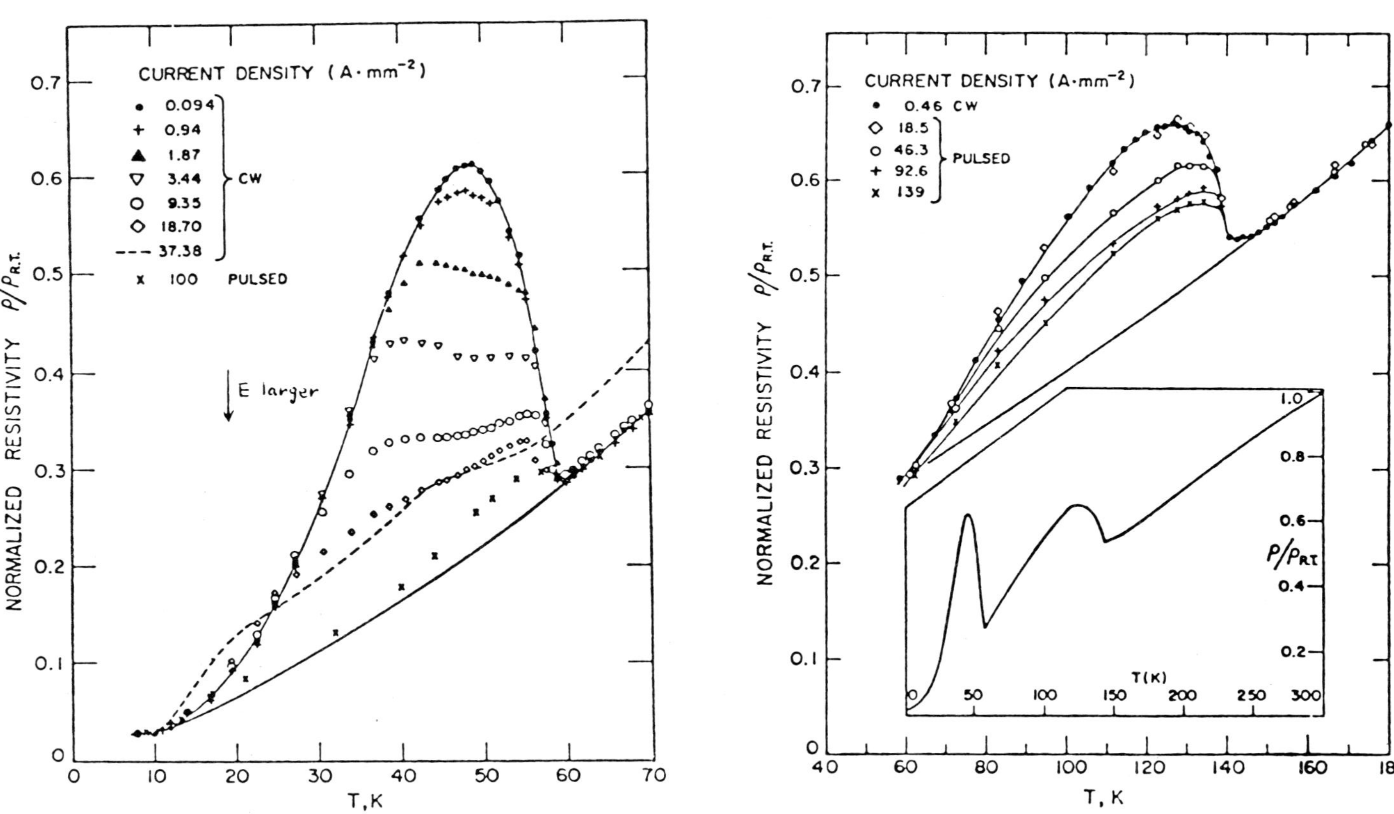

Fig. 10. Temperature and electric current dependences of upper and lower resistivity peaks of $NbSe_3$ [17]. 'CW' stands for measurements with d.c. current. The heavy line represents the limiting resistivity with available electric fields.

For nonohmic conduction, there exists a threshold electric field E_{th}; in other words, E_{th} corresponds to the upper limit in the electric field for ohmic conduction. The peaks of the resistivity curve also disappear under an a.c. field [63–65, 76]. In addition to these phenomena, quasi-periodic noise is observed in the nonohmic region [22, 66–75]. These anomalous properties are considered to result from depinning the ICDW (incommensurate charge density wave) or Fröhlich sliding mode. Several theories to explain the nonohmicity and quasi-periodic noise have been proposed so far. These are depinning theories [82–86], tunneling theories [87–89], soliton models [90–93], phase locking models [96, 97], etc. [94, 95, 98], but none of them has succeeded in explaining the phenomenon completely and consistently.

1.3.3 *CDW Transitions and Band Calculation*

The microscopic mechanisms of CDW transition, as well as of the accompanied anomalous transport properties, are strongly dependent on an electronic structure. In a case of a CDW transition, its nesting vector and the disappearance of a Fermi surface are determined mainly from the geometry and shape of the Fermi surface. Of course, the phonon structure of a crystal is also an important factor. The dynamics of ICDW or Fröhlich sliding mode which is considered to cause the nonohmic conductivity of $NbSe_3$ is also dependent on the electronic structure, i.e. an effective mass and location of ICDW or sliding mode. Thus in order to understand these CDW characteristics, it is necessary to calculate the precise band structure, especially for such a material as $NbSe_3$.

As described previously, electrons are expected to transfer from Nb atoms to Se atoms in $NbSe_3$. The charge distribution greatly affects the relative positions and shapes of the energy bands through the Coulomb potential. As explained in Section 1.2, the shapes, number, and positions of Fermi surfaces in a Brillouin zone depend on the relative positions of the conduction bands. Thus, a self-consistent calculation is essential to determine the Fermi surfaces and to study the CDW transitions.

So far, three band structure calculations of $NbSe_3$ have been reported. D. W. Bullett [103, 104, 106] calculated the band structure by a chemical pseudo-potential method using neutral atomic potentials of Nb and Se atoms. Because the effects of charge redistribution are not considered exactly in this calculation, interchain interactions and the relative location of the σ_B^* band to that of the d_{z^2} bands are not very accurately determined. Hoffmann *et al.* [105] also calculated the band structure of $NbSe_3$ by the extended Hückel method using semi-empirical parameters. Because they have considered all chains in a unit cell of $NbSe_3$ as type A, the existence of a σ_B^* band is ignored. Thus the number of conduction electrons in their result is incorrect. Further, the effects of interchain interactions and of charge redistribution are not determined correctly. Both calculations are not *ab-initio* self-consistent with regard to a charge distribution in a crystal, and important factors described in Section 1.2, which are especially important in $NbSe_3$, are not treated perfectly. In this view, one of the present authors (N. Shima), using the relativistic LCAO-Xα-SCF (linear combination of atomic orbitals – local Xα exchange correlation potential – self-consistent field) method, has recently calculated the electronic structure and the Fermi surfaces of $NbSe_3$. In the following we describe this method in detail.

2. Methodology of a Band Calculation

2.1. Formalism of a Band Calculation

In $NbSe_3$ the conduction bands are expected to be made of d_{z^2} orbitals of Nb atoms which are supposed to form narrow bands. Thus, in the band calcualtion it is suitable to perform an LCAO (linear combination of atomic orbitals) calculation. From the Herman–Skillman table of atomic levels of Mo [112], it can be seen that the relativistic effects on $4d$ levels of the Nb atom cannot be neglected. Therefore, it may be necessary to include them in a band calculation of $NbSe_3$. From this standpoint Shima [108] has performed a relativistic band calculation, including mass–velocity, Darwin, and spin–orbit coupling effects.

This relativistc band calculation is carried out by a numerical LCAO-Xα-SCF method [113, 114]. The state of each electron in a solid is approximately described by the one-electron Hamiltonian $\mathscr{H}$. Here we describe the local density functional formalism [115]. In this method, a problem of solving the ground state of systems is reduced to a problem of solving the states of one electron in a mean field. The exchange–correlation effect is described by a local potential determined by the electron density in a solid. This local density procedure is justified for the ground state of a system and has been successfully applied to many band structure calculations. Though a relativistic effect cannot be completely related with the interaction between many electrons, it can be approximately described by the Dirac-Slater form [114] as will be presented in Section 2.2.

Energy bands which are formed by one-electron energy levels of the one-electron Hamiltonian $\mathscr{H}$ are calculated by a familiar variational procedure using linear combinations of atomic orbitals.

As a variational basis set, we take Bloch functions,

$$\phi_{j,\mathbf{k}}(\mathbf{r}) = \frac{1}{\sqrt{N}}\sum_{\mathbf{l}} e^{i\mathbf{k}\cdot\mathbf{l}}\varphi_j(\mathbf{r} - \mathbf{l}) \tag{1}$$

where $\mathbf{r}$ is a space vector, $\mathbf{l}$ is a lattice vector, $\mathbf{k}$ is a crystal momentum, N is the number of atoms in a cyclic period, j is an orbital index of an atom in a unit cell, and φ_j is the jth atomic orbital function. Using these variational bases, the usual eigenvalue matrix equation is obtained for the αth band dispersion of $\mathscr{H}$:

$$\sum_j H_{ij}(\mathbf{k})C_{\alpha,j}(\mathbf{k}) = E_\alpha(\mathbf{k})\sum_j S_{ij}(\mathbf{k})C_{\alpha j}(\mathbf{k}), \tag{2}$$

where $E_\alpha(\mathbf{k})$ is the αth band dispersion and

$$H_{ij}(\mathbf{k}) = \langle \phi_{i,\mathbf{k}}|\mathscr{H}|\phi_{j,\mathbf{k}}\rangle$$

$$S_{ij}(\mathbf{k}) = \langle \phi_{i,\mathbf{k}}|\phi_{j,\mathbf{k}}\rangle$$

$$\langle f|g\rangle = \int f^* g\,\mathrm{d}\mathbf{r}$$

An αth band's eigenfunction is approximated by

$$\psi_{\alpha,\mathbf{k}}(\mathbf{r}) = \sum_j C_{\alpha j}(\mathbf{k})\phi_{j,\mathbf{k}}(\mathbf{r}). \tag{3}$$

The steps of a band calculation are: to construct a reasonable one-electron Hamiltonian $\mathscr{H}$; to choose appropriate basis functions; to calculate matrix elements H_{ij} and S_{ij} and to solve a secular equation. Actual procedures are described briefly below.

2.2. Hamiltonian of a System

A total one-electron Hamiltonian $\mathscr{H}$ is expressed as the sum of the nonrelativistic and relativistic parts:

$$\mathscr{H} = \mathscr{H}_{\mathrm{NR}} + \mathscr{H}_{\mathrm{Rel}}. \tag{4}$$

First, the nonrelativistic part of Hamiltonian is expressed as

$$\mathscr{H}_{\mathrm{NR}} = T_{\mathrm{kin}} + V_{\mathrm{Coul}} + V_{\mathrm{ex}}. \tag{5}$$

Here T_{kin} is a kinetic energy of an electron which is expressed in atomic unit as $T_{\mathrm{kin}} = -\Delta$, where Δ is the Laplacian operator (we shall use the atomic unit in this article); and V_{Coul} is the Coulomb potential due to nuclei and electrons in a solid:

$$V_{\mathrm{Coul}}(\mathbf{r}) = \sum_{j,\mathbf{l}} v_j(\mathbf{r} - \mathbf{l}), \tag{6}$$

where $\mathbf{l}$ is the lattice vector, j is the atomic index in a unit cell, and $v_j(\mathbf{r})$ is the Coulomb potential at the jth atom.

$$v_j(\mathbf{r}) = \int \frac{\rho_j(\mathbf{r}') - Z_j\,\delta(\mathbf{r}')}{|\mathbf{r} - \mathbf{r}'|}\,\mathrm{d}\mathbf{r}', \tag{7}$$

where $\rho_j(\mathbf{r})$ is the charge distribution of the electrons in the jth atom; $Z_j/2$ the atomic number of jth atom; and $\delta(\mathbf{r})$ the delta function, $v_j(\mathbf{r})$ can be divided into a short-range and a long-range part,

$$\begin{aligned} v_j(\mathbf{r}) &= v_j^s(\mathbf{r}) + v_j^l(\mathbf{r}) \\ v_j^s(\mathbf{r}) &= \int \frac{\rho_j(\mathbf{r}') - Z_j\,\delta(\mathbf{r}') - \rho_j'(\mathbf{r}')}{|\mathbf{r} - \mathbf{r}'|}\,\mathrm{d}\mathbf{r}' \\ v_j^l(\mathbf{r}) &= \int \frac{\rho_j'(\mathbf{r}')}{|\mathbf{r} - \mathbf{r}'|}\,\mathrm{d}\mathbf{r}'. \end{aligned} \tag{8}$$

where $v_j^s(\mathbf{r})$ is the short-range part, $v_j^l(\mathbf{r})$ the long-range part, and $\rho_j'(\mathbf{r})$ an excess multipole moment of charge distribution in the jth atom. Any form can be used for $\rho_j'(\mathbf{r})$ only if it satisfies the following relation:

$$\int r^{\mu} Y_{\mu\nu} \rho_j'(\mathbf{r})\,\mathrm{d}\mathbf{r} = \int r^{\mu} Y_{\mu\nu} [\rho_j(\mathbf{r}) - Z_j\,\delta(\mathbf{r})]\,\mathrm{d}\mathbf{r}, \tag{9}$$

where $Y_{\mu\nu}$ is a spherical harmonics function and $\mu = 0, 1, 2, \ldots, \rho_j(\mathbf{r})$ damps rapidly as $|\mathbf{r}| \to \infty$. Then the total Coulomb potential $V_{\mathrm{Coul}}(\mathbf{r})$ is

$$V_{\mathrm{Coul}}(\mathbf{r}) = \sum_{j,\mathbf{l}} v_j^s(\mathbf{r} - \mathbf{l}) + \sum_{j,\mathbf{l}} v_j^l(\mathbf{r} - \mathbf{l}). \tag{10}$$

The value of the Coulomb potential is needed only in a unit cell. As $v_j^s(\mathbf{r})$ is a short-range potential, the summation over j and $\mathbf{l}$ can be restricted only in a small region around a unit cell. As regards the summation for the long-range potential $v_j^l(\mathbf{r})$, the Ewalt's technique is used. If $\rho_j^{\prime}(\mathbf{r})$ has the form

$$\rho_j^{\prime}(\mathbf{r}) = \sum_{\mu\nu} \rho_{j\mu}^{\prime}(r)\, Y_{\mu\nu}(\theta, \phi), \tag{11}$$

where θ and ϕ represent angular part of $\mathbf{r}$, then

$$\begin{aligned} \sum_{j\mathbf{l}} v_j^l(\mathbf{r} - \mathbf{l}) \; & \sum_{\substack{j\mu \\ \mathbf{K}_n \neq 0}} \frac{(4\pi)^2}{\Omega \mathbf{K}_n^2}\, i^{\mu} Y_{\mu\nu}(\theta_{\mathbf{K}_n}, \phi_{\mathbf{K}_n}) \\ & \times \left[\int_0^{\infty} \mathrm{d}r' r'^2 \rho_{j\mu}^{\prime}(r')\, j_{\mu}(|\mathbf{K}_n| r') \right] \mathrm{e}^{-i\mathbf{K}_n \cdot \mathbf{r}}, \end{aligned} \tag{12}$$

where $\mathbf{K}_n$ is a reciprocal lattice vector, Ω the volume of a unit cell, $(\theta_{\mathbf{K}_n}, \phi_{\mathbf{K}_n})$ the angular part of $\mathbf{K}_n$, and $j_{\mu}(r)$ the μth sperical Bessel function. $\rho_{j\mu}^{\prime}(r)$ has, for example, the form

$$\rho_{j\mu}^{\prime}(r) = A_{j\mu} r^{\mu} \mathrm{e}^{-a_{j\mu} r^2}, \tag{13}$$

where $A_{j\mu}$ and $a_{j\mu}$ are constants.

$V_{\mathrm{ex}}(\mathbf{r})$ represents the exchange–correlation potential, which consists of the exchange interaction v_{exch}, the self-interaction correction of electrons v_{self}, and the correlation energy v_{corr}.

$$V_{\mathrm{ex}} = v_{\mathrm{exch}} - v_{\mathrm{self}} + v_{\mathrm{corr}}. \tag{14}$$

In the LDF (local density functional) formalism, V_{ex} can be expressed in a local potential form [115, 116]. If an electronic charge in a solid $\rho(\mathbf{r})$ is varying slowly enough, we have

$$\begin{aligned} v_{\mathrm{exch}} - v_{\mathrm{self}} &= -2\left[\frac{3}{\pi}\rho(\mathbf{r})\right]^{1/3} \\ v_{\mathrm{corr}} &\simeq -0.18 \ln[1 + 33.9\rho^{1/3}(\mathbf{r})]. \end{aligned} \tag{15}$$

In the X–α method [117]

$$\begin{aligned} V_{\mathrm{ex}} &= v_{\mathrm{X}\alpha} \\ &= -3\alpha\left[\frac{3}{\pi}\rho(\mathbf{r})\right]^{1/3}, \end{aligned} \tag{16}$$

where α is a parameter which is normally chosen as 0.7. This choice of α works well in many calculations of molecules [118]. The Xα potential with $\alpha = 0.7$ is considered to include the correlation effect as well as v_{exch} and v_{self}. In a real case the difference between the LDF potential and the Xα potential is nearly a small constant. The Xα potential with $\alpha = 0.7$ is used in the following calculation.

Second, we would like to explain the relativistic part $\mathscr{H}_{\mathrm{Rel}}$. Relativistic effects derived from the Dirac equation are added to the nonrelativistic part of one-electron Hamiltonian. After expanding the operator and wave functions of the Dirac equation in terms of α_F (a fine structure constant) up to α_F^2, two components Hamiltonian and wave function Φ whose upper component represents up spin state and lower part represents down spin state are derived:

$$H\Phi = E\Phi$$

$$H = -\Delta - V - \tfrac{1}{4}\alpha_F^2\Delta^2 + \tfrac{1}{4}\alpha_F^2(-\Delta V) + \tfrac{1}{4}\alpha_F^2\boldsymbol{\sigma}\cdot\left(-\nabla V \times \frac{\nabla}{i}\right), \tag{17}$$

where E is an eigenvalue, $\sigma_{x,y,z}$ are 2×2 Pauli matrices, and V is a local scalar potential. The last three terms of H represent the relativistic effects. Then the relativistic part $\mathscr{H}_{\mathrm{Rel}}$ is expressed as

$$\mathscr{H}_{\mathrm{Rel}} = H_V + H_{\mathrm{D}} + H_{\mathrm{SO}}, \tag{18}$$

where

$$H_V = -\tfrac{1}{4}\alpha_F^2\Delta^2 \tag{19}$$

and it represents a mass–velocity effect. Further, H_{D} represents a Darwin effect and is given by

$$H_{\mathrm{D}} = \pi\alpha_F^2\rho_{\mathrm{Tot}}, \tag{20}$$

with the following approximation,

$$\Delta V \simeq -4\pi\rho_{\mathrm{Tot}}, \tag{21}$$

where V is the crystal potential and ρ_{Tot} is the total charge distribution in a crystal. H_{SO} represents the spin–orbit effect and is given by

$$H_{\mathrm{SO}} = \frac{\alpha_F^2}{4}\sum_{j\mathbf{l}} \zeta_j(|\mathbf{r} - \boldsymbol{\mu}_j - \mathbf{l}|)(\boldsymbol{\sigma}\cdot\mathbf{L})[\mathbf{r} - \boldsymbol{\mu}_j - \mathbf{l}], \tag{22}$$

where $\mathbf{l}$ is the lattice vector, $\boldsymbol{\mu}_j$ a site vector of the jth atom, and $\mathbf{L}$ an angular momentum operator. $\zeta_j(\mathbf{r})$ is a function of $\mathbf{r}$, defined by

$$\zeta_j(\mathbf{r}) = \frac{1}{r}\left\{\frac{\mathrm{d}}{\mathrm{d}r}[v_j(r)] - \alpha\left(\frac{3}{\pi}\right)^{1/3}\rho_j^{-2/3}(r)\frac{\mathrm{d}}{\mathrm{d}r}[\rho_j(r)]\right\}, \tag{23}$$

where the crystal Coulomb potential and the crystal electronic charge are approximated by the Coulomb potential of the jth atom, $v_j(r)$, and the electronic charge of the jth atom, $\rho_j(r)$, for ease of calculation. The approximations used in deriving the terms H_{D} and H_{SO} are not serious because these terms have little effect on the real calculation.

From the Herman–Skillman table [112], relativistic effects on atomic levels can be seen. For $4d$ orbitals of Mo^{42}, $E_V = -0.018$ Ry, $E_{\mathrm{D}} = 0.0005$ Ry, and $E_{\mathrm{SO}} = 0.004$ Ry. Thus the important relativistic effect for the $4d$ orbitals of Nb is expected to come from H_V term. The effects of H_{D} and H_{SO} terms are expected to be small.

Because H_V and H_{D} are spin-independent operators, they can be treated together with the nonrelativistic part $\mathscr{H}_{NR}$. As regards the spin-dependent operator H_{SO} whose effect is expected to be small, we treat it as a perturbation to the result of SCF calculation of $\mathscr{H}_0$, where

$$\mathscr{H}_0 = \mathscr{H}_{NR} + H_V + H_D. \tag{24}$$

2.3. Basis Functions and Matrix Element

By analogy with the electronic structure of MX_2, such as shown in the introduction, it is supposed that conduction bands of $NbSe_3$ are mainly composed of $4d$ orbitals of Nb atoms. From the atomic energy levels of neutral Nb atom and neutral Se atom in the Herman–Skillman table [112], it can be seen that the energy levels of $4p$ orbital of Nb atom and $3d$ orbital of Se atom are deep enough so that they do not contribute to forming upper bands near the conduction bands of $NbSe_3$. For variational basis functions of valence bands, we use thc Se $4s$ and $4p$ orbitals and Nb $4d$ and $5s$ orbitals. These orbitals are orthogonal to the core orbitals at their own atomic sites and further the tails of these orbitals are modified to shrink so as to orthogonalize the core orbitals of the neighboring atoms by adding a suitable well potential to the atomic potential.

Once a Hamiltonian and basis set are determined, one can calculate the matrix elements of a secular equation. For this purpose it is necessary to get an expectation value of an operator $\mathscr{O}$ in a unit cell, which is calculated numerically here using the Painter–Ellis random sampling method [119, 120],

$$\int_\Omega \varphi_i^*(\mathbf{r})\,\mathscr{O}(\mathbf{r})\,\varphi_j(\mathbf{r})\,d\mathbf{r} \simeq \sum_{\substack{p=1\\ \mathbf{r}_p\in\Omega}}^{N} \varphi_i^*(\mathbf{r}_p)\,\mathscr{O}(\mathbf{r}_p)\,\varphi_j(\mathbf{r}_p)\,\mathscr{W}(\mathbf{r}_p), \tag{25}$$

where $\mathbf{r}_p$ is a sampling point in a unit cell and N is a number of sampling points. Further, $\mathscr{W}(\mathbf{r})$ is a weight function. A unit cell is divided into two regions, the intra-atomic region and the interatomic region using muffin-tin like division of the APW method. The inverse of $\mathscr{W}(\mathbf{r})$, $[\mathscr{W}(\mathbf{r})]^{-1}$, which means density of sampling points is defined in the two regions,

$$[\mathscr{W}(\mathbf{r})]^{-1} = \begin{cases} \dfrac{A}{r^2}\dfrac{1}{1+e^{\beta(r-R)^2}} & \text{for the intra-atomic region} \\ \text{const.} & \text{for the interatomic region} \end{cases} \tag{26}$$

where A, β, and R are constants. In a unit cell of $NbSe_3$, 9600 points are chosen for sampling points.

There are differential operators Δ and Δ^2 in a total Hamiltonian $\mathscr{H}$. Their matrix elements are calculated by the following equations.

$$\langle\varphi_i|\,\Delta\,|\varphi_j\rangle_\Omega = \tfrac{1}{2}\langle\varphi_i|\,(\epsilon_i - v_i) + (\epsilon_j - v_j)\,|\varphi_j\rangle_\Omega \tag{27}$$

$$\langle\varphi_i|\,\Delta^2\,|\varphi_j\rangle_\Omega = \langle\varphi_i|\,(\epsilon_i - v_i)(\epsilon_j - v_j)\,|\varphi_j\rangle_\Omega, \tag{28}$$

where ϵ_i and ϵ_j are orbital energies of i and jth atomic sites, and v_i and v_j are the potential energies of ith and jth atomic sites.

A spin–orbit interaction term $\mathscr{H}_1 = H_{SO}$ is taken into account as a perturbation to $\mathscr{H}_0 = \mathscr{H}_{NR} + H_V + H_D$. A perturbation calculation is done using the unperturbed spin-up and spin-down energy band wave functions. Because of the inversion symmetry of $NbSe_3$ all bands keep double degeneracy, namely a state at every $\mathbf{k}$ point has the spin degeneracy even under the spin–orbit perturbation H_{SO}. For the simplicity of a calculation, two- or three-center matrix elements of H_{SO} are all neglected. This one-center approximation is justified by the fact that $\zeta(\mathbf{r})$ decreases rapidly with the r dependence faster than $1/r^3$ as r becomes large. Then the spherical symmetric terms in the H_{SO} determines nonzero matrix elements. It is easily seen that the bands which are composed of d_{z^2} orbitals do not mix with each other through H_{SO} perturbation. As the conduction bands of $NbSe_3$ are mainly composed of d_{z^2} orbitals, the effect of H_{SO} is expected to be small. This is confirmed from the resulting band structure. In a real calculation, ten unperturbed valence bands (from 67th band to 76th band) are used to calculate the perturbation effects of H_{SO} [108].

2.4. Charge Distribution

From a given charge distribution, a Hamiltonian is constructed. Resulting band wave functions from the Hamiltonian give a new charge distribution that is used in a next iteration of SCF band calculation. In this procedure, it is necessary to construct a charge distribution from calculated band wave functions, $\{\Psi_{\alpha\mathbf{k}}\}$. In the following, the procedure is explained for a simple case of a single atom in a unit cell.

A crystal charge density is given by

$$\begin{aligned} \rho(\mathbf{r}) &= \sum_{\substack{\alpha \\ E_\alpha(\mathbf{k}) \leq E_F}} \rho_{\alpha,\mathbf{k}}, \\ \rho_{\alpha,\mathbf{k}} &= |\Psi_{\alpha,\mathbf{k}}|^2, \end{aligned} \tag{29}$$

where α is a band index and the summation is taken over the occupied bands below Fermi energy E_F. $\rho_{\alpha\mathbf{k}}$ is divided into two parts, one-center part $\rho^{(1)}_{\alpha\mathbf{k}}$ and two-center part $\rho^{(2)}_{\alpha\mathbf{k}}$ as follows;

$$\begin{aligned} \rho^{(1)}_{\alpha,\mathbf{k}} &= \frac{1}{N}\sum_{st} C^*_{\alpha s}(\mathbf{k})\,C_{\alpha t}(\mathbf{k})s_{st;0}\sum_{\mathbf{l}} \varphi^2_t(\mathbf{r} - \mathbf{l}) \\ &\quad + \frac{1}{N}\sum_{s\neq t} C^*_{\alpha s}(\mathbf{k})\,C_{\alpha t}(\mathbf{k})\sum_{\mathbf{l}} \varphi_s\varphi_t(\mathbf{r} - \mathbf{l}), \end{aligned} \tag{30}$$

and

$$\rho^{(2)}_{\alpha,\mathbf{k}} = \frac{1}{N}\sum_{\substack{st \\ \mathbf{l}\neq\mathbf{l}'}} \mathrm{e}^{-i\mathbf{k}\cdot(\mathbf{l}-\mathbf{l}')}C^*_{\alpha s}(\mathbf{k})\,C_{\alpha t}(\mathbf{k})\varphi_s(\mathbf{r} - \mathbf{l})\varphi_t(\mathbf{r} - \mathbf{l}'), \tag{31}$$

where

$$s_{st;(\mathbf{l}-\mathbf{l}')} = \langle \varphi_s(\mathbf{r} - \mathbf{l})\,|\,\varphi_t(\mathbf{r} - \mathbf{l}')\rangle. \tag{32}$$

A difficulty in calculating a Coulomb potential occurs from the two-center charge distribution $\rho^{(2)}_{\alpha\mathbf{k}}$. Though one may expand $\rho^{(2)}_{\alpha\mathbf{k}}$ by Fourier series or one-center orthogonal sets and calculate a Coulomb potential, these procedures are very tedious. Here, using Mulliken's population analysis [121], $\rho^{(2)}_{\alpha\mathbf{k}}$ is approximated by one-center function as follows.

$$\rho^{(2)}_{\alpha,\mathbf{k}} \cong \sum_{st} C^*_{\alpha s}(\mathbf{k}) C_{\alpha t}(\mathbf{k}) [S_{st}(\mathbf{k}) - s_{st;0}] \sum_{\mathbf{l}} \varphi_s^2(\mathbf{r} - \mathbf{l}), \tag{33}$$

where

$$S_{st}(\mathbf{k}) = \sum_{\mathbf{l}} \mathrm{e}^{-i\mathbf{k}\cdot\mathbf{l}} s_{st;\mathbf{l}} \tag{34}$$

is an overlap matrix element. Then the total $\rho_{\alpha\mathbf{k}}$ is approximated by one-center functions;

$$\begin{aligned} \rho_{\alpha,\mathbf{k}} \cong & \frac{1}{N} \sum_{st} C^*_{\alpha s}(\mathbf{k}) S_{st}(\mathbf{k}) C_{\alpha t}(\mathbf{k}) \sum_{\mathbf{l}} \varphi_t^2(\mathbf{r} - \mathbf{l}) \\ & + \frac{1}{N} \sum_{s \neq t} C^*_{\alpha s}(\mathbf{k}) C_{\alpha t}(\mathbf{k}) \sum_{\mathbf{l}} \varphi_s \varphi_t(\mathbf{r} - \mathbf{l}). \end{aligned} \tag{35}$$

It is easy to calculate a Coulomb potential from a one-center function as shown before. The second term in $\rho_{\alpha\mathbf{k}}$ which produces short-range multipole Coulomb potentials is omitted in the band calculation of $NbSe_3$ because of its complex crystal structure. After all the crystal charge density becomes

$$\rho(\mathbf{r}) \cong \frac{1}{N} \sum_{st} \Big[\sum_{\substack{\alpha \\ E_\alpha(\mathbf{k}) \leq E_F}} C^*_{\alpha s}(\mathbf{k}) S_{st}(\mathbf{k}) C_{\alpha t}(\mathbf{k}) \Big] \sum_{\mathbf{l}} \varphi_t^2(\mathbf{r} - \mathbf{l}). \tag{36}$$

Though the above approximation for a two-center function makes the calculation of a charge distribution and a Coulomb potential in a unit cell easier, it also has some demerits which we shall mention briefly. This approximation cannot describe a very localized overlapping charge distribution. It has a tendency to spread a charge distribution. Nodes of an overlapping charge are neglected. Unless all $\{\varphi_i\}$ are s orbitals, the rotational invariance of the calculation is not maintained in this approximation.

In spite of these demerits, this approximation is useful and does not result in very serious difficulties in many cases of energy level calculations in large molecules, perhaps because the energy levels and Coulomb potentials are mainly determined from gross population of charge and may not be sensitive to local change of charge distribution.

2.5. Density of states and Brillouin zone integral

An energy density of states of a resulting band structure is calculated by the following expression,

$$D(\epsilon) = \sum_{\alpha} \frac{2\Omega}{(2\pi)^3} \int_{E_\alpha(\mathbf{k})=\epsilon} \mathrm{d}\mathbf{S} \frac{1}{|\nabla_{\mathbf{k}} E_\alpha(\mathbf{k})|} \text{ states} \cdot \text{molecule}^{-1} \cdot \text{Ry}^{-1} \tag{37}$$

where Ω is the volume of a unit cell, $\int \mathrm{d}\mathbf{S}$ denotes a surface integral in $\mathbf{k}$ space, $E_\alpha(\mathbf{k})$ is the energy dispersion of the αth band, and $\nabla_{\mathbf{k}}$ is a nabla operator in $\mathbf{k}$ space. To investigate the contribution of the φ_j orbital to the energy band dispersion, the partial density of states $n_j(\epsilon)$ at the jth atomic site is calculated using Mulliken's population analysis, as follows;

$$n_j(\epsilon) = \sum_{\alpha} \frac{2\Omega}{(2\pi)^3} \int_{E_\alpha(\mathbf{k})=\epsilon} \mathrm{d}\mathbf{S} \frac{1}{|\nabla_{\mathbf{k}} E_\alpha(\mathbf{k})|} \sum_{t} C^*_{\alpha j}(\mathbf{k}) S_{jt}(\mathbf{k}) C_{\alpha t}(\mathbf{k}). \tag{38}$$

Of course, $n_j(\epsilon)$ satisfies the following relation:

$$D(\epsilon) = \sum_{j} n_j(\epsilon). \tag{39}$$

It is necessary to carry out some $\mathbf{k}$-space integrals, in order to get a crystal charge distribution, a density of states, and a partial density of states. These $\mathbf{k}$-space integrals are easily performed by the tetrahedron method of Lehmann *et al.* [122, 123]. A Brillouin zone is divided into many small tetrahedron regions, where energy dispersion and integrand functions are approximated by $\mathbf{k}$-linear functions. Then it is easy to carry out a Brillouin zone integral. The first Brillouin zone of $NbSe_3$ is divided into about 2600 small tetrahedron regions in the present calculation.

2.6. Summary

We summarize the procedure of a relativistic LCAO-Xα-SCF band calculation in Figure 11. From an input charge, a crystal potential is constructed. Atomic orbitals $\{\varphi_i\}$ are calculated using atomic potentials $\{v_i\}$ which are obtained by taking the average of the crystal potentials locally. By taking the linear combination of atomic orbitals $\{\varphi_i\}$, the basis functions $\{\phi_{i\mathbf{k}}\}$ are formed. With these basis functions, a secular equation for $\mathscr{H}_0$ is derived and then diagonalized. The band energy dispersion and band wave functions $\{E_\alpha(\mathbf{k}), \Psi_{\alpha\mathbf{k}}(\mathbf{r})\}$ are then obtained. The $\{\Psi_{\alpha\mathbf{k}}(\mathbf{r})\}$ of the occupied states yield an output charge. Using this output charge as a new input charge, another iterative calculation is performed until the difference between the input charge and the output charge becomes acceptably small. The spin–orbit interaction H_{SO} is then included as a perturbation effect to the result of the SCF calculation. Spin–orbital functions $\{\Psi^\sigma_{\alpha\mathbf{k}}(\mathbf{r})\}$ are constructed from a part of $\{\Psi_{\alpha\mathbf{k}}(\mathbf{r})\}$. After solving the secular equation of the perturbation, we get the relativistic SCF band structure, Fermi surfaces, and density of states, etc.

3. Results of the Band Calculation

3.1. Charge distribution in a unit cell

In this section we describe the results of the energy band calculations done by Shima using the method mentioned in previous sections. First, the charge distribution at

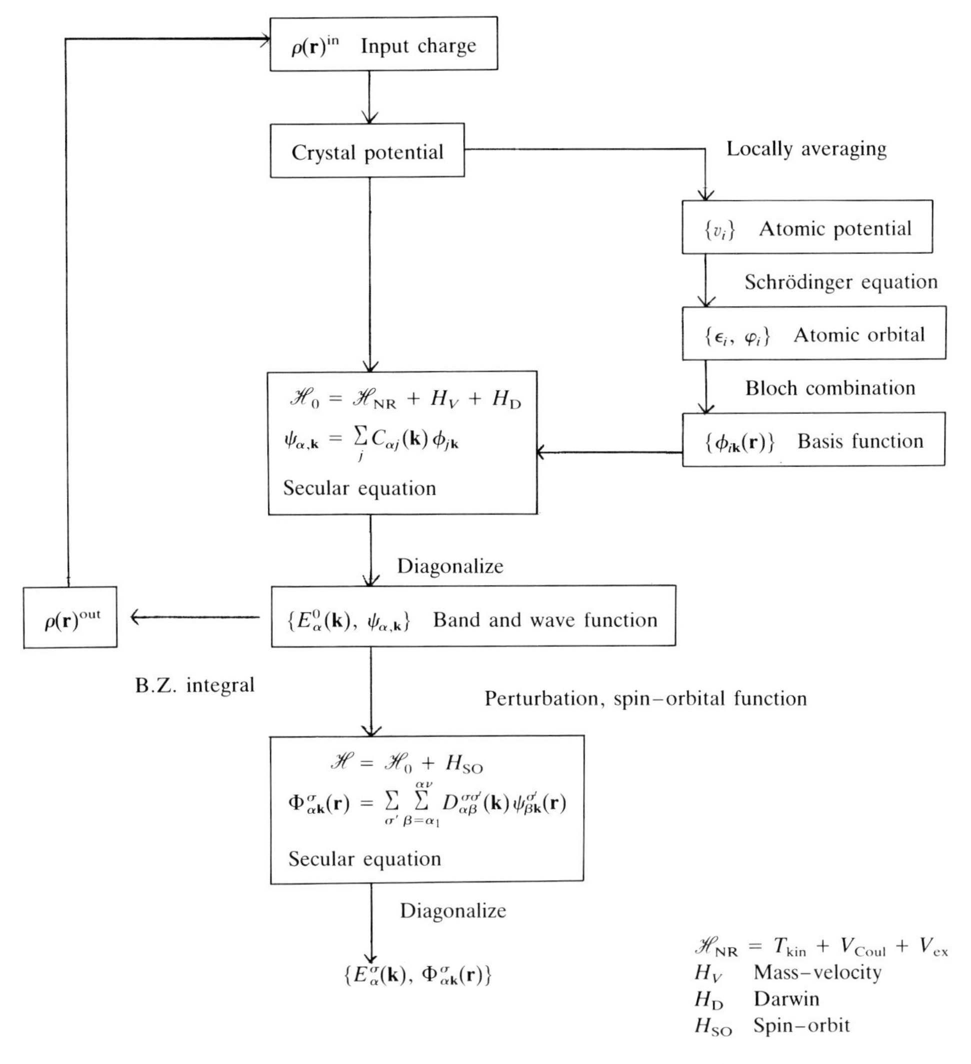

Fig. 11. Relativistic LCAO-Xα-SCF band calculation scheme.

each atomic site is given in Figure 12. This is determined by Mulliken's population analysis after seven SCF iterations. In the following, an atom in a unit cell is specified by designating the chain (I, II, III, I′, II′, III′) and the atomic site (Nb, $Se_{①}$, $Se_{②}$, $Se_{③}$) where the atoms in a unit cell are arranged as shown in Figure 13. For instance, the Se atom which is indicated by an arrow in the figure is specified as $Se_{②}$ in chain, I or $Se^{I}_{②}$. Chain I and chain I′ are transferred into each other by a screw symmetric operation.

It is seen from Figure 12 that the more Nb atoms surrounding a Se atom, the more does the ionicity of that Se atom become negative. For example, $Se^{I}_{③}$, which is strongly negative among the Se atoms, is surrounded by three Nb atoms of Nb^{I},

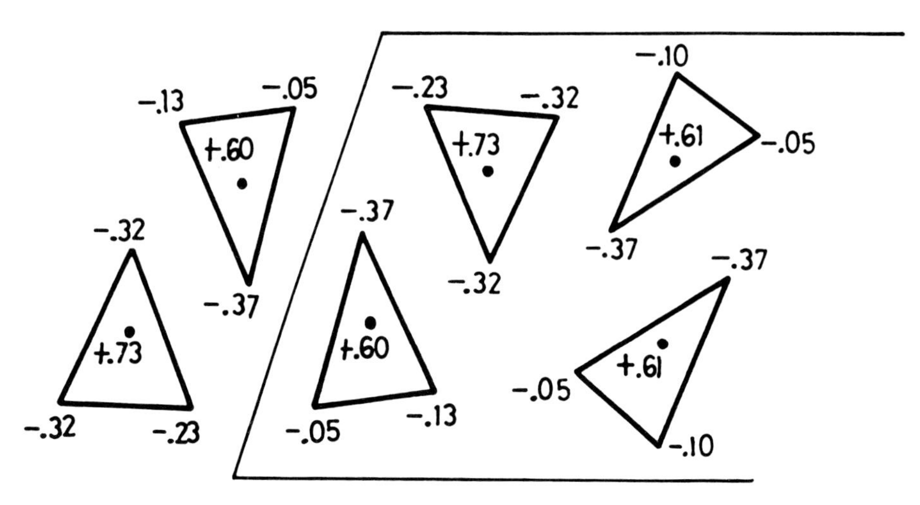

Fig. 12. Charge distribution at each atomic site determined by Mulliken's population analysis. Each numeral represents the change of charge of each atom from a neutral state.

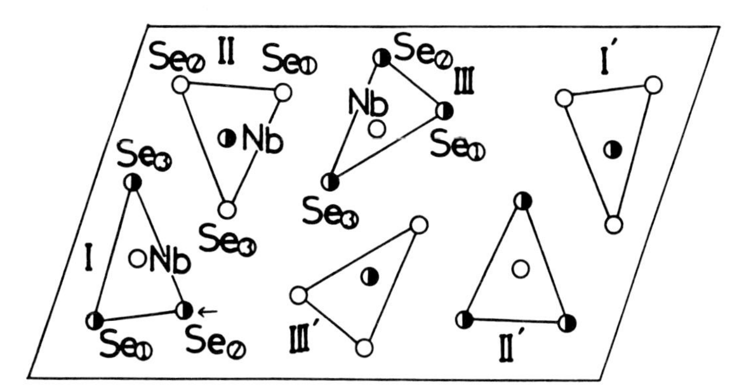

Fig. 13. Labeling of atoms in a unit cell of $NbSe_3$. Chains I and I′ locate at mutually screw symmetric sites, for example.

$\mathrm{Nb}^{\mathrm{I}'}$, and $\mathrm{Nb}^{\mathrm{II}}$. $\mathrm{Se}^{\mathrm{III}}_{(3)}$, which is surrounded by $\mathrm{Nb}^{\mathrm{III}}$, $\mathrm{Nb}^{\mathrm{III}'}$ and $\mathrm{Nb}^{\mathrm{II}}$, is also strongly negative. $\mathrm{Se}^{\mathrm{II}}_{(3)}$, which is surrounded by $\mathrm{Nb}^{\mathrm{II}}$ and Nb^{I}, is less negative. This can be understood by considering the existence of chemical bonds (or charge distribution) between Se atoms and neighboring Nb atoms.

On the other hand, it is possible to explain that the types of chain which were described in Section 1.2 reflect the charge distribution. The charge distributions of chains I and III, both of which are type A, are very similar. Chain II, which is type B, becomes more ionic than chain I or III.

According to the calculated results, about 0.6 electrons are transferred from a Nb atom in chain I or III to neighboring Se atoms, and about 0.7 electrons are transferred from a Nb atom in chain II to neighboring Se atoms. On the other hand, from an experimental measurement of the chemical shift of the core level of Nb atoms in

$NbSe_3$ [40], it is estimated that about 0.2 electrons are transferred from a Nb atom to Se atoms. Thus, the calculated values of the charge transfer are about three times as large as the experimental value. However, there is an ambiguity in the estimation of a charge transfer from the chemical shift and in that of Mulliken's population analysis. Thus, we think that this discrepancy of the charge transfer from a Nb atom to Se atoms between the calculated and experimental values is not so serious.

3.2. Energy dispersion

A nonrelativistic self-consistent field band calculation was also carried out by Shima in addition to the relativistic one [108]. First, a nonrelativistic band structure is shown and then a relativistic one is shown. The differences between relativistic and nonrelativistic calculations are then investigated.

A Brillouin zone of $NbSe_3$ is shown in Figure 14. In Figure 15 the nonrelativistic band structure is shown. A vertical axis represents energy in Rydberg unit. A horizontal axis represents crystal momentum **k** in a Brillouin zone along the line **G–B–A–Y–G–Z–D–E–C–Z**, namely along the edges of a central plane and a top plane of a Brillouin zone and the **G–Z** line. The **G–Y**, **G–Z**, and **G–B** lines are

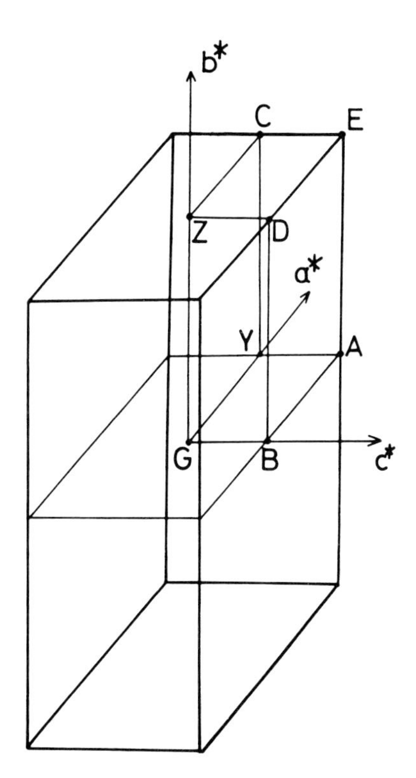

Fig. 14. Brillouin zone of $NbSe_3$ and its special points (**G**, **B**, **A**, **Y**, **Z**, **D**, **E**, **C**). The crystal momentum **k** at point **Z** is ($0.5\mathbf{b}^*$, 0), for example.

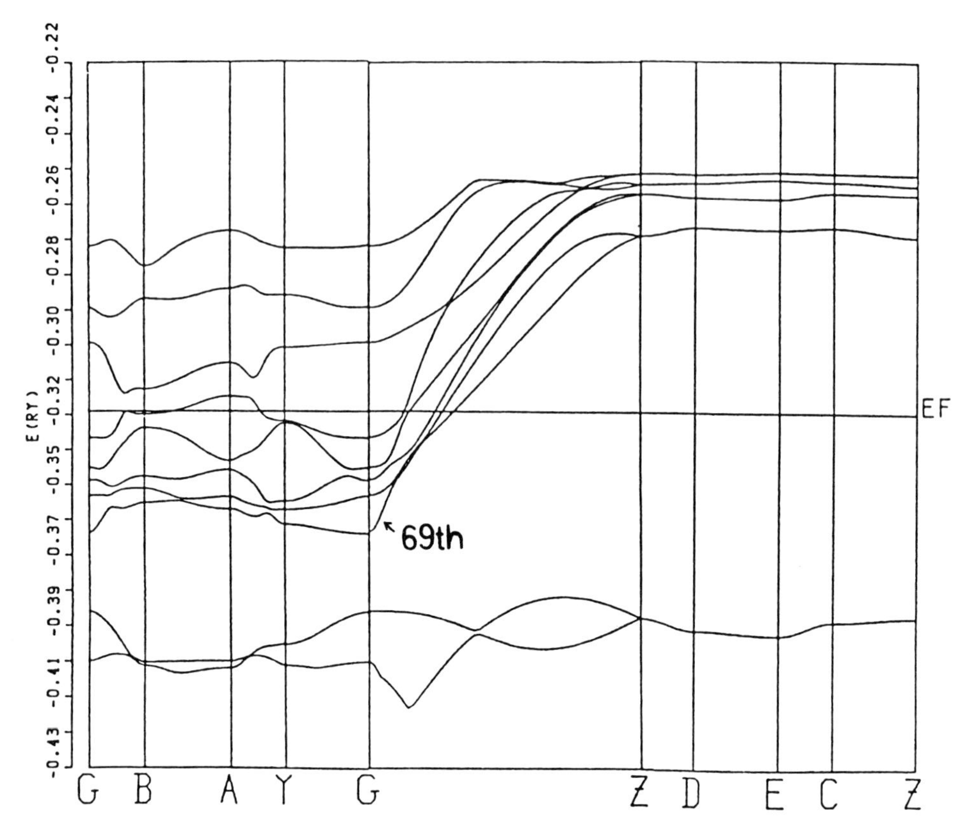

Fig. 15. Nonrelativistic band structure of $NbSe_3$. The 67th band to 76th band are shown. EF represents the Fermi level.

parallel to the **a***, **b***, and **c*** reciprocal lattice vectors, respectively. In Figure 15, 'EF' represents 'Fermi level'.

Because of screw axes of $NbSe_3$, all bands are doubly degenerate on the top and bottom planes of a Brillouin zone, namely the **Z**–**D**–**E**–**C**–**Z** plane. There are five conduction bands of the 69th to 73th band above the 67th and 68th bands which are made from $4p$ orbitals of Se atoms and correspond to the σ_B^* band discussed in Section 1.2. (Here the number of the band is counted from the lowest $4s$ band of Se.) Around the point **G** in the Brillouin zone these five conduction bands are mainly made from $4d_{z^2}$ ($z \| \mathbf{b}$-axis) orbitals of Nb atoms. Thus they have larger dispersions in the **b***-direction than in the **a*** or **c*** directions. It can be said that the conduction bands of $NbSe_3$ are $4d_{z^2}$ bands having quasi-one-dimensional characters. There is an energy gap between the 69th and 68th bands. As the Fermi level EF crosses five d_{z^2} bands, there are five pairs of Fermi surfaces corresponding to each conduction band. Because the 73rd band has a hole around the point **A**, the Fermi surface of this band, which is known here as Fermi surface 5, is partially closed. Four other pairs of Fermi surfaces, known as Fermi surface 1 to 4 belonging to the

69th to the 72nd conduction bands, are rather flat. They have plane-like shapes and have no closed orbit in the Brillouin zone. Since several Fermi surfaces have plane-like shapes, we can expect the existence of some nesting vectors.

As the energy differences between conduction bands are not very large, the relativistic effects may not be neglected. The relativistic band structure is shown in Figure 16. Because of the inversion symmetry of $NbSe_3$, each band retains a two-fold spin degeneracy at least. The main relativistic effect is due to the mass–velocity term. Other two effects are rather small. The mass–velocity effect gives a constant shift of energy to the energy bands which have the same orbital character as a whole. Thus the main relative structure of the bands is not different from the nonrelativistic one. The 69th band shifts downwards around the point **G** in the relativistic calculation, and the energy gap which has existed between the 69th and 68th bands, in the nonrelativistic case disappears. The 73rd and 74th bands are mixed and cross each other. Then the shape of the Fermi surface 5 becomes more sensitive to the position of the Fermi level. A tiny electron pocket may result from the 74th band around **B** in a Brillouin zone. In the following description of the electronic structure of $NbSe_3$, Shima neglects this tiny pocket. Except for the above

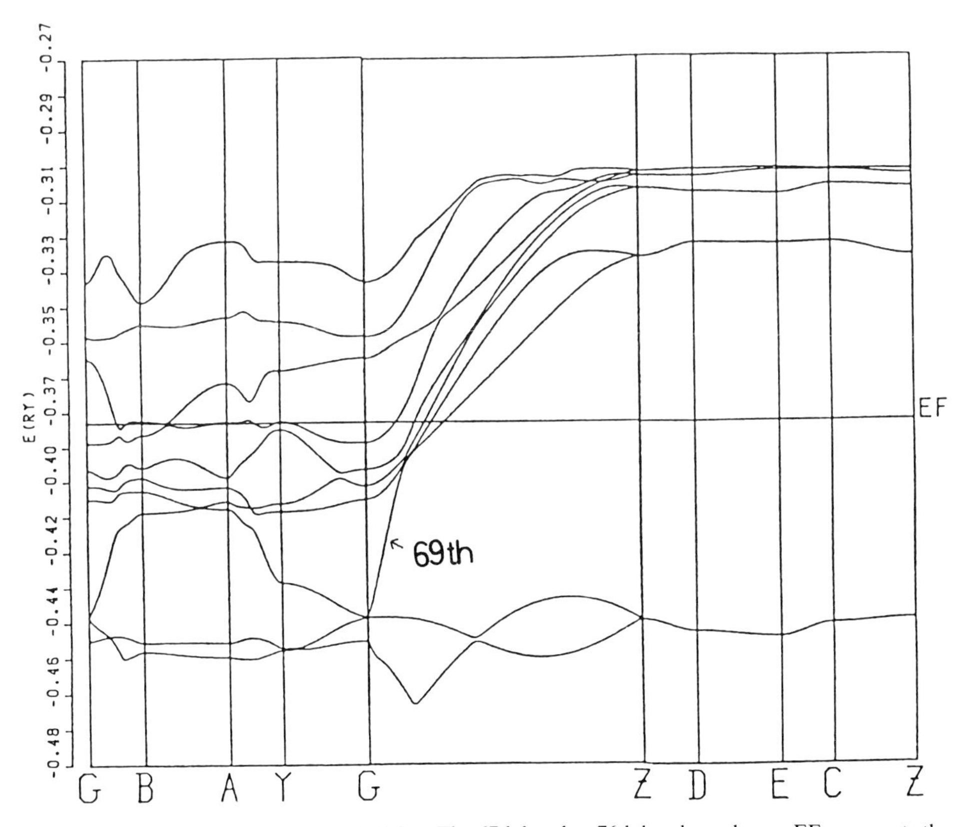

Fig. 16. Relativistic band structure of $NbSe_3$. The 67th band to 76th band are shown. EF represents the Fermi level.

points, the main features of relativistic conduction bands and Fermi surfaces are not different from those of nonrelativistic bands.

The global structure and character of the conduction bands obtained by Shima are not remarkably different from the results calculated by Bullett [103, 104, 106] and Hoffman *et al.* [105]. However, because of the lack of charge self-consistency or of neglecting the existence of both types of A and B chains in $NbSe_3$ in the calculations of latter groups, the number of conduction bands, their relative positions, and the curvatures are not correct. Since these features determine the number and shape of Fermi surfaces which have influence on the transport properties, and especially on the CDW transitions, a precise self-consistent calculation is essential.

3.3. Density of States

The energy density of states is calculated from the relativistic band structure. In Figure 17 the calculated result is shown by a solid line. The density of states can be divided into four regions. In the energy region from −1.7 Ry to −1.35 Ry states are made from Se 4*s* orbitals (the first region). The density of states extending from −0.95 Ry to −0.45 Ry corresponds to Se 4*p* orbitals and Nb 4*d* orbitals (the second region). The third region in the density of states appears near −0.45 Ry, which is made from Se 4*p* orbitals. Finally, the fourth region appears in the density of states

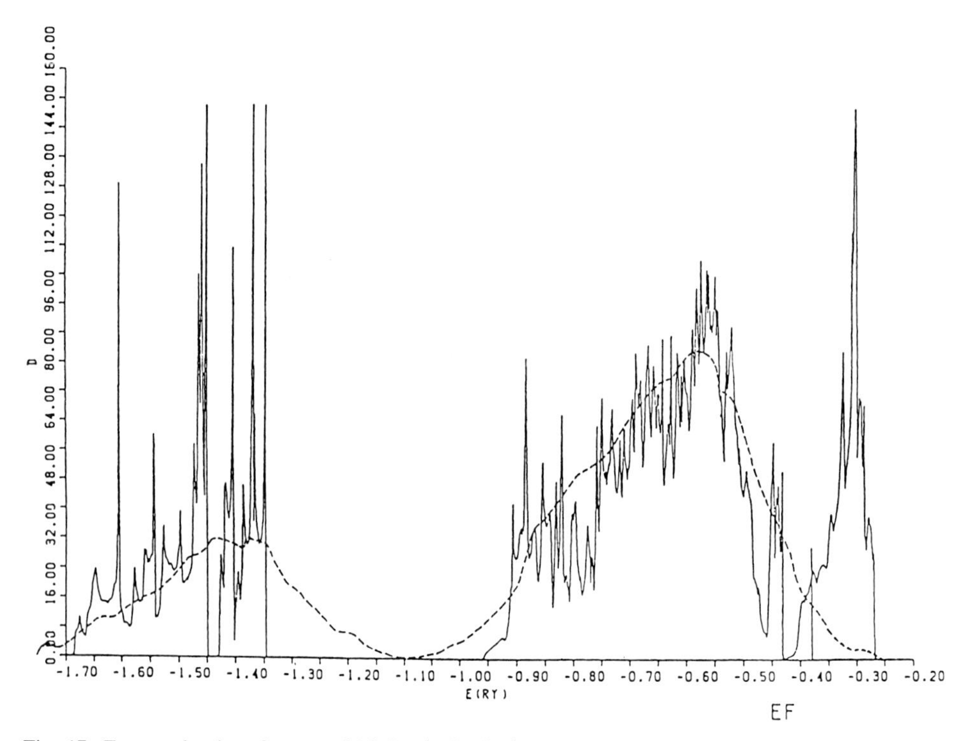

Fig. 17. Energy density of states of $NbSe_3$ (solid line) and experimental value of XPS spectrum [40] (dotted line).

above −0.4 Ry. This region corresponds to the states made mainly from Nb 4*d* orbitals. More precise components are described later. Here, the first region is called a 4*s* band, the second region a bonding band, the third region an isolated band, and the fourth region a *d* band. Up to the isolated band, there are 68 bands. The Fermi level is located at the lower part of the *d* band.

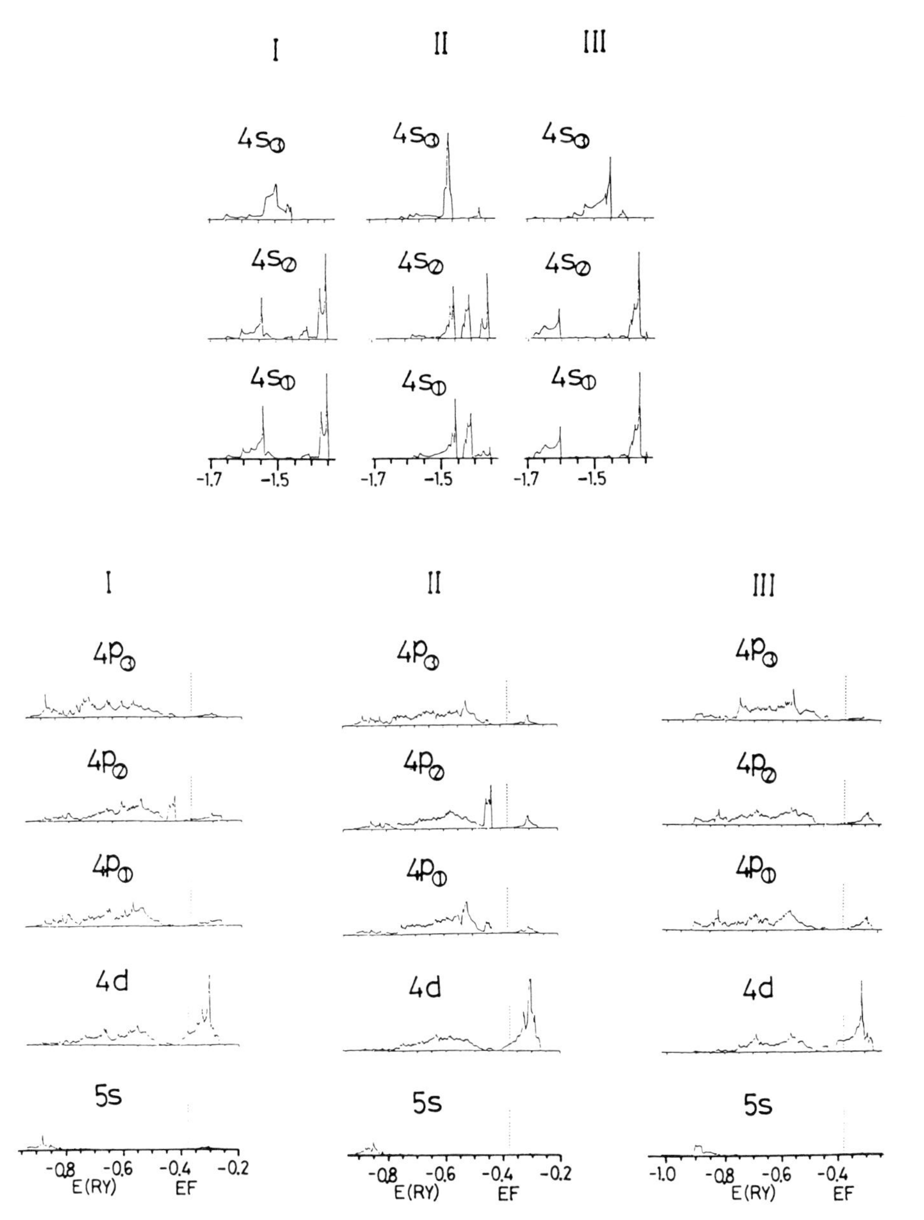

Fig. 18. Partial energy density of states for each orbital component of Nb and Se atoms.

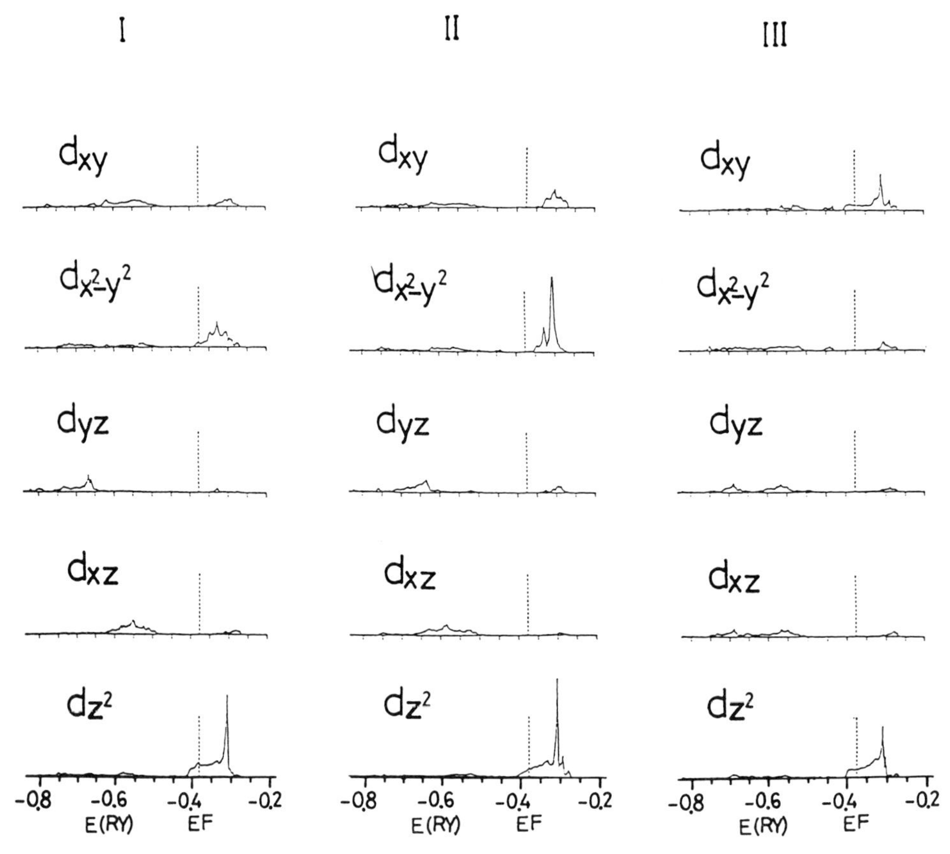

Fig. 19. Partial energy density of states for d orbital components of Nb atoms. From this figure it is obvious that the main component of the conduction bands are d_{z^2} orbitals.

In the same figure, the density of states obtained from observed XPS spectrum at room temperature [40] whose resolution is about 0.04 Ry is shown by a dotted line, where the unit is arbitrary. The part of the XPS spectrum corresponding to the 4*s* band is rather broad and shifts slightly to higher energy as compared with the calculated 4*s* band region. The isolated band region is not seen clearly but appears as a shoulder in the XPS intensity. Qualitative features, for example, as for peak positions and peak heights, coincide with the calculated density of states.

The calculated density of states at the Fermi level is 17.2 states/Nb-atom Ry, which can be compared with the observed value, of 21.7 states/Nb-atom Ry, estimated from NMR measurement [42]. This corresponds to the value before two CDW transitions occur. Because of the occurrence of the gap in the Fermi surfaces being accompanied with CDW transitions, the density of states at the Fermi level decreases after two CDW transitions have occurred. Assuming the disappearance of Fermi surfaces 1–4, which we shall explain in detail later, the density of states at low temperature becomes 5.6 states/Nb-atom Ry, which is close to the experimental values of 4.7 states/Nb-atom Ry (specific heat [27]), and 3.8 states/Nb-atom Ry (Knight shift [42]).

In order to investigate the character of orbitals contributing to the density of states, and also to understand the bonding nature in $NbSe_3$, some partial densities of states corresponding to different orbitals are calcualted. Partial densities of states for five d orbitals (d_{z^2}, d_{xz}, d_{yz}, $d_{x^2-y^2}$, d_{xy}) of chains I, II, and III are also given. The results are shown in Figures 18 and 19. The symbols I, II, and III represent the chain locations in a unit cell and ①, ②, and ③ represent the locations of Se atom in a chain (see Figure 13). Because of the inversion symmetry of the system, the partial densities of states for chains I′, II′, and III′ are the same as those for chains I, II, and III, respectively.

As for the 4s band, it is understood from the partial density of states that bonding and antibonding orbitals between two Se atoms, which are close to each other in the trigonal chains, are composed of bonding and antibonding band parts in the 4s band. The nonbonding 4s orbitals of Se atoms, each of which is isolated from two other Se atoms in the trigonal chains, are also composed of nonbonding parts in the 4s band. The energy splitting between bonding and antibonding parts reflects the distance between the two Se atoms for each case.

The bonding band is mainly composed of Se 4p orbitals and Nb e_g (d_{xz}, d_{yz}) and two t_{2g} (d_{zy}, $d_{x^2-y^2}$) orbitals. Thus, bonds between Nb atoms and Se atoms have not only an ionic character but also a covalent character, as discussed in Section 1.2. The mixture of Nb e_g orbitals and of Se 4p orbitals corresponds to covalent bonds between the Nb atom and Se atoms in a chain. The mixture of two t_{2g} orbitals and of Se 4p orbitals, which is relatively small compared with that of e_g and 4p orbitals, corresponds to covalent bonds between the Nb atom in a chain and Se atoms in adjacent chains.

The main component of the isolated band is $4p_{①}$ and $4p_{②}$ orbitals of chain II and $4p_{②}$ of chain I, as shown in Figure 18. This band is considered to correspond to σ_B^*, as explained in Section 1.2. This result is different from the simple picture described in Section 1.2 in which the $4p_{②}$ orbital of chains I and I′, i.e. type A chains, contribute to the isolated band.

The main character of the 4d band is 4d orbitals of Nb atoms. The lower part of the 4d band is composed of d_{z^2} orbitals of Nb atoms. The d band also contains a small amount of Se 4p orbital components. It is obvious from Figure 19 that the main character of the conduction bands are d_{z^2} orbitals. In addition to these components, the $d_{x^2-y^2}$ orbital of chain I and the d_{xy} orbital of chain III are also components of the conduction bands. The mixing of $d_{x^2-y^2}$ and d_{xy} orbital components to the d_{z^2} conduction bands influences the shapes of Fermi surfaces and characters of CDWs. Orbital characters of Fermi surfaces are discussed in the next section.

3.4. Fermi Surfaces

In Figure 20 five pairs of Fermi surfaces, corresponding to the five conduction bands from the 69th to the 73rd band, are pictured [107]. As before, we call the Fermi

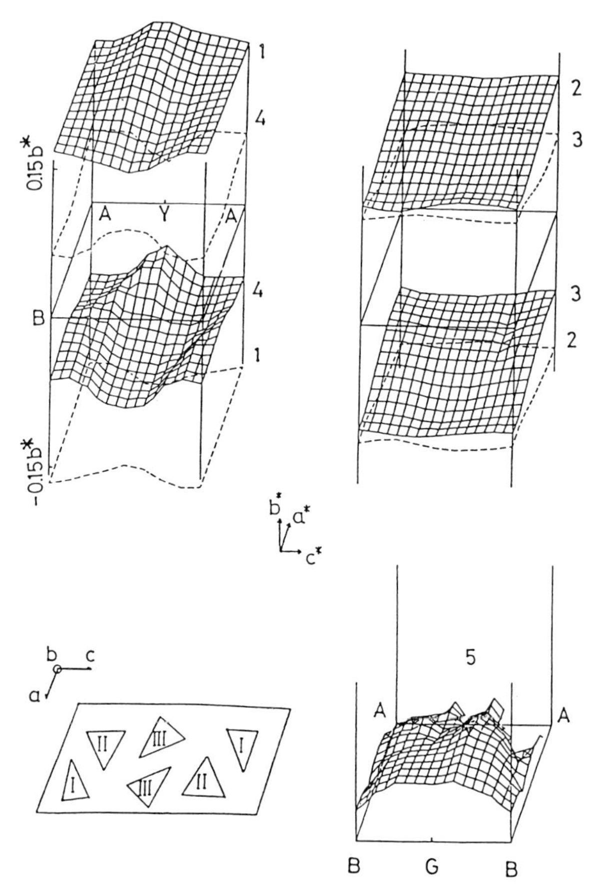

Fig. 20. Fermi surfaces corresponding to each conduction band and the chain location in a unit cell of $NbSe_3$. The nesting pairs, 1–4 and 2–3, are pictured in the half region of the Brillouin zone. Each of the nesting pieces of the Fermi surfaces is drawn in the solid and dashed lines. Fermi surface 5 is pictured in the quarter of the Brillouin zone.

surfaces corresponding to the 69th band Fermi surface 1, the Fermi surfaces corresponding to the 70th band, Fermi surface 2, and so on. Fermi surfaces 1–4 are plane-like. They do not have a closed orbit. This indicates a one-dimensional character. On the other hand, Fermi surface 5 is partially closed, because the 73rd band has a hole around the point **A** in the Brillouin zone. Partial densities of states for d orbitals are calculated for each Fermi surface. By examining the orbital densities, each Fermi surface can be assigned to different chains in the unit cell (see Figure 20). States which contribute to Fermi surfaces 1 and 4 are composed of d_{z^2} orbitals of chains I,I′ and II,II′. States involved in Fermi surfaces 2 and 3 are mainly composed of d_{z^2} orbitals and d_{xy} orbitals of chain III,III′. The warped Fermi surface 5 is made of d_{z^2} orbitals of chains I,I′ and II,II′ and $d_{x^2-y^2}$ orbitals of chain I,I′. Thus Fermi surface 1 is assiged to chains I,I′ and II,II′; 2 and 3 to III,III′; 4 and 5 to I,I′ and II,II′.

As Fermi surfaces 1–4 are plane-like, it is expected that they nest each other and disappear after the two CDW transitions occur. (We describe the CDW transitions later.) It is also expected that at low temperatures below the two transition temperatures there exists only Fermi surface 5, which is rather warped and partially closed. As was shown in Section 3.3, the calculated value of density of states for Fermi surface 5 explains the experimental facts at low temperatures.

Shubnikov–de Haas oscillations [31] in magnetoresistance are observed at low temperatures, as shown in Figure 9. From the shape of Fermi surface 5, a model Fermi surface on the basis of which Shubnikov–de Haas oscillations at low temperatures are easily calculated can be proposed. As shown in Figure 21, this is partially closed. This model Fermi surface can explain the behavior of the fundamental branch of the quantum oscillation qualitatively. The experimental and calculated extreme cross-sections of Fermi surface along the $\mathbf{a}^*$ and $\mathbf{c}^*$ axes are presented in Table II. As seen from this table, quantitative agreement is not good. This means that the Shubnikov–de Haas oscillations are very sensitive to the position of the Fermi level.

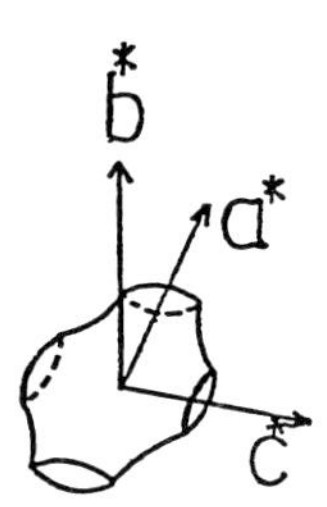

Fig. 21. A model Fermi surface to explain the Shubnikov–de Haas oscillations [31].

TABLE II

Cross-sections of the Fermi surface 5

	Experiment	Calculated
S_{a^*}	0.0041	0.019 (A^{-2})
S_{c^*}	0.0030	0.020

4. Discussion

4.1. Fermi Surfaces and CDW Transitions

As seen in Section 1.3, $NbSe_3$ shows two independent CDW phase transitions at $T_1 = 142$ K and $T_2 = 58$ K which are accompanied with two independent superstructure vectors, $\mathbf{q}_1 = (0.0, 0.243\mathbf{b}^*, 0.0)$ and $\mathbf{q}_2 = (0.5\mathbf{a}^*, 0.263\mathbf{b}^*, 0.5\mathbf{c}^*)$. We discuss in this section the relation between these two phase transitions and the Fermi surfaces of $NbSe_3$ derived by Shima.

In a previous section, it has been shown that there are four pairs of flat Fermi surfaces and one partially closed Fermi surface in the Brillouin zone. Because of flatness. Fermi surfaces 1–4 may nest at low temperatures. In order to understand a nesting process and to determine nesting vectors, we may apply the Chan–Heine criterion [124]. This criterion needs information on an electron–phonon coupling constant, precise phonon structure and electron–electron interactions in addition to a calculation of a generalized susceptibility. For lack of such information, we here discuss a possibility of nesting in $NbSe_3$ qualitatively.

From the curvatures of these Fermi surfaces (see Figure 20), it can be said that Fermi surface 2, in a region of a Brillouin zone where the $\mathbf{b}^*$ component of a crystal momentum is positive, has a possibility of nesting with Fermi surface 3 in a region of a Brillouin zone where the $\mathbf{b}^*$ component of a crystal momentum is negative, and vice versa. Similarly, Fermi surface 1 has a possibility of nesting with Fermi surface 4. Each of the nesting vectors has only a $\mathbf{b}^*$ component. At the point **G** in the Brillouin zone, the distances of the Fermi surfaces between 1 and 4 and between 2 and 3 are $0.23\mathbf{b}^*$ and $0.22\mathbf{b}^*$, respectively. The nesting vector of the 1–4 pair of Fermi surfaces is slightly longer than that of the 2–3 pair. As Fermi surfaces 2 and 3 are flatter and have a more parallel region than Fermi surfaces 1 and 4, the generalized susceptibility of the former is larger than that of the latter. From these considerations, it can be said that two pairs of Fermi surfaces, 2–3 and 1–4, which can be assigned to the different chains III and I,II, nest at T_1 and T_2 respectively (see Section 3.4), and thus different distortions occur in different chains at T_1 and T_2. From NMR experiments, Devreux [41] and Wada *et al.* [42] also concluded that the two CDW transitions occurred independently at different sites in a unit cell. They assigned the first transition to chain III,III′ and the second one to chain I,I′, which are slightly different conclusion from that by Shima.

Apparently the 1–4 pair nesting does not cause $\mathbf{a}^*$ and $\mathbf{c}^*$ components for the superstructure vector $\mathbf{q}_2$. These components can be considered to result from phase ordering of CDWs because the inter-CDWs Coulomb interaction in chains I and II plays an important role, just in the same way as in the case of TTF-TCNQ [134]. The 1–4 pair's CDW is localized on four chains I, I′, II, and II′ in the unit cell. If the phases of CDWs in adjacent unit cells along the directions of $\mathbf{a}^*$ and $\mathbf{c}^*$ differ by π, the size of a new unit cell is twice that of a unit cell before the transition. Then new superstructure reciprocal vectors of $(\pm 0.5\mathbf{a}^*, 0.0, \pm 0.5\mathbf{c}^*)$ are obtained.

As the innermost Fermi surface 5 is rather warped and partially closed, only this Fermi surface survives after the two transitions. (It should be re-emphasized that the existence of this Fermi surface can explain the observed results of specific heat and Knight shift quantitatively, and that of Shubnikov–de Haas oscillations qualitatively, as we described in Section 3.4.)

4.2. CDW TRANSITIONS AND TRANSPORT PROPERTIES

A large area of a Fermi surface disappears after a CDW transition. This disappearance of a Fermi surface has great influence on some transport properties. A Hall constant and an electric resistance can be calculated from the Fermi surfaces described in previous sections, by making some assumptions. In this section we compare the theoretical and experimental results in order to investigate the nesting processes of $NbSe_3$.

From the ordinary linearized Boltzmann equation under the relaxation time approximation, a conductivity tensor can be easily derived using Chamber's expression [126, 127] up to linear field term:

$$\sigma_{ij} = \frac{2q^2}{(2\pi)^2\hbar^2}\left\langle\frac{\tau\epsilon_i\epsilon_j}{\sqrt{\epsilon^k\epsilon_k}}\right\rangle_{\text{F.S.}} - \frac{2q^2}{(2\pi)^3\hbar^3}qE^\nu\left\langle\frac{\tau^2\epsilon_i\epsilon_{j\nu}}{\sqrt{\epsilon^k\epsilon_k}}\right\rangle_{\text{F.S.}} - \frac{2q^2}{(2\pi)^3\hbar^4}\frac{qH^\mu}{c}\mathscr{E}_\mu^{\ \nu\lambda}\left\langle\frac{\tau^2\epsilon_i\epsilon_{j\nu}\epsilon_\lambda}{\sqrt{\epsilon^k\epsilon_k}}\right\rangle_{\text{F.S.}}, \tag{40}$$

where c is the velocity of light, $-q$ the electric charge, $A^kB_k = \Sigma_k A_k B_k$, $\mathscr{E}_{\mu\nu\lambda}$ an antisymmetric tensor, ϵ the energy dispersion, $\epsilon_x = (\partial\epsilon/\partial k_x)$, τ the relaxation time, E_ν and H_μ electric and magnetic fields, and $\langle\ \rangle_{\text{F.S.}}$ an integral over a Fermi surface. Generally, the relaxation time τ is dependent on crystal momentum, band index and temperature.

4.2.1. *Hall constant*

If we assume that τ is constant over the Fermi surfaces, we can obtain a very simple form for a zero-field-limit Hall constant under a magnetic field perpendicular to a b–c plane and an electric field along the **b**-axis [107],

$$R_{\text{H}}(0) = \frac{(2\pi)^3}{2qc}\frac{\left\langle\frac{\epsilon_{bb}\epsilon_c^2 - \epsilon_b\epsilon_c\epsilon_{bc}}{\sqrt{\epsilon^k\epsilon_k}}\right\rangle_{\text{F.S.}}}{\left\langle\frac{\epsilon_b^2}{\sqrt{\epsilon^k\epsilon_k}}\right\rangle_{\text{F.S.}}\left\langle\frac{\epsilon_c^2}{\sqrt{\epsilon^k\epsilon_k}}\right\rangle_{\text{F.S.}}}, \tag{41}$$

where suffixes b and c represent the directions along the **b** and **c** axes, and $\epsilon_b = \partial\epsilon(\mathbf{k})/\partial k_b$, etc. In calculating $R_{\text{H}}(0)$, using the above expression, the energy dispersion of the αth band near the Fermi surface is expressed by the following analytical form

$$E_\alpha(\tilde{\mathbf{k}}, k_b) = \sum_n k_b^n \sum_{\tilde{\mathbf{m}}} A_{\tilde{\mathbf{m}},n,\alpha}\, e^{i\tilde{\mathbf{k}}\cdot\tilde{\mathbf{m}}}, \tag{42}$$

where $\tilde{\mathbf{k}}$ means (k_a, k_c), and the coefficients $A_{\tilde{\mathbf{m}},n,\alpha}$ are determined so as to reproduce the αth band near the Fermi energy well.

Complete disappearance of Fermi surfaces 2–3 and 1–4 are assumed below the transition temperatures T_1 and T_2, respectively. $R_{\text{H}}(0)$ is calculated numerically using the above analytical expansion of energy bands. The result is shown by the dotted lines in Figure 22, where the experimental values are also shown by solid lines. Above T_2, the calculated Hall constant can explain the experiment qualitatively. Below T_2, the observed Hall constant changes its sign from negative to positive at about 15 K, while the present results always give a negative sign.

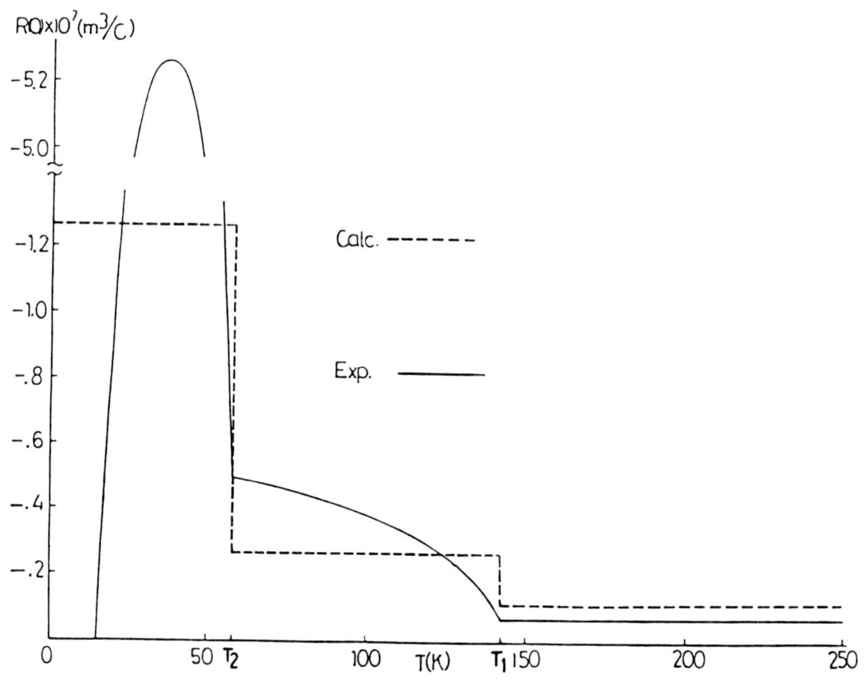

Fig. 22. Hall constant $R_H(0)$ of $NbSe_3$, calculated (dashed line) and experimental [33] (solid line) values.

Probably this sign change can be explained by taking account of the temperature and valley dependences of relaxation time, which are not considered here.

4.2.2. *Resistivity*

Resistivity along the **b**-axis has also been calculated. From the conductivity tensor calcualted in Section 4.2.1., the resistivity is obtained as:

$$\rho_b = \frac{(2\pi)^3}{2q^2}\hbar^2\tau^{-1}\left\langle\frac{\epsilon_b^2}{\sqrt{\epsilon^k\epsilon_k}}\right\rangle_{\text{F.S.}}^{-1} + \rho_{\text{imp}}, \tag{43}$$

where ρ_{imp} is a residual impurity resistance. The above expression contains a relaxation time constant τ explicitly. Although the exact form of τ is not known, the Grüneisen's form for τ is assumed for simplicity, and then ρ_b becomes

$$\rho_b = BT^5\int_0^{\Theta_D/T}\frac{x^5}{(1-e^{-x})(e^x-1)}\mathrm{d}x\left\langle\frac{\epsilon_b^2}{\sqrt{\epsilon^k\epsilon_k}}\right\rangle_{\text{F.S.}}^{-1} + \rho_{\text{imp}}, \tag{44}$$

where B is a constant and Θ_D a Debye temperature which is roughly estimated from the resistivity at low temperatures.

The constant B is also estimated from the resistivity at room temperature. Like the case of a Hall constant $R_H(0)$, complete disappearance of Fermi surfaces 2–3 and 1–4 are assumed below the transition temperatures T_1 and T_2.

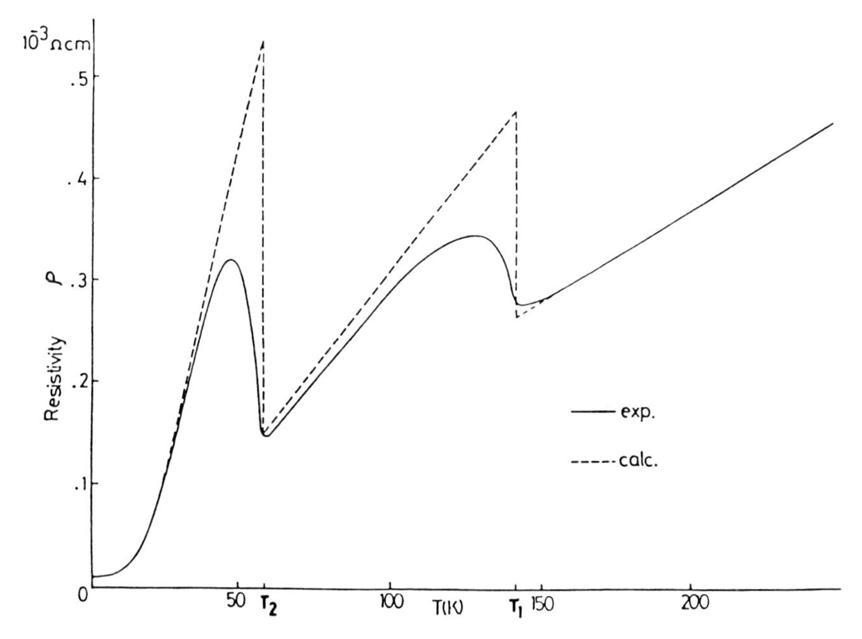

Fig. 23. Resistivity ρ of $NbSe_3$, calculated (dashed line) and experimental [11] (solid line) values.

The calculated result is shown in Figure 23, together with the experimental result. An approximate calculation of the resistivity ratio ρ_c/ρ_b is carried out using the same procedure. Its result is given in Table III with experimental values [34]. Though quantitative agreement is not always good, these results can explain the transport properties of $NbSe_3$ fairly well in a qualitative aspect. These qualitative agreements between theory and experiment in various transport properties support the picture of the disappearance of Fermi surfaces 2–3 and 1–4 below the transition temperatures.

Strictly speaking, one must take account of valley and crystal momentum dependences of τ, electron–electron interactions, temperature variation of the

TABLE III

Calculated and experimental [34] resistivity ratio ρ_c/ρ_b of $NbSe_3$ in three temperature regions

	Resistivity ratio		
ρ_c/ρ_b	$T < T_2$	$T_2 < T < T_1$	$T_1 < T$
Calculated	5.7	9.1	16.5
Experiment	10 ~ 22	16 ~ 22	16 ~ 20
	$T_1 = 142\,K$	$T_2 = 58\,K$	

nesting area of Fermi surfaces, and electron–phonon interaction, etc., all of which have not been considered so far.

4.3. Concluding Remarks

Reflecting the chain structure, there are five d_{z^2} conduction bands in $NbSe_3$, and from these five bands there appear four pairs of Fermi surfaces which are rather flat, and one partially closed Fermi surface. Fermi surfaces 2 and 3 are assigned to chains III and III′. Fermi surfaces 1 and 4 are assigned to chains I,II and I′,II′. Fermi surface 5, which is partially closed, is assigned to chains I,I′ and II,II′. These four flat Fermi surfaces tend to nest at two transition temperatures and then cause two CDW phases. As the pairs of the Fermi surface 2–3 and 1–4 belong to different chains in a unit cell of $NbSe_3$, the nesting of each Fermi surface pair at T_1 and T_2 is independent of each other, resulting in two independent CDW phases.

Assuming complete disappearance of Fermi surfaces 2–3 and 1–4 at T_1 and T_2, respectively, the Hall constant $R_H(0)$ and resistivity ρ_b have been calculated and compared with the experiments. Though continuous changes of $R_H(0)$ and ρ_b through the transition temperatures are not reproduced here, qualitative behaviors of these transport properties through the transition temperatures can be understood by the two independent nestings of Fermi surfaces. Of course, there may be other mechanisms to cause the anomalous transport behaviors, for example, a mechanism due to strongly temperature-dependent scattering of carriers. Although such a possibility cannot be excluded, it seems rather natural to ascribe anomalous behaviors to the losses of carriers due to CDW transitions.

As mentioned previously, the second nesting vector corresponding to Fermi surface pair 1–4 does not have $\mathbf{a}^*$ and $\mathbf{c}^*$ components. We have suggested that the appearance of these $\mathbf{a}^*$ and $\mathbf{c}^*$ components in the superstructure is due to the phase ordering of CDWs by the Coulomb interaction. More precise theoretical investigation must be performed for exploring this possibility.

The band structure calculations have revealed that two independent CDWs localize around different sites in a unit cell of $NbSe_3$. The CDW associated with the first phase transition at T_1 localizes around two chains III and III′. The CDW associated with the second transition at T_2 localizes around four chains I, II, I′, II′. At two different sites of the two-chain area and the four-chain area, the effects of impurity potential on sliding modes [20, 21] which cause the non-Ohmic conduction may be different. Moreover, as explained in Section 3.4, the two independent CDWs have a different orbital character. The CDW of the second phase transition at T_2 localizing around chains I,I′ and II,II′ has a d_{z^2} character only. On the other hand, the CDW of the first phase transition at T_1 localizing around chain III,III′ has not only a d_{z^2} character but also a d_{xy} character. From these facts it is expected that an amplitude of CDW of the latter in the direction perpendicular to the chain is larger than that of the former. When the amplitude of CDW is large, the pinning energy of the CDW by charged impurities becomes large. Thus, the latter CDW is supposed to interact with charged impurities strongly. The threshold and activation electric fields of the non-Ohmic conductions at transition temperatures T_1 and T_2

have different magnitudes experimentally; that is, the CDW at T_1 has larger threshold and activation electric fields. According to the above-mentioned argument, this suggests a stronger Coulomb interaction between CDW and impurities for the CDW of the first transition temperature.

The possible existence of two kinds of CDW in a unit cell of $NbSe_3$ will be useful for investigating a macroscopic mechanism of CDW transition and dynamics of Fröhlich sliding mode or ICDW (incommensurate charge density wave).

Acknowledgements

We are very grateful to Professors Masaru Tsukada, Yasutada Uemura, Kenji Nakao, Takashi Sambongi, Masayuki Ido, and Kazuhiko Yamaya for their stimulating conversation and discussions on the subject reviewed in this article. We are also indebted to Professors Ryozo Aoki and Shinji Wada, and Doctors Kazuhiro Endo and Kazushige Kawabata for valuable discussions and suggestions.

References

1. S. Barišić, A. Bjeliš, J. R. Cooper, and B. Leontić (eds), *Lecture Notes in Phys.* **95** (1981), Springer-Verlag, Berlin.
2. J. S. Miller (ed.), *Extended Linear Chain Compounds*, **I**, Plenum, New York (1982).
3. R. M. Fleming, *Solid State Science* **23**, 253 (1981).
4. N. P. Ong, *Can. J. Phys.* **60**, 757 (1982).
5. G. Grüner, *Comm. Sol. St. Phys.* **10**, 183 (1983).
6. H. Fröhlich, *Proc. R. Soc. London* **A223**, 296 (1954).
7. P. A. Lee, T. M. Rice, and P. W. Anderson, *Sol. St. Commun.* **14**, 703 (1974).
8. P. Haen, P. Monceau, B. Tissier, G. Waysand, A. Meerschaut, P. Molinie, and J. Rouxel, *Proc. 14th Int. Conf. Low Temp. Phys.* **5**, 445 (1975).
9. A. Meerschaut and J. Rouxel, *J. Less-Common Metals* **39**, 197 (1975).
10. P. Haen, G. Waysard, G. Boch, A. Waintal, P. Monceau, N. P. Ong, and R. B. Portis, *J. Physique* **C4**, 179 (1976).
11. J. Chaussy, P. Haen, J. C. Lasjaunias, P. Monceau, G. Waysand, A. Waintal, A. Meerschaut, P. Moline, and J. Rouxel, *Sol. St. Commun.* **20**, 759 (1976).
12. K. Tsutsumi, T. Takagi, M. Yamamoto, Y. Shiozaki, M. Ido, T. Sambongi, K. Yamaya, and Y. Abe, *Phys. Rev. Lett.* **39**, 1675 (1977).
13. S. Nakamura and R. Aoki, *Sol. St. Commun.* **27**, 151 (1978).
14. J. L. Hodeau, M. Marezio, C. Rouceau, R. Ayroles, A. Meerschaut, J. Rouxel, and P. Monceau, *J. Phys.* **C11**, 4117 (1978).
15. R. M. Fleming, D. E. Moncton, and D. B. McWhan, *Phys. Rev.* **B18**, 5560 (1978).
16. K. Tsutsumi and T. Sambongi, *J. Phys. Soc. Japan* **49**, 1859 (1980).
17. P. Monceau, N. P. Ong, A. M. Portis, A. Meerschaut, and J. Rouxel, *Phys. Rev. Lett.* **37**, 602 (1976).
18. N. P. Ong and P. Monceau, *Phys. Rev.* **B16**, 3443 (1977).
19. N. P. Ong, *Phys. Rev.* **B17**, 3243 (1978).
20. N. P. Ong, J. W. Brill, J. C. Eckert, J. W. Savage, S. K. Khanna, and R. B. Somoano, *Phys. Rev. Lett.* **42**, 811 (1979).
21. J. W. Brill, N. P. Ong, J. C. Eckert, J. W. Savage, S. K. Khanna, and R. B. Somoano, *Phys. Rev.* **B23**, 1517 (1981).

22. R. M. Fleming and C. C. Grimes, *Phys. Rev. Lett.* **42**, 1423 (1979).
23. R. M. Fleming, *Phys. Rev.* **B22**, 5606 (1980).
24. J. Richard and P. Monceau, *Sol. St. Commun.* **33**, 635 (1980).
25. S. Tomić, K. Biljaković, D. Djurek, J. R. Cooper, P. Monceau, and A. Meerschaut, *Sol. St. Commun.* **38**, 109 (1981).
26. S. Tomić, *Sol. St. Commun.* **40**, 321 (1981).
27. J. C. Lasjaunias and P. Monceau, *Sol. St. Commun.* **41**, 911 (1982).
28. J. D. Kulick and J. C. Scott, *Sol. St. Commun.* **32**, 217 (1979).
29. F. J. DiSalvo, J. V. Waszczak, and K. Yamaya, *J. Phys. Chem. Solids* **41**, 1311 (1980).
30. P. Monceau, *Sol. St. Commun.* **24**, 331 (1977).
31. P. Monceau and A. Briggs, *J. Phys.* **C11**, L465 (1978).
32. R. M. Fleming, J. A. Polo, and R. V. Coleman, *Phys. Rev.* **B17**, 1634 (1978).
33. N. P. Ong and P. Monceau, *Sol. St. Commun.* **26**, 487 (1978).
34. N. P. Ong and J. W. Brill, *Phys. Rev.* **B18**, 5265 (1978).
35. K. Kawabata, M. Ido, and T. Sambongi, *J. Phys. Soc. Japan* **50**, 739 (1981).
36. K. Kawabata, M. Ido, and T. Sambongi, *J. Phys. Soc. Japan* **50**, 1992 (1981).
37. G. X. Tessema and N. P. Ong, *Phys. Rev.* **B23**, 5607 (1981).
38. T. Takagi, M. Ido, and T. Sambongi, *J. Phys. Soc. Japan* **45**, 2039 (1978).
39. R. H. Dee, P. M. Chaikin, and N. P. Ong, *Phys. Rev. Lett.* **42**, 1234 (1979).
40. K. Endo, H. Ihara, S. Gonda, and K. Watanabe, *Phisica* **105B**, 159 (1981).
41. F. Devreux, *J. Physique* **43**, 1489 (1982).
42. S. Wada, M. Sasakura, R. Aoki, and O. Fujita (submitted to *J. Phys.* **F** (1983)).
43. P. M. Chaikin, W. W. Fuller, R. Lacoe, J. F. Kwak, R. L. Greene, J. C. Eckert, and N. P. Ong, *Sol. St. Commun.* **39**, 553 (1981).
44. P. Monceau, J. Peyrard, J. Richard, and P. Molinie, *Phys. Rev. Lett.* **39**, 161 (1977).
45. P. Haen, F. Lapierre, P. Monceau, M. N. Regueiro, and J. Richard, *Sol. St. Commun.* **26**, 725 (1978).
46. W. W. Fuller, P. M. Chaikin, N. P. Ong, *Sol. St. Commun.* **30**, 689 (1979).
47. K. Nishida, T. Sambongi, and M. Ido, *J. Phys. Soc. Japan* **48**, 331 (1980).
48. A. Briggs, P. Monceau, M. N. Regueiro, J. Peyrard, M. Ribault, and J. Richard, *J. Phys.* **C13**, 2117 (1980).
49. C. M. Basstuscheck, C. M. Burman, R. A. Kulick, and J. C. Scott, *Sol. St. Commun.* **36**, 983 (1980).
50. W. W. Fuller, P. M. Chaikin, and N. P. Ong, *Phys. Rev.* **B24**, 1333 (1981).
51. K. Yamaya and G. Oomi, *J. Phys. Soc. Japan* **52**, 1886 (1983).
52. J. W. Brill and N. P. Ong, *Sol. St. Commun.* **25**, 1075 (1978).
53. J. W. Brill, C. P. Tzou, G. Verma, and N. P. Ong, *Sol. St. Commun.* **39**, 233 (1981).
54. R. J. Wagner and N. P. Ong, *Sol. St. Commun.* **46**, 491 (1983).
55. N. P. Ong and C. M. Gould, *Sol. St. Commun.* **37**, 25 (1980).
56. J. C. Gill, *J. Phys.* **F10**, L81 (1980).
57. J. C. Gill, *Sol. St. Commun.* **39**, 1203 (1981).
58. G. Grüner, W. G. Clark, and A. M. Portis, *Phys. Rev.* **B24**, 3641 (1981).
59. A. Zettel and G. Grüner, *Phys. Rev.* **B26**, 2298 (1982).
60. J. C. Gill, *Sol. St. Commun.* **44**, 1041 (1982).
61. R. M. Fleming, *Sol. St. Commun.* **43**, 167 (1982).
62. A. Zettel and G. Grüner, *Sol. St. Commun.* **46**, 501 (1983).
63. G. Grüner, L. C. Tippie, J. Sanny, W. G. Clark, and N. P. Ong, *Phys. Rev. Lett.* **45**, 935 (1980).
64. J. C. Gill, *Sol. St. Commun.* **37**, 459 (1981).
65. P. Bak, *Phys. Rev. Lett.* **48**, 692 (1982).
66. G. X. Tessema and N. P. Ong, *Phys. Rev.* **B27**, 1417 (1983).
67. M. Weger, G. Grüner, and W. G. Clark, *Sol. St. Commun.* **35**, 243 (1980).
68. P. Monceau, J. Richard, and M. Renard, *Phys. Rev. Lett.* **45**, 43 (1980).
69. G. Grüner, A. Zawadowski, and P. M. Chaikin, *Phys. Rev. Lett.* **46**, 511 (1981).

70. M. Weger and B. Horovitz, *Sol. St. Commun.* **43**, 583 (1982).
71. J. Bardeen, E. Ben-Jacob, A. Zettel, and G. Grüner, *Phys. Rev. Lett.* **49**, 493 (1982).
72. P. Monceau, J. Richard, and M. Renard, *Phys. Rev.* **B25**, 931 (1982).
73. P. Monceau, J. Richard, and M. Renard, *Phys. Rev.* **B25**, 948 (1982).
74. P. Monceau, J. Richard, and M. Renard, *Sol. St. Commun.* **41**, 609 (1982).
75. N. P. Ong and G. Varma, *Phys. Rev.* **B27**, 4495 (1983).
76. G. Grüner, A. Zettel, W. G. Clark, and J. Bardeen, *Phys. Rev.* **B24**, 7427 (1981).
77. K. Seeger, W. Mayr, and A. Philipp, *Sol. St. Commun.* **43**, 113 (1982).
78. D. Djurek, M. Prester, and S. Tomić, *Sol. St. Commun.* **42**, 807 (1982).
79. K. K. Fung and J. W. Steeds, *Phys. Rev. Lett.* **45**, 1696 (1980).
80. C. H. Chen, R. M. Fleming, and P. M. Petroff, *Phys. Rev.* **B27**, 4459 (1983).
81. W. W. Fuller, P. M. Chaikin, and N. P. Ong, *Sol. St. Commun.* **39**, 547 (1981).
82. W. W. Fuller, G. Grüner, P. M. Chaikin, and N. P. Ong, *Phys. Rev.* **B23**, 6259 (1981).
83. P. A. Lee and T. M. Rice, *Phys. Rev.* **B19**, 3970 (1979).
84. L. P. Gor'kov and E. N. Dolgov, *Sov. Phys. JETP* **50**, 203 (1979).
85. J. B. Sokoloff, *Phys. Rev.* **B23**, 1992 (1981).
86. L. P. Gor'kov and E. N. Dolgov, *J. Low Temp. Phys.* **42**, 101 (1981).
87. J. Bardeen, *Phys. Rev. Lett.* **42**, 1498 (1979).
88. J. Bardeen, *Phys. Rev. Lett.* **45**, 1978 (1980).
89. J. R. Tucker, J. H. Miller, K. Seeger, and J. Bardeen, *Phys. Rev.* **B25**, 2979 (1982).
90. K. Maki, *Phys. Rev. Lett.* **39**, 46 (1977).
91. K. Maki, *Phys. Rev.* **B18**, 1641 (1978).
92. K. Maki, *Phys. Lett.* **70A**, 449 (1979).
93. S. E. Burkov and V. L. Pokrovsky, *Sol. St. Commun.* **46**, 609 (1983).
94. M. Papoular, *Phys. Rev.* **B25**, 7856 (1982).
95. L. Sneddon, M. C. Cross, and D. S. Fisher, *Phys. Rev. Lett.* **49**, 292 (1982).
96. V. J. Emery and D. Mukamel, *J. Phys.* **C12**, L677 (1979).
97. R. Bruinsma and S. E. Trullinger, *Phys. Rev.* **B22**, 4543 (1980).
98. B. A. Huberman and J. P. Crutchfield, *Phys. Rev. Lett.* **43**, 1743 (1979).
99. M. L. Boriack, *Phys. Rev.* **B21**, 4478 (1980).
100. N. P. Ong, *Phys. Rev.* **B18**, 5272 (1978).
101. J. A. Wilson, *Phys. Rev.* **B19**, 6456 (1979).
102. M. Danino and M. Weger, *Phys. Rev.* **B26**, 7035 (1982).
103. D. W. Bullett, *Sol. St. Commun.* **26**, 563 (1978).
104. D. W. Bullett, *J. Phys.* **C12**, 277 (1979).
105. R. Hoffmann, S. Shaik, J. C. Scott, M. H. Whangbo, and M. J. Foshee, *J. Sol. St. Chem.* **34**, 263 (1980).
106. D. W. Bullett, *J. Phys.* **C15**, 3069 (1982).
107. N. Shima, *J. Phys. Soc. Japan* **51**, 11 (1982).
108. N. Shima, *J. Phys. Soc. Japan* **52**, 578 (1983).
109. J. A. Wilson and A. D. Yoffe, *Advin. Physics* **18**, 193 (1969).
110. L. F. Mattheiss, *Phys. Rev.* **B8**, 3719 (1973).
111. N. Shima, Master Thesis (Japanese), Tokyo Univ., Tokyo (1978).
112. F. Herman and S. Skillman, *Atomic Structure Calculations*, Prentice-Hall, (1963).
113. A. Zunger and A. J. Freeman, *Phys. Rev.* **B15**, 4716 (1977).
114. A. Rosen and D. E. Ellis, *J. Chem. Phys.* **62**, 3039 (1975).
115. W. Kohn and L. J. Shan, *Phys. Rev.* **140**, A1133 (1965).
116. L. Hedin and B. I. Lundqvist, *J. Phys.* **C4**, 2064 (1971).
117. J. C. Slater, *Phys. Rev.* **81**, 385 (1951).
118. E. J. Baerends and P. Ros, *Chem. Phys.* **2**, 52 (1973).
119. G. S. Painter and D. E. Ellis, *Phys. Rev.* **B1**, 4747 (1970).
120. F. W. Averill and D. E. Ellis, *J. Chem. Phys.* **59**, 6412 (1973).
121. R. S. Mulliken, *J. Chem. Phys.* **23**, 1833 (1955).

122. G. Lehmann and M. Taut, *Phys. Status Solidi* **54**, 469 (1972).
123. J. Rath and A. J. Freeman, *Phys. Rev.* **B11**, 2109 (1975).
124. S. K. Chan and V. Heine, *J. Phys.* **F3**, 795 (1973).
125. T. D. Schultz and S. Etemad, *Phys. Rev.* **B13**, 4928 (1976).
126. R. Chambers, *Proc. Phys. Soc. London* **A65**, 458 (1952).
127. C. Kittel, *Quantum Theory of Solids*, Wiley, New York (1963).

INDEX OF NAMES

INDEX OF SUBJECTS